SYMPOSIA OF THE ZOOLOGICAL SOCIETY OF LONDON NUMBER 61

# The Biology of Large African Mammals in their Environment

SYMPOSIA OF THE ZOOLOGICAL SOCIETY OF LONDON 61

# The Biology of Large African Mammals in their Environment

The Proceedings of a Symposium
held at the Zoological Society of London
on 19th and 20th May 1988

*Edited by* **P. A. JEWELL**

Research Group in Mammalian Ecology and Reproduction,
University of Cambridge

*and* **G. M. O. MALOIY**

Department of Animal Physiology, University
of Nairobi, Kenya

*Published for* THE ZOOLOGICAL SOCIETY OF LONDON

*by* CLARENDON PRESS · OXFORD

1989

Oxford University Press, Walton Street, Oxford OX2 6DP
Oxford New York Toronto
Delhi Bombay Calcutta Madras Karachi
Nairobi Dar es Salaam Cape Town
Melbourne Auckland
and associated companies in
Berlin Ibadan

Oxford is a trade mark of Oxford University Press

Published in the United States
by Oxford University Press, New York

British Library Cataloguing in Publication Data
The biology of large African Mammals in their
environment.
1. Africa. Mammals
I. Zoological Society of London II. Jewell, Peter A.
(Peter Arundell, 1925–  III. Maloiy, G. M. O. IV.
Series
599.096
ISBN 0–19–854009–4

Library of Congress Cataloging in Publication Data
The Biology of large African mammals in their environment:
proceedings of a symposium held at the Zoological Society of London
on 19th and 20th May, 1988/edited by P. A. Jewell and G. M. O.
Maloiy.
p. cm.—(Symposia of the Zoological Society of London; no.
61)
Includes bibliographies.
ISBN 0–19–854009–4
1. Mammals—Africa—Congresses. I. Jewell, Peter Arundel.
II. Maloiy, G. M. O. III. Zoological Society of London.
IV. Series.
QL 1.Z733 no. 61
[QL731.A1]
5991 s—dc20
[599.096]    89–3333   CIP

Typeset by Cambrian Typesetters, Frimley, Surrey
Printed in Great Britain by
Bookcraft (Bath) Ltd
Midsomer Norton, Avon

# Dedication

## Richard Maitland Laws, CBE, PhD, FIBiol, FRS

This symposium is dedicated to Richard Maitland Laws, who has done so much in the development of our understanding of the biology of large mammals and their habitats.

Dick Laws was born in 1926. His earlier education was marked by a series of scholarships, culminating in one to St. Catherine's College, Cambridge, where he obtained a first class honours degree in zoology, in both parts of the natural sciences tripos. This led on to research in the Antarctic into the reproduction of the southern elephant seal, *Mirounga leonina* Linn., for his Ph.D., which paved the way for the extensive studies of the biology of the large marine mammals for which he is so well known. These continued at the National Institute of Oceanography, and after an eight-year period in Africa, he returned to this research. He spent almost 20 years until his retirement in April, 1987, with the British Antarctic Survey; for the last 14 of these he was Director. From 1977 he was also Director of the Sea Mammal Research Unit.

During the middle of his career he moved to the terrestrial tropics in Africa—a dramatic contrast from the Antarctic marine environment! First he was the Director of the Nuffield Unit of Tropical Animal Ecology in Uganda, and then the Director of the Tsavo Research Project in Kenya, both of which he set up. Our knowledge of some of the large African mammals (notably the elephant and hippopotamus) was greatly enhanced by the wealth of experience and the expertise that he was able to contribute. During his stay in NUTAE, he supervised a number of research students, many of whom have themselves made notable contributions to our knowledge of the biology of large African mammals.

Recognition of his work on the ecology and physiology of large mammals has come from many sources—the Bruce Memorial Medal for Antarctic Work (1954), the Scientific Medal of the Zoological Society of London (1966), fellowship of the Institute of Biology (1973) and of the Royal Society of London (1980). He served on the Council of the Zoological Society of London from 1982 to 1988, being made Vice President in 1983 and elected to the key position of Secretary in 1984. He has been deeply involved with many facets of the work of the International Scientific Committee on Antarctic Research.

From the outset of his scientific career he has shown an interest in, and a remarkable grasp of, the essential parameters required for an understanding

R.M.Laws.

of population dynamics, which is a *sine qua non* for the management of animal populations. These parameters include the time and rate at which various biological processes occur, neither of which can be determined without the ability to estimate an animal's age. This poses a real problem in large wild mammals, because of their intractable nature and relatively long life-span. He has greatly assisted the studies of large marine and terrestrial mammals, by elucidating techniques for the ageing of elephants and hippopotamuses, as well as seals and whales.

Inevitably the growth of individuals and populations (in biomass as well as numbers) has been an essential component of these studies, and again techniques appropriate to animals of exceptional size have been developed.

The problem that has claimed as much of his attention as any other has been the estimation of the levels of the reproductive parameters that affect the population dynamics of a species. In Africa we are particularly indebted to him for the light he has thrown on the dynamics of elephant and hippopotamus populations.

His interest in population dynamics has expanded to encompass the ecology of the biological systems with which the particular species of his concern are associated. Our knowledge of the ecology of elephants and hippopotamuses is still primarily dependent upon his work in Uganda and Kenya between 1961 and 1967.

In summary, the investigations by Dick Laws and his associates in the 1960s:–

1. Led to formulations of management plans for dealing with large mammal populations and habitat deterioration within National Parks and Game Reserves in Uganda and Kenya.
2. Provided a clear insight into aspects of the population and ecology and reproductive biology of such African mammals as elephants, hippopotamus, buffalo and warthog. He also developed techniques for determining age and growth rates in these mammals.
3. Elucidated the feeding behaviour and strategy of elephants, buffalo and waterbuck.
4. Provided basic information on the ecology of other large mammals such as buffalo, waterbuck, Uganda kob and topi.
5. Raised important biological and ecological questions which are still being actively investigated by students interested in the biology of African mammals.

Dick Laws supported both basic and applied research carried out inside East African National Parks and Game Reserves. He strongly maintained, however, that research should attempt to provide answers that are appropriate and applicable to both management and conservation of game animals and National Parks. At that time, although some conservationists in

East Africa accepted this approach, others greatly opposed it. It is now fully
appreciated, thanks to Dick's foresight, that research is indeed an essential
tool in the management of our resources in Africa.

In conclusion, this is an appropriate and very pleasant opportunity to
wish him well and continued success in the future.

May 1988                                                      G. M. O. Maloiy

# Preface

The Zoological Society of London is a highly appropriate place in which to hold this Symposium. Large mammals from Africa have been a mainstay of the Society's exhibits since its foundation in 1826. Moreover the Society has always enjoyed an association with fieldwork in Africa[1]. The Society even attempted to find out something about the life of the okapi soon after its 'discovery' by Johnston by supporting an expedition to the Congo. The attempt was abortive and the okapi has remained an inscrutable species. Now, at last, we shall learn something about its life-style at this Symposium.

The Zoological Society of London co-operated with zoologists at Cambridge to set up the Nuffield Unit of Tropical Animal Ecology in Uganda in 1960. This was the first research institute of its kind in East Africa and has produced a continuous flow of important work since that time. We are pleased to say that all three scientists who have been Directors of NUTAE are at this Symposium: Dr R. M. Laws, who set it up, Dr S. K. Eltringham, who saw the transition to the Uganda Institute of Ecology, and Dr E. Edroma, the present Director. We have high expectations for the continued work of this Institute.

Long after we had thought of an appropriate title for this Symposium we came across a paper by Dr Leo Harrison Matthews, the former Scientific Director of the Zoological Society, published in *Nature* in 1954. It was simply entitled 'Research on the mammals of Africa'[2]. In fact it deplored the lack of research on African mammals and lamented the waste of a wealth of biological material that could have been salvaged from profligate hunting and ill-conceived eradication schemes. Matthews wrote: 'Little if anything is known of the carrying capacity per acre, in terms of game animals, of any African territory, nothing exact about the numbers of animals normally present or their fluctuations, about breeding-cycles, food preferences and requirements, or the effects of game on the environment, about the relationships of prey to predators, the incidence of parasites, and innumerable allied subjects.' One particular comment of Matthews makes chilling reading today: 'It is greatly regretted that when the clearance of the country for the abortive ground-nuts scheme was undertaken no zoologist was informed of the material that might have become available for scientific examination. As a consequence, more than nine hundred black rhinoceroses were slaughtered and left to the hyaenas and vultures —a timely postcard to

[1] Jewell, P. A. (1976). The contribution of the Zoological Society of London to field studies and prospects for the future. *Symp. zool. Soc. Lond.* No. 40: 269–81.
[2] Matthews, L. H. (1954). Research on the mammals of Africa. *Nature, Lond.* 174: 670–1.

the right quarter could have ensured that a zoologist would have been enabled to take advantage of this great destruction and have turned it to some use in elucidating many of the unanswered questions about the life-cycle of this interesting species. It is hoped that arrangements can be made to take advantage of a 'kill-out' of black rhinoceros on a much smaller scale (about fifty animals) that is shortly to be undertaken in an area of Uganda.'

Even at that time it could hardly have been anticipated that 35 years later, in 1989, the black rhinoceros would be facing extinction. A local population of black rhinoceros of 50 individuals is probably the largest of any that survive in East Africa today. One such population is on the Ol Ari Nyiro Ranch (in Laikipia, Kenya), which is dedicated to the conservation ideals of the Gallmann Memorial Foundation, and where the Zoological Society has mounted an important study of the rhinoceros, under the supervision of the Director of Zoos, Mr David Jones. We shall hear more about this at this Symposium.

In Africa large wild mammals have a profound impact on human activities and it is impossible to give full consideration to their ecology without recognizing the confrontation that exists. None is more stark than that between elephants and people and this interaction will be explored in the Symposium. A more complex interaction occurs between wild mammals and domestic livestock. Sometimes they can live in harmony, as in areas exploited by nomadic pastoralists, in other places the wild mammals have been regarded as vermin making poor use of the environment. It has been held that more intensive exploitation of such areas is justified if it produces more food for people. But that increased production may be short-lived and it can hardly be held to be moral to pass on a degraded environment to future generations who will have to pay the penalty for our mismanagement and face food shortages of an ever more destructive nature. It is with these problems in mind that we have used the broad term 'environment' in the title of our Symposium, and we have thought it appropriate to include new work on domestic livestock amongst the papers. We hope the claims of all systems of land use will continue to be explored and that these will range from intensive beef production, through pastoralism, and wildlife ranching, to wildlife viewing and recreation. If these systems are to be well managed we need a great deal more knowledge about African large mammals.

We think that Harrison Matthews would have enjoyed this Symposium and would have been impressed to see how much detailed physiological, ecological, ethological and anatomical research has indeed been initiated on a wide range of African mammals. We cannot pause, however, and Matthew's words remain apt: 'Time is short, practically everything still remains to be done.'

G. M. O. Maloiy
P. A. Jewell

24 October 1988

# Acknowledgements

The Zoological Society of London gratefully acknowledges the grants and donations towards the costs of holding the Symposium made by the following organizations:

The African Biosciences Network
Anglia Television Ltd.
Barclays Bank PLC
The Commonwealth Foundation
The Norwegian Agency for International Development
The Royal Society
The Society for the Study of Fertility
The Wellcome Trust.

The grant from the Commonwealth Foundation was specifically awarded to cover the cost of participation by a contributor from each of three East African countries, Kenya, Tanzania and Uganda. The grants from the Norwegian Agency for International Development and from the Society for the Study of Fertility were also each made to assist an individual contributor from Africa.

The Society also thanks the Ciba Foundation for providing accommodation for contributors.

Without the generous help of all these organizations the meeting could not have taken place.

# Contents

## The response of tropical vegetation to grazing and browsing in Queen Elizabeth National Park, Uganda

ERIC L. EDROMA

## Interactions of plants of the field layer with large herbivores

S. J. McNAUGHTON

## Ranging and feeding behaviour of okapi (*Okapia johnstoni*) in the Ituri Forest of Zaire: food limitation in a rain-forest herbivore?

JOHN A. HART & TERESE B. HART

## Buffalo and their food resources: the exploitation of Kariba lakeshore pastures

R. D. TAYLOR

## Strategies for water economy amongst cattle pastoralists and in wild ruminants

P. A. JEWELL & M. J. NICHOLSON

# The ecology of female behaviour and male mating success in the Grevy's zebra

JOSHUA R. GINSBERG

# Elephant mate searching: group dynamics and vocal and olfactory communication

JOYCE H. POOLE & CYNTHIA J. MOSS

# Ontogeny of female dominance in the spotted hyaena: perspectives from nature and captivity

LAURENCE G. FRANK, STEPHEN E. GLICKMAN & CYNTHIA J. ZABEL

# Assessment of reproductive status of the black rhinoceros (*Diceros bicornis*) in the wild

R. A. BRETT, J. K. HODGES & E. WANJOHI

# Contents

## African trypanosomiasis in wild and domestic ungulates: the problem and its control
MAX MURRAY & A. R. NJOGU

## Men, elephants and competition
I. S. C. PARKER & A. D. GRAHAM

## A survey of wildlife populations in Tanzania and their potential for research
K. N. HIRJI

Contents

# Development of research on large mammals in East Africa
RICHARD M. LAWS

# Contributors

ALEXANDER, R.McN., Department of Pure & Applied Biology, University of Leeds, Leeds LS2 9JT, UK.

BRETT, R. A., The Gallmann Memorial Foundation, P.O. Box 45593, Nairobi, Kenya; and Institute of Zoology, The Zoological Society of London, Regent's Park, London NW1 4RY, UK.

EDROMA, E. L., Uganda Institute of Ecology, P.O. Box 3530, Kampala, Uganda.

FRANK, L. G., Psychology Department, University of California, Berkeley, California 94720, USA.

GINSBERG, J. R., Department of Biology, Princeton University, Princeton, New Jersey 08544, USA; present address Department of Zoology, University of Oxford, South Parks Road, Oxford OX1 3PS, UK.

GLICKMAN, S. E., Psychology Department, University of California, Berkeley, California 94720, USA.

GRAHAM, A. D., B.P. 1444, Bangui, Central African Republic.

HART, B. L., School of Veterinary Medicine, Department of Physiological Sciences, University of California, Davis, California 95616, USA.

HART, J. A., Wildlife Conservation International, New York Zoological Society, Bronx, New York 10460, USA; and Epulu via Mambasa, P.O. Box 21285, Nairobi, Kenya.

HART, L. A., School of Veterinary Medicine, Human–Animal Program, University of California, Davis, California 95616, USA.

HART, T. B., Wildlife Conservation International, New York Zoological Society, Bronx, New York 10460, USA; and Epulu via Mambasa, P.O. Box 21285, Nairobi, Kenya.

HIRJI, K. N., Serengeti Wildlife Research Institute, P.O. Box 661, Arusha, Tanzania.

HODGES, J. K., Institute of Zoology, The Zoological Society of London, Regent's Park, London NW1 4RY, UK.

JEWELL, P. A., Research Group in Mammalian Ecology and Reproduction, University of Cambridge, Physiological Laboratory, Downing Street, Cambridge CB2 3EG, UK.

KAYANJA, F. I. B., Makerere University, P.O. Box 7062, Kampala, Uganda.

LAWS, R. M., St Edmund's College, Mount Pleasant, Cambridge CB3 0BN, UK.

MAINA, J. N., Department of Veterinary Anatomy, University of Nairobi, P.O. Box 30197, Nairobi, Kenya.

MALOIY, G. M. O., Department of Animal Physiology, University of Nairobi, P.O. Box 30197, Nairobi, Kenya.

McNAUGHTON, S. J., Biological Research Laboratories, Syracuse University, Syracuse, New York 13244–1220, USA.

MOSS, C. J., Amboseli Elephant Project, African Wildlife Foundation, P.O. Box 48177, Nairobi, Kenya.

MURRAY, M., Department of Veterinary Medicine, Veterinary School, University of Glasgow, Glasgow G61 1HQ, UK.

NICHOLSON, M. J., International Livestock Centre for Africa, P.O. Box 568a, Addis Ababa, Ethiopia.

NJOGU, A. R., Kenya Trypanosomiasis Research Institute, P.O. Box 362, Kikuyu, Kenya.

PARKER, I. S. C., P.O. Box 15093, Nairobi, Kenya.

POOLE, J. H., Amboseli Elephant Project, African Wildlife Foundation, P.O. Box 48177, Nairobi, Kenya; *and* Biology Department, Princeton University, Princeton, New Jersey 08544, USA.

TAYLOR, R. D., Department of National Parks & Wild Life Management, P. Bag 2003, Kariba, Zimbabwe; *present address* WWF Multispecies Project, P.O. Box 8437, Causeway, Harare, Zimbabwe.

WANJOHI, E., Research Section, Wildlife Conservation & Management Department, P.O. Box 40241, Nairobi, Kenya; *present address* Institute of Zoology, The Zoological Society of London, Regent's Park, London NW1 4RY, UK.

ZABEL, C. J., Psychology Department, University of California, Berkeley, California 94720, USA.

# Organizers of symposium

P. A. Jewell, Research Group in Mammalian Ecology and Reproduction, University of Cambridge, Physiological Laboratory, Downing Street, Cambridge CB2 3EG, UK.

G. M. O. Maloiy, Department of Animal Physiology, University of Nairobi, P.O. Box 30197, Nairobi, Kenya.


# Chairmen of sessions

E. L. Edroma, Uganda Institute of Ecology, P.O. Box 3530, Kampala, Uganda.

S. K. Eltringham, Department of Applied Biology, University of Cambridge, Pembroke Street, Cambridge CB2 3DX, U.K.

K. N. Hirji, Serengeti Wildlife Research Institute, P.O. Box 661, Arusha, Tanzania.

D. M. Jones, Director of Zoos, The Zoological Society of London, Regent's Park, London NW1 4RY, UK.

G. M. O. Maloiy, Department of Animal Physiology, University of Nairobi, P.O. Box 30197, Nairobi, Kenya.

The following pages show the chairmen of the Symposium and the contributors who attended the meeting. Above: the organizers, Professor G.M.O. Maloiy (left) and Professor P.A. Jewell (right) in the foyer of the Zoological Society's Meeting Rooms, with Dame Elisabeth Frink's sculpture of the former President of the Society, Professor Lord Zuckerman. Below, left to right: Professor F.I.B. Kayanja, Professor K.N. Hirji, Dr E.L. Edroma.

Above, left to right: Professor R. McNeill Alexander, FRS, talking to Dr S.K. Eltringham; Sir William Henderson, FRS. Centre, left to right: Dr T.B. Hart, Dr R.A. Brett, Dr J.A. Hart. Bottom, left to right: Mr I. Parker, Dr J.H. Poole, Dr M.J. Nicholson.

Above, left to right: Professor M. Murray, Dr B.L. Hart, Dr L.A. Hart. Centre, left to right: Professor S.J. McNaughton, Miss E. Wanjohi, Dr J.H. Ginsberg. Bottom, left to right: Professor A.P.F. Flint (Director of Science at the Zoological Society of London), Dr J.K. Hodges, Dr C.J. Moss.

Above, left to right: Dr R.D. Taylor, Professor J.N. Maina, Dr R.M. Laws, FRS. Centre, left to right: Dr A.R. Njogu, Dr L.G. Frank, Mr D.M. Jones. Bottom: speakers and audience in conversation during an interval.

Symp. zool. Soc. Lond. (1989) No. 61: 1–13

# The response of tropical vegetation to grazing and browsing in Queen Elizabeth National Park, Uganda

Eric L. EDROMA

*Uganda Institute of Ecology*
*P.O. Box 3530*
*Kampala, Uganda*

## Synopsis

The Queen Elizabeth National Park in south-western Uganda, which supported the largest mammalian biomass in the world up to the mid 1970s, has today 10–15% of its previous populations of large game. The effects of the changing numbers on the vegetation and the resulting response of two types of grassland, bushy thickets and *Acacia* woodland, are presented. When their numbers are too high, hippopotamus degrade their habitat to mosaic short grassland. However, moderate grazing and fire in adequate frequency and intensity clear dead smothering cover and reduce competition, and therefore promote tillering, cover, overall production of dry matter and increase in species richness. Species repeatedly and selectively grazed become severely suppressed and are eventually killed. Species of lower nutritive value replace them.

Before 1973, high elephant densities plus annual fires had suppressed regeneration of woody species, and these factors maintained an open grassland. But the catastrophic decline in the numbers of large mammals during the past 16 years of political unrest in the country has resulted in widespread regeneration of woody species, thus reconverting what is believed to be derived grassland into woodland. The regeneration particularly of *Acacia* species is largely by vegetative growth. Despite profuse flowering and seed production, only up to 3% of seeds of *Acacia sieberiana* that germinate are able to develop into mature trees. Data are given to account for the poor recruitment.

## Introduction

The Queen Elizabeth National Park (QENP), with an area of 1978 km$^2$, stands within the western arm of the Great East African Rift Valley astride the equator in south-western Uganda. The Park enjoys a bimodal rainfall pattern of climate with two wet seasons (March–May, September–November) and two dry seasons (June–August, December–February). It receives between 700 and 1500 mm of rainfall a year and experiences moderate temperatures (16°–18.7°C minimum and 26.8–29.0°C maximum). The vegetation of the Park is diverse, ranging from aquatic communities through varied grassland types interspersed with thickets and dry woodlands to remnants of dense tropical rain forest. Descriptions of the communities were given in general by Langdale-Brown (1960) and Edroma (1975) and in greater detail by Lock (1977).

The variety of vegetation types provides important habitat for a diversity of animals comprising at least 66 species of mammals and 546 species of birds and a wide range of insects and microfauna. In terms of numbers and biomass the grasslands of the QENP had until the mid 1970s supported the highest mammalian biomass in the world (Bourlière 1965; Petrides & Swank 1965; Stewart 1967; Talbot & Talbot 1963; Field 1968). Both directly and indirectly the animals exert a controlling influence on the character, structure, distribution and performance of the vegetation types (Lock 1967, 1972; Laws 1968; Strugnell & Pigott 1978; Edroma 1975; Sabiiti 1986).

The different animal species affect the individual plants and the vegetation stands in the Park in different ways. The feeding behaviour of the hippopotamus *Hippopotamus amphibius* is such that some plant species are totally uprooted and destroyed (Field 1970; Lock 1972). The elephant *Loxodonta africana*, on the other hand, debarks trees which eventually die through drying up and fire (Buechner & Dawkins 1961; Laws, Parker & Johnstone 1975; Wyatt & Eltringham 1974). However, the majority of the grazers and browsers inflict damage on the plants through removing parts when feeding, without killing the whole plant. The performance of the vegetation is further influenced by various other types of animal activities, for example, trampling, rubbing, defaecation and soil compaction. The impact of these activities on the plants may be modified further by other equally important environmental factors (e.g. fire, climatic variations, topographic irregularities, changes in numbers of animals, human influences, etc.). This paper reviews the responses of the major vegetation types (*Sporobolus pyramidalis* grassland, *Hyparrhenia filipendula* and *Themeda triandra* grassland, *Capparis tomentosa* and *Securinega virosa* thickets and *Acacia sieberiana* woodlands) to changing grazing and browsing pressures in the QENP.

## Disruptive forces in grazing and browsing ecosystems

The vegetation communities of savanna ecosystems are typically subject to many disruptive forces. In Murchison Falls National Park, Uganda, several researchers, including Buechner & Dawkins (1961) and Laws *et al.* (1975), described how high elephant densities, combined with fire, progressively converted the woodlands of the Park into grassland. During the late 1950s too many hippopotamus caused overgrazing, soil erosion, increased rainfall run-off, deterioration of the habitat, creation of mosaic short grassland (Lock 1972) and reduction of the carrying capacity of the QENP. A management decision to reduce a significant component of the biomass of large mammals through a programme of hippopotamus cropping in QENP resulted in significant increases of the density and biomass of other large mammals, of 54% and 20% respectively (Eltringham 1974), of diversity, from a factor of 1.26 to 1.78, and in production of dry matter and cover (Thornton 1971). This management practice achieved the desired goal of establishing an ecological equilibrium between the vegetation and the herbivorous animal populations. This equilibrium was later described by Laws (1968) as not only delicate but temporary, prevailing only as long as the populations of the elephant, buffalo (*Syncerus caffer*) and hippopotamus remained at around 2800, 20 000 and 12 000 respectively. Similar relationships were observed elsewhere by Croze (1974), Agnew (1968), Corfield (1973) and Western & Van Praet (1973) for the Serengeti, Tsavo and Amboseli National Parks respectively.

## Grazing

The influence of large mammals on their habitat is complex. The response of an individual plant to herbivory inevitably depends on a number of factors including (1) the plant parts removed, (2) the position above ground of the apical primordia, (3) the intensity (amount), frequency (how often) and season of use, (4) the developmental growth of the plant, (5) the genetic make-up of the species, (6) the history of the plant (particularly the time since offtake by herbivores or fire last occurred), and (7) the modifying effects of environmental factors such as rainfall, drought, type of soil, number of other animals, human influences, diseases, and so on.

The total plant response to grazing as reported here was measured by monthly harvesting of the above-ground herbage in 1 m² quadrats randomly laid and replicated five times from grazed and ungrazed *Hyparrhenia/Themeda* grassland type. The increase (Table 1) and biomass of shoots closely reflected the rainfall pattern (Table 2) and was at a much higher rate in the ungrazed than in the grazed swards. Leaf production was proportionately higher but declined as flowering intensified six weeks after the fire. The total biomass a year from burning was twice as much in the ungrazed as in the grazed plots, emphasizing the concentration of grazers on the fresh regrowth. Throughout the one-year period, the apparent rate of production in the grazed area continued to be half that in the protected area. The difference was assumed to be a result of grazing which removed half of the production on the grazed plot. The differences in dry weight of the crop

Eric L. Edroma

**Table 1.** Changes in above-ground dry matter ($g/m^2$) in *Hyparrhenia/Themeda* grassland in 1972/73.

| Month | Grazed | Ungrazed |
|-------|--------|----------|
| March | 10.8 | 21.6 |
| April | 32.1 | 67.1 |
| May | 60.9 | 125.4 |
| June | 118.2 | 208.3 |
| July | 120.7 | 240.7 |
| August | 130.4 | 243.8 |
| September | 172.6 | 288.6 |
| October | 205.3 | 411.8 |
| November | 241.8 | 438.0 |
| December | 266.3 | 520.3 |
| January | 270.8 | 524.1 |
| February | 281.4 | 518.6 |

Whole area burnt 15 February 1972.

**Table 2.** Rainfall (mm) for the study area in *Hyparrhenia/Themeda* grassland from March 1972 to February 1973 and mean (mm) for QENP during 1964–73.

| Month | M | A | M | J | J | A | S | O | N | D | J | F |
|-------|---|---|---|---|---|---|---|---|---|---|---|---|
| Rainfall, study area | 101 | 124 | 82 | 49 | 36 | 67 | 74 | 102 | 116 | 62 | 37 | 56 |
| Mean rainfall, QENP | 92 | 120 | 87 | 44 | 32 | 71 | 83 | 108 | 104 | 57 | 35 | 46 |

harvested from the grazed and ungrazed plots were wide during the wet seasons, demonstrating the importance of the amount and reliability of rainfall in determining plant response to grazing.

In separate studies designed to simulate grazing, the first two to three clippings at variable heights above 5 cm and at intervals greater than 2 weeks stimulated tillering and dry matter production (Edroma 1985). Subsequent repeated clippings at all heights and frequencies caused continuous decline in yield. Production in grazed plots maintained a decreasing trend virtually until the plants died.

Plants receiving selective clipping (simulating selective grazing) produced more and lived longer under simulated grazing than in protected conditions (Edroma 1981a). The response to the treatment was most pronounced in the tall species such as *Themeda triandra*, *Cymbopogon afronardus* and *Hyparrhenia filipendula*, but least in *Sporobolus pyramidalis* and *Imperata cylindrica*. As the individual vigour and numbers of the dominant tall species were reduced by selective clipping they were gradually suppressed and eventually replaced by short grasses or species lower in the successional hierarchy. The cumulative effects of repeated selective clipping (grazing) resulted in replacement of the plants receiving the treatment by weedy annuals and certain perennials such as *Sporobolus stapfianus, S. festivus, Aristida adoensis* and *Microchloa kunthii.* These are species of lower nutritive value, recorded elsewhere by Rattray (1954) and Heady (1966) as species of overgrazed or disturbed areas.

The maintenance of one species below the general vegetation cover of the sward by selective clipping significantly reduced the ability of the clipped plants to yield and live longer. Selective clipping of the dominant species reduced competition for environmental resources in the swards, and the unclipped associate plants were given a chance to develop and increase their leaf area indices. The plants repeatedly clipped or grazed (*Themeda triandra, Hyparrhenia filipendula, Brachiaria platynota*) progressively lost a greater portion of their photosynthetic organs, became shaded by the associated plants, received reduced light intensity, and experienced increased competitive stress, loss of vigour and productivity potential. Furthermore, they decreased in number and basal area and became increasingly unable to resist environmental stress, and consequently died earlier than those plants in the quadrats receiving total clipping.

Clipping at 5 cm height simulated close grazing and created conditions favourable for low-growing species to become established. Smith (1940) has shown that continuous overgrazing depletes plant carbohydrate reserves, eventually killing the affected species. The areas along the shores of the Kazinga Channel must have evolved from the rich tall grassland reported by Lugard (1896) to the present degraded mosaic habitat through a similar process. Additionally, erosion has removed the rich topsoil (Bishop 1962). This negative response of the tall nutritious and palatable species to grazing had resulted in a reduced number of ungulates and a lowering of the carrying capacity of the mosaic grassland until a programme of cropping hippopotamus was carried out in the mid-1960s.

In contrast, all species in the swards receiving total clipping continued to live and produce in both grazed and protected areas. By keeping the canopy short this treatment reduced competition between the plants so that more species and plants were able to co-exist at the same place. In the absence of grazing, tall grasses overgrew and shaded out shorter species to the point of exclusion. Normally the tall grasslands get burnt once every one to three years (Eltringham 1976). This allows the short species to maintain vigour. The fires clear the dead and nutritionally poor smothering cover and therefore allow those short species otherwise intolerant of shade to persist in small quantities. Also, by removing herbaceous cover, grass competition and plant litter, the trampling of grazing animals exposes the minerally-rich seed-beds, leading to germination and establishment of dense forbs and shrubs. Grazing and fire, in adequate frequency and intensity, play important roles in maintaining the species diversity of the grasslands (Edroma 1981b). In the absence of grazing and fire, the grasses gradually become replaced by shrubs. In all the grassland types studied, grass production gradually declined with increasing numbers of harvests, and with decreasing distance to the shores of Kazinga Channel or to the lakes where the concentration of hippopotamus activity was heaviest (Laws 1968; Lock 1972). Similar gradients of grazing intensity were attributed to hippopotamus by Field & Laws (1970) and Lock (1972) in the QENP, and by Olivier & Laurie (1974) in Serengeti National Park.

By removing the over-mature standing crop and litter (Hadley 1970; Lemon 1968) grazing (like fire) stimulates sprouting primary production, and it maintains the grassland in a sub-mature productive successional phase of growth and palatability (Hulbert 1969; Vesey-Fitzgerald 1974; Edroma 1975). Grazing has also been said to act as the main pathway for recycling mature herbage, but the effect of this on the structure and chemistry of soils is not fully known in the QENP. However, in Murchison Falls National Park, Hatton & Smart (1984) reported significant increase in the soil nutrient status following the exclusion of large mammals for 20 years.

There is a consensus of opinion that moderate densities of grazing animals exert beneficial influences on the vegetation. McNaughton (1976) showed an immediate increase in tall growing grass in Serengeti when the grazing pressure was relaxed. Optimal defoliation will tend to stimulate compensatory growth, delay tiller senescence, redirect substrates from inflorescences to leaf and shoot, maximize net assimilation rate and increase the rate of nutrient cycling. In QENP rainfall is high and the soil is rich with considerable reserve of weatherable materials (Harrop 1960); so there exists the high biomass of grass species (Bell 1982). On the other hand, large concentrations of herbivores often exert harmful effects on the plants because of selectivity and overgrazing. The grazing habits of some animals encourage certain grasses and eliminate others. Different intensities of grazing have definite effects on plant succession and on the local distribution of species. Grazing intensity can affect production and botanical composition by eliminating grasses if grazed too frequently or at susceptible stages in their seasonal growth. Heavily defoliated plants become unable to produce healthy roots and ultimately deteriorate and die (Edroma 1981a).

The damaging effects of defoliation and the theories to account for the responses of plants to defoliation have been reviewed by Alcock (1964) and Milthorpe & Davidson (1965). Defoliation reduced dry weights of roots of *Brachiaria platynota* and *Themeda triandra* in laboratory experiments

(Edroma 1985). Severe defoliation has been known to reduce the rate of root respiration (Sukurai 1960), root weight, length, diameter and sometimes specific gravity (Bhaskaran & Chokvaborty, 1965), photosynthetic activity of the shoot (Richardson 1953), reserve assimilates, survivorship of tiller primordia and of roots, and plant performance.

Other grasses like *S. pyramidalis* which are resistant to the various direct and indirect influences of grazing animals come to dominance. Mott, Bridge & Arnt (1979) demonstrated reduced productivity due to heavy grazing in Australia. Harrington (1974) recorded a switch from dominance by tall *Themeda* and *Hyparrhenia* grasses to stoloniferous *Brachiaria* species under a regime of heavy grazing in western Uganda.

## Browsing

Over the past 16 years of political upheaval in Uganda the QENP has experienced a massive reduction in the numbers of large mammals (Eltringham & Malpas 1980) resulting in dramatic changes in the vegetation. Profuse regeneration of woody species has become widespread in all the Uganda National Parks (Lock 1985; Sabiiti 1986), reconverting what is believed to be derived grassland into woodland. Although the capacity of savanna ecosystems to absorb change and disturbance is said to be great, making them highly resilient (Holling 1973; Noy-Meir 1975), the QENP is still adjusting itself to the effects of the massive perturbation.

Profuse *Acacia* regeneration in QENP is one of the dramatic responses of woody species to a reduction in browsing by large game. *Acacia* trees are an important component of the vegetation in national parks and game reserves of East and Central Africa. In Serengeti 60% of the woodlands are *Acacia* (Pellew 1983). Nearly 80% of all the trees in QENP outside Maramagambo Forest are *Acacia* (Sabiiti 1986). *Acacia* trees provide important browse for wildlife especially in dry seasons (Field 1971; Field & Ross 1976; Taylor & Walker 1978). Owing to a continuing decline in large animal populations (Table 3), *Acacia* regeneration is changing large areas of QENP into woody green mantle.

Before 1972 grazing and browsing plus regular fires suppressed regeneration of *Acacia* and other woody species, thus maintaining an open grassland habitat (Buechner & Dawkins 1961; Laws 1970; Field 1971; Spence & Angus 1971; Harrington & Ross 1974; Lock 1977). Elephants browse and uproot *Acacia* seedlings and saplings (Field & Ross 1976; Ross, Field & Harrington 1976; Barnes 1983; Pellew 1983). Death of most mature *Acacia* trees is caused by elephant damage (Glover 1963; Field 1971; Croze 1974; Harrington & Ross 1974; Laws *et al.* 1975; Caughley 1976; Barnes 1983; Pellew 1983; Norton-Griffiths 1979; Cumming 1982). In Kidepo Valley National Park, Field & Ross (1976) showed that 23% of *A. gerrardii* was

**Table 3.** Changes in numbers of major mammalian species in the QENP

| Year | Elephant | Buffalo | Hippopotamus | Antelope |
|------|----------|---------|--------------|----------|
| 1971 | 2847 | 20 000 | 15 000 | 30 000 |
| 1972 | 2809 | | | |
| 1973 | 2400 | | | |
| 1974 | 1700 | | | |
| 1975 | 1048 | | | |
| 1976 | 807 | 15 400 | 11 500 | |
| 1977 | 724 | | | |
| 1978 | 420 | | | |
| 1979 | | | | |
| 1980 | 152 | 8 000 | 4 500 | 10 000 |
| 1981 | 206 | | | |
| 1982 | 426 | | | |
| 1983 | 500 | 9 250 | 6 000 | 14 300 |
| 1984 | 630 | | | |
| 1985 | 710 | | | |
| 1986 | 760 | 10 500 | 7 000 | 16 800 |
| 1987 | 225 | 4 359 | 3 115 | |

killed by elephant in 3 years. Elephants push over trees, and remove bark, thus exposing the wood to attack by fungi and boring insects.

In recent years the vegetation of the QENP has changed because pressure from browsing and grazing has been lower. Extensive areas have become dominated by densely regenerating saplings of *A. sieberiana*, *A. gerrardii*, *A. hockii*, *Capparis tomentosa*, *Securinega virosa* and *Euphorbia* trees. Several other terrestrial and shoreline species which had been suppressed by large mammals before 1972 are also regenerating to significant proportions. *Acacia sieberiana* generates by seed and by vegetative propagation. It flowers profusely so as to offset flower destruction by a wide range of animals including baboons, insects and birds. Nearly 84% of the flowers the author marked produced pods, 78% of which developed seeds. Half (50%) of the seeds became bored by bruchid beetles and died; the remainder were viable. Some seeds were ingested by animals, while others entered the soil seed bank. Seeds in QENP required a minimum of 6 months' dormancy before germination (Sabiiti 1986). In Tanzania *Acacia* seeds required fire or digestion by an animal (Lamprey, Halvey & Makacha 1974) to break dormancy. In QENP nearly 90% of the seeds the author collected germinated under laboratory conditions, but only a 66% germination success was achieved under field conditions. It is assumed that some of the seeds were burnt to ashes by the fierce hot fires resulting from increased loads of ungrazed herbage. Not only do digestive fluids break seed dormancy, but they also kill bruchid larvae. The catastrophic decline of large herbivorous animals has necessarily deprived *Acacia* seeds of those advantages, and has resulted in the accumulation of fuelwood that has, in turn, led to hot fires that destroy not only seeds but also seedlings.

The author recorded high rates of mortality in seedlings of *A. sieberiana* in the *Hyparrhenia/Themeda* grassland type in QENP. Some 60–70% died by the first year, 90% by the second year, and 97% by the end of the third year. Only 2–3% reached maturity. The mortality was attributed to damping off, intra- and inter-specific competition, pathogens, rodents, trampling by large animals and fire. A growth rate of nearly 1 m per year was recorded in the seedlings under field conditions. This rapid growth is interpreted as an adaptation to avoid destruction by fire. Mortality through fire was particularly high during the first three years after germination. By the fourth year, nearly all the seedlings were safe from destructive fires and they survived.

The success of regeneration by vegetative growth depends on the interaction between the influences of animals and fire. Table 4 demonstrates the growth in height of 100 marked *A. sieberiana* trees (25 in each of grazed burnt, grazed unburnt, ungrazed burnt and ungrazed unburnt plots) in the southern sector (Ishasha) of QENP during the period 1973–1987. The effect of browsing was pronounced up to 1975, when populations of the large ungulates were high and the impact of grazing was significant. Fire further decreased tree heights in the browsed plots. From 1976 onwards fire became the dominant factor suppressing regeneration. Decreasing grazing and browsing pressure in the 1970s permitted a greater production of grass litter which fuelled severe fires. These in turn stimulated greater germination and a subsequent increase in the density of *Acacia* woodland.

**Table 4.** Changes in height (inches) of *Acacia sieberiana* in relation to the presence of grazing animals and the occurrence of fire (+, present or occurring; −, absent).

| Year | Mean height (in) of 25 trees in plot | | | |
| | − Animals − Fire | − Animals + Fire | + Animals − Fire | + Animals + Fire |
|---|---|---|---|---|
| 1973 | 126 | 121 | 131 | 128 |
| 1974 | 158 | 107 | 134 | 94 |
| 1975 | 230 | 113 | 145 | 107 |
| 1976 | 312 | 127 | 150 | 116 |
| 1977 | 350 | 146 | 179 | 127 |
| 1978 | 367 | 152 | 205 | 132 |
| 1979 | | | | |
| 1980 | | | | |
| 1981 | 402 | 236 | 362 | 159 |
| 1982 | 410 | 245 | 269 | 166 |
| 1983 | 415 | 266 | 289 | 163 |
| 1984 | 423 | 282 | 303 | 148 |
| 1985 | 426 | 295 | 307 | 157 |
| 1986 | 426 | 306 | 315 | 161 |
| 1987 | 437 | 319 | 330 | 171 |

## Acknowledgements

Thanks to Drs Robert C. D. Olivier and Mike J. Lock for useful comments on an earlier draft of the paper, Mr Yasent Kariba for assistance during field studies and to Mrs Ben Kawesi for typing the paper.

## References

Agnew, A.D.Q. (1968). Observations on the changing vegetation of Tsavo National Park (East). *E. Afr. Wildl. J.* **6**:75–80.

Alcock, M.B. (1964). The physiological significance of defoliation on the subsequent regrowth of grass–clover mixtures and cereals. In *Grazing in terrestrial and marine environments*: 25–41. (Ed. Crisp. D.J.). Blackwell Scientific Publications, Oxford. (*Symp. Br. ecol. Soc.* No. 4.)

Barnes, R.F.W. (1983). Effects of elephant browsing on woodlands in a Tanzanian national park: measurements, models and management. *J. appl. Ecol.* **20**:521–39.

Bell, R.H.V. (1982). The effect of soil nutrient availability on community structure in African ecosystems. In *Ecology of tropical savannas*: 193–216. (Eds Huntley, B.J. & Walker, B.H.). Springer Verlag, Berlin. (*Ecol. Stud. Anal. Synth.* **42**.)

Bhaskaran, A.R. & Chokvaborty, D.C.A. (1965). A preliminary study on the variation in the soil binding capacity of some grass roots. *Indian J. Agron.* **10**:326–30.

Bishop, W.W. (1962). Gully erosion in Queen Elizabeth National Park. *Uganda J.* **26**:161–5.

Bourlière, F. (1965). Densities and biomasses of some ungulate populations in Eastern Congo and Rwanda, with notes on population structure and lion/ungulate ratios. *Zool. afr.* **1**:199–207.

Buechner, H.K. & Dawkins, H.C. (1961). Vegetation change induced by elephants and fire in Murchison Falls National Park, Uganda. *Ecology* **42**:752–66.

Caughley, G. (1976). The elephant problem—an alternative hypothesis. *E. Afr. Wildl. J.* **14**:265–83.

Corfield, T.F. (1973). Elephant mortality in Tsavo National Park, Kenya. *E. Afr. Wildl. J.* **11**:339–68.

Croze, H. (1974). The Seronera bull problem. II. The trees. *E. Afr. Wildl. J.* **12**:29–47.

Cumming, D.H.M. (1982). The influence of large herbivores on savanna structure in Africa. In *Ecology of tropical savannas*: 231–45. (Eds Huntley, B.J. & Walker, B.H.). Springer-Verlag, Berlin. (*Ecol. Stud. Anal. Synth.* **42**.)

Edroma, E.L. (1975). *Influences of burning and grazing on the productivity and dynamics of grasslands in Rwenzori National Park, Uganda*. Unpubl. Ph.D. thesis: University of Giessen.

Edroma, E.L. (1981a). Some effects of grazing on the productivity of grassland in Rwenzori National Park, Uganda. *Afr. J. Ecol.* **19**:313–26.

Edroma, E.L. (1981b). The role of grazing in maintaining high species-composition in *Imperata* grassland in Rwenzori National Park, Uganda. *Afr. J. Ecol.* **19**:215–33.

Edroma, E.L. (1985). Effects of clipping on *Themeda triandra* and *Brachiaria platynota* in Queen Elizabeth National Park, Uganda. *Afr. J.Ecol.* 23:45–51.

Eltringham, S.K. (1974). Changes in the large mammal community of Mweya Peninsula, Rwenzori National Park, following removal of hippopotamus. *J. appl. Ecol.* 11:855–65.

Eltringham, S.K. (1976). The frequency and extent of uncontrolled grass fires in the Rwenzori National Park, Uganda. *E. Afr. Wildl. J.* 14:215–22.

Eltringham, S.K. & Malpas, R.C. (1980). The decline in elephant numbers in Rwenzori and Kabalega Falls National Parks, Uganda. *Afr. J. Ecol.* 18:73–86.

Field, C.R. (1968). *The food habits of some ungulates in Uganda.* Unpubl. Ph.D. thesis: University of Cambridge.

Field, C.R. (1970). A study of the feeding habits of the hippopotamus (*Hippopotamus amphibius* Linn.) in the Queen Elizabeth National Park, Uganda, with some management implications. *Zool. afr.* 5:71–86.

Field, C.R. (1971). Elephant ecology in the Queen Elizabeth National Park, Uganda. *E. Afr. Wildl. J.* 9:99–123.

Field, C.R. & Laws, R.M. (1970). The distribution of the larger herbivores in the Queen Elizabeth National Park, Uganda. *J. appl. Ecol.* 7:273–94.

Field, C.R. & Ross, I.C. (1976). The savanna ecology of Kidepo Valley National Park. II. Feeding ecology of elephant and giraffe. *E. Afr. Wildl. J.* 14:1–15.

Glover, J. (1963). The elephant problem at Tsavo. *E. Afr. Wildl. J.* 1:30–9.

Hadley, E.B. (1970). Net productivity and burning responses of native eastern North Dakota prairie communities. *Am. Midl. Nat.* 84:121–35.

Harrington, G.N. (1974). Fire effects on a Ugandan savanna grassland. *Trop. Grassld* 8:87–105.

Harrington, G.N. & Ross, I.C. (1974). The savanna ecology of Kidepo Valley National Park. I. The effects of burning and browsing on the vegetation. *E. Afr. Wildl. J.* 12:93–105.

Harrop, J.F. (1960). The soils of the Western Province of Uganda. *Mem. Res. Div. Dep. Agric. Uganda* (1) No. 6:1–106.

Hatton, J.C. & Smart, N.O.E. (1984). The effect of long-term exclusion of large herbivores on soil nutrient status in Murchison Falls National Park, Uganda. *Afr. J. Ecol.* 22:23–30.

Heady, H.F. (1966). Influence of grazing on the composition of *Themeda triandra* grassland, East Africa. *J. Ecol.* 54:705–27.

Holling, C.S. (1973). Resilience and stability of ecological systems. *A. Rev. Ecol. Syst.* 4:1–23.

Hulbert, L.C. (1969). Fire and litter effects in undisturbed bluestem prairie in Kansas. *Ecology* 50:874–7.

Lamprey, H.F., Halevy, G. & Makacha, S. (1974). Interactions between *Acacia*, bruchid seed beetles and large herbivores. *E. Afr. Wildl. J.* 12:81–5.

Langdale-Brown, I. (1960). The vegetation of the Western Province of Uganda. *Mem. Res. Div. Dep. Agric. Uganda* (2) No.4:1–111.

Laws, R.M. (1968). Interactions between elephant and hippopotamus populations and their environments. *E. Afr. agric. For. J.* 33 (Spec. issue): 140–7.

Laws, R.M. (1970). Elephants as agents of habitat and landscape change in East Africa. *Oikos* 21:1–15.

Laws, R.M., Parker, I.S.C. & Johnstone, R.C.B. (1975). *Elephants and their habitats: the ecology of elephants in North Bunyoro, Uganda.* Clarendon, Oxford.

Lemon, P.C. (1968). Effects of fire on an African plateau grassland. *Ecology* **49**:316–22.

Lock, J.M. (1967). *The vegetation in relation to grazing and soils in Queen Elizabeth National Park, Uganda.* Unpubl. Ph.D. thesis: University of Cambridge.

Lock, J.M. (1972). The effects of hippopotamus grazing on grasslands. *J. Ecol.* **60**:445–67.

Lock, J.M. (1977). Preliminary results from fire and elephant exclusion plots in Kabalega Falls National Park, Uganda. *E. Afr. Wildl. J.* **15**:229–32.

Lock, J.M. (1985). Recent changes in the vegetation of Queen Elizabeth National Park, Uganda. *Afr. J. Ecol.* **23**:63–5.

Lugard, F.D. (1896). *The rise of our East African Empire.* Frank Cass & Co. Ltd., London.

McNaughton, S.J. (1976). Serengeti migratory wildebeest: facilitation of energy flow by grazing. *Science, N.Y.* **193**:92–4.

Milthorpe, F.L. & Davidson, J.L. (1965). Physiological aspects of regrowth following defoliation. In *The growth of cereals and grasses*: 241–55. (Eds Milthorpe, F.L. & Ivins, J.D.). Butterworths, London.

Mott, J.J., Bridge, B.J. & Arndt, W. (1979). Soil seals in tropical tallgrass pastures of Northern Australia. *Aust. J. Soil Res.* **30**:483–94.

Norton-Griffiths, M. (1979). The influence of grazing, browsing, and fire on the vegetation dynamics of the Serengeti. In *Serengeti: dynamics of an ecosystem*: 310–52. (Eds Sinclair, A.R.E. & Norton-Griffiths, M.). Univ. Chicago Press, Chicago and London.

Noy-Meir, I. (1975). Stability of grazing systems: an application of predator-prey graphs. *J. Ecol.* **63**:459–81.

Olivier, R.C.D. & Laurie, W.A. (1974). Habitat utilization by hippopotamus in the Mara River. *E. Afr. Wildl. J.* **12**:249–71.

Pellew, R.A.P. (1983). The impacts of elephant, giraffe and fire upon the *Acacia tortilis* woodlands of the Serengeti. *Afr. J. Ecol.* **21**:41–74.

Petrides, G.A. & Swank, W.G. (1965). Population densities and the range carrying capacity for large mammals in Queen Elizabeth National Park, Uganda. *Zool. afr.* **1**:209–25.

Rattray, J.M. (1954). Some plant indicators in southern Rhodesia. *Rhodesia agric. J.* **51**:176–86.

Richardson, S.D. (1953). Studies of root growth in *Acer saccherinum.* I. The relation between root growth and photosynthesis. *Proc. K. ned. Akad. Wet.* (C) **56**:185–93.

Ross, I.C., Field, C.R. & Harrington, G.N. (1976). The savanna ecology of Kidepo Valley National Park, Uganda. III. Animal populations and park management recommendations. *E. Afr. Wildl. J.* **14**:35–48.

Sabiiti, E.N. (1986). *Fire effects on* Acacia *regeneration.* Unpubl. Ph.D. thesis: Univ. New Brunswick.

Smith, C.C. (1940). The effect of overgrazing and erosion upon the biota of the mixed grass prairie of Oklahoma. *Ecology* **21**:381–97.

Spence, D.H.N. & Angus, A. (1971). African grassland management: burning and grazing in Murchison Falls National Park, Uganda. In *Scientific management of animal and plant communities for conservation*: 319–31. (Eds Duffey, E. & Wyatt, A.S.). Blackwell Scientific Publications, Oxford. (*Symp. Br. ecol. Soc.* No. 11.)

Stewart, D.R.M. (1967). Analysis of plant epidermis in faeces: A technique for studying the food preferences of grazing herbivores. *J. appl. Ecol.* 4:83–111.

Strugnell, R.G. & Pigott, C.D. (1978). Biomass, shoot-production and grazing of two grasslands in the Rwenzori National Park, Uganda. *J. Ecol.* **66**:73–96.

Sukurai, M. (1960). Physiological and ecological studies on the mechanisms of regrowth of grasses. I. Changes in respiration of root of orchard grass, Italian ryegrass and ladino clover by cutting. *Proc. Crop Sci. Soc. Japan* **28**:311–2.

Talbot, L.M. & Talbot, M.H. (1963). The high biomass of wild ungulates on East African savanna. *Trans N. Am. Wildl. Conf.* **28**:465–76.

Taylor, R.D. & Walker, B.H. (1978). Comparisons of vegetation use and herbivore biomass on a Rhodesian game and cattle ranch. *J. appl. Ecol.* **15**:565–81.

Thornton, D.D. (1971). The effect of complete removal of hippopotamus on grassland in the Queen Elizabeth National Park, Uganda. *E. Afr. Wildl. J.* **9**:47–55.

Vesey-Fitzgerald, D.F. (1974). Utilization of the grazing resources by buffaloes in the Arusha National Park, Tanzania. *E.Afr. Wildl. J.* **12**:107–34.

Western, D. & Van Praet, P.C. (1973). Cyclical changes in the habitat and climate of an East African ecosystem. *Nature, Lond.* **241**:104–6.

Wyatt, J.R. & Eltringham, S.K. (1974). The daily activity of the elephant in the Rwenzori National Park, Uganda. *E. Afr. Wildl. J.* **12**:273–89.

Symp. zool. Soc. Lond. (1989) No. 61: 15–29

# Interactions of plants of the field layer with large herbivores

S.J. McNAUGHTON

*Biological Research Laboratories
Syracuse University
Syracuse, NY 13244–1220, USA*

## Synopsis

Grasslands with large grazing mammals as dominant members of the fauna were one of Earth's major ecosystem types just a century and a half ago, but most disappeared in the last half of the 19th century. Africa is the continent that to the greatest extent preserves significant remnants of these once widespread ecosystems. Large grazing ungulates are related to their environments along complicated temporal scales, ranging from seconds to years, and spatial scales, ranging from a single bite of food to a vast geographic region. A hierarchical approach to those scales indicates that animals encounter substantial forage mineral heterogeneity at all levels, with nutritional heterogeneity varying from the level of bites to the level of landscape regions. Thus, rather than being a homogeneous food source, grass swards are highly variable in both space and time. Evolution of large mammals, therefore, took place amidst a background of substantial nutritional heterogeneity.

## Introduction

Grass is the forgiveness of nature—her constant benediction. Fields trampled with battle, saturated with blood, torn with the ruts of the cannon, grow green again with grass, and carnage is forgotten. Forests decay, harvests perish, flowers vanish, but grass is immortal.

John J. Ingalls (1904), *On Bluegrass*

Grasslands are prosaic places. They are seemingly simple of structure, stretching featureless to the horizon, boring even, so that explorers, trappers, and settlers on the North American Great Plains in the 18th and 19th centuries greeted topographic anomalies, a bluff, a butte, or any prominence, with pleasure and fanciful names (Morison 1965). On the flat or gently rolling landscapes that grasslands typically occupy, riverine lowlands, escarpments and rocky outcrops where tree seedlings can escape the prevalent fires of grasslands are the only habitats with trees in a seemingly endless expanse of grass (Wells 1965). During grassland

ZOOLOGICAL SYMPOSIUM No. 61
ISBN 0–19–854009–4

colonization, these natural firebreaks became favoured locations for camps and then homes, protecting them from the ravages of recurrent grass fires (Sauer 1952).

Geologically, grasslands are a comparatively recent phenomenon, the earliest grass fossils being from Eocene deposits in Patagonia (Frenguelli 1930; Teruggi 1955). However, geological evidence of grasses does not become abundant until the Miocene, by which time the family was fully differentiated, suggesting significant prior evolution in rare, local habitats not well-represented in fossil strata (Stebbins 1981). Appearing simultaneously with the Poaceae were mammals with dentition indicating that they fed upon abrasive plant tissues, and documentation of the evolution of hypsodonty (Kowalevsky 1873–4) was one of the earliest and most thorough bodies of evidence supporting Darwinism. Orogeny, creating rainshadows, and increasing global aridity were associated with greatly increasing expanses of treeless grasslands and open woodlands. The climatic trends were accompanied by explosive adaptive radiations of the Poaceae and large, hyposodont mammals from the Miocene into the Pleistocene (Scott 1937; Webb 1977, 1978).

Grazing ecosystems with grasses dominating the field layer and Artiodactyla, Perissodactyla and Proboscidea or, in Australia, Macropodidae, dominating the fauna, occupied a third to a half of Earth's terrestrial surface as recently as a century and a half ago (Barnard & Frankel 1964; Williams, Allred, DeNio & Paulsen 1968). Because primary production had outstripped decomposition for millennia, grassland soils became highly fertile repositories of organic matter (Jenny 1930). But available tillage techniques made these heavy soils resistant to agriculture until perfection of the all-steel plough in the 1830s (Higgs, Fussell, Nair, Rasmussen, Gray & Ordish 1978). Within six decades, most of Earth's native grazing ecosystems were engulfed by agricultural grain belts.

The subject of this symposium paper is the interactions of plants of the field layer with mammals in those remnants of grazing ecosystems that are still to be found in Africa, where a semblance of the great herds has survived. Its principal theme is that the seemingly monotonous, uniform physiognomy of grasslands cloaks a diverse, highly dynamic, spatially and temporally heterogeneous complex of ecological relationships. Those relationships belie the view that grasses constitute a uniform food source exploited in a simple, uniform fashion by mammals with large, unselective mouths.

## Grazing: a landscape perpective

Accompanying the evolution of hypsodonty was the evolution of cursorial limb and body forms and of stomachs or hindguts modified for fermenting cellulose. Limbs and feet, then hooves, were progressively modified to allow animals to travel long distances with increasing energetic efficiency and reduced damage to locomotory organs contacting terrestrial surfaces that were often hard and destructive. These trends indicate that increasing aridity was associated with poorer-quality, more cellulosic and progressively less predictable food sources, with primary production and, therefore, food availability often highly stochastic in both space and time (McNaughton 1979, 1983, 1985).

Large, grazing ungulates, therefore, are related to their environments along complicated spatial and temporal scales (Laca & Demment 1986; Senft, Coughenour, Bailey, Rittenhouse, Sala & Swift 1987; Demment & Greenwood 1988). Such a herbivore consumes perhaps $10^7$ bites per year (Chacon, Stobbs & Saldland 1976) and devotes much more of its time-budget to feeding and digestion than to any other activities (Leuthold 1977). Some species occupy comparatively circumscribed, well-defined home ranges; others, including those that commonly dominate the biomass in arid to subhumid climates (Bell 1982), are often nomadic, moving annually over huge geographic regions (Fryxell, Greever & Sinclair 1988). Resident animals are generally confined to localities with annual rainfall regimes providing periodic showers throughout the year, permanent water supplies (McNaughton & Georgiadis 1986) and habitat heterogeneity duplicating many features of the ranges of landscapes and vegetations that highly mobile species encounter along their huge, annual migratory routes (McNaughton 1983, 1985).

On a localized, subregional scale, the movement of grazers along topographic and soil catena gradients (Vesey-Fitzgerald 1960; Bell 1970) is a means of partially stabilizing food quantity and quality on an interseasonal basis. Resident grazers commonly concentrate on upper catenas during the wet season, with a tendency for small-bodied animals to occupy drier sites than larger-bodied species. This is exactly equivalent to the wet-season concentration of migratory species in the regions of their ranges with the lowest annual rainfall and the tendency for the smaller-bodied species to occupy the driest end of the gradient available (Maddock 1979). As the dry season begins and progresses, resident animals move topographically downward and nomads move up an increasing rainfall gradient. Although food quality and quantity certainly decline as the dry season progresses (Sinclair 1977), these mobility patterns partially stabilize intake and dietary quality (McNaughton 1985).

## Grassland landscape heterogeneity: a Serengeti mineral perspective

The Serengeti ecosystem encompasses some 25 000km$^2$ in Tanzania and Kenya, defined (Grzimek & Grzimek 1960; Talbot & Talbot 1963; Watson

1967) by the movements of 1.4 million wildebeest (*Connochaetes taurinus*) and 0.25 million zebra (*Equus burchelli*) (Sinclair & Norton-Griffiths 1982) over their annual ranges. These animals, therefore, encounter the entire scale of spatial and temporal heterogeneity that ungulates encounter (Laca & Demment 1986; Senft *et al.* 1987; Demment & Greenwood 1988), from the second-to-second scale of individual bites to the annual scale of a vast geographic region (Fig. 1). An ecosystem consideration of the biology of large African mammals in their environment must, therefore, take a hierarchical approach, recognizing and integrating phenomena that operate at time scales from real to evolutionary time and spatial scales ranging from a bite to a vast region.

Studies of the qualty of forage available to, and consumed by, large ungulates in the Serengeti, and throughout Africa, have concentrated on nitrogen and, to a lesser extent, digestibility (McNaughton 1987). However, the mineral elements in tropical forages commonly are so insufficient to meet animal requirements (McDowell 1985) that full mineral supplementation has been advocated as a standard practice in the husbandry of domestic livestock in the tropics (McDowell, Conrad, Ellis & Loosli 1983). It is possible that wild ungulates may have evolved different requirements for minerals under the selective force of chronic mineral deficiency (McNaughton 1987). But there is good evidence that different ungulates differ little when genetic size-scaling is taken into account (Taylor & Murray 1987), and the African veterinary literature has frequently reported evidence of mineral deficiencies, 'forage diseases', in wild ungulates (McNaughton & Georgiadis 1986). Therefore, a hierarchically constructed, landscape-directed sampling of Serengeti forages was started in 1986 to begin integrating ungulates, grasses, and soil properties into a nutritional, ecosystem context. The content of these studies is not animal and plant nutrition *per se* but an understanding of how nutrition can influence, and be influenced by, the ecosystem processes of nutrient cycling and energy flow. This paper presents initial data from those studies relevant to the symposium theme of the biology of large African mammals in their environment, providing, perhaps, a new, mineral-based perspective on that theme.

The samples from which data were derived were collected during the wet season from the youngest blades of actively growing grasses, which should stabilize many potential variables related to developmental state and environment (Martin & Matocha 1973; Chapin 1980; Mengel & Kirkby 1982). Samples contacted only plastic containers after collection. They were sun-dried in mesh bags, transported to Syracuse University, washed in double-distilled water to remove surface soil contamination, redried, and prepared and analysed with a Leeman Labs PlasmaSpec 2.5 Inductively Coupled Plasma Spectrometer following standard protocols and quality

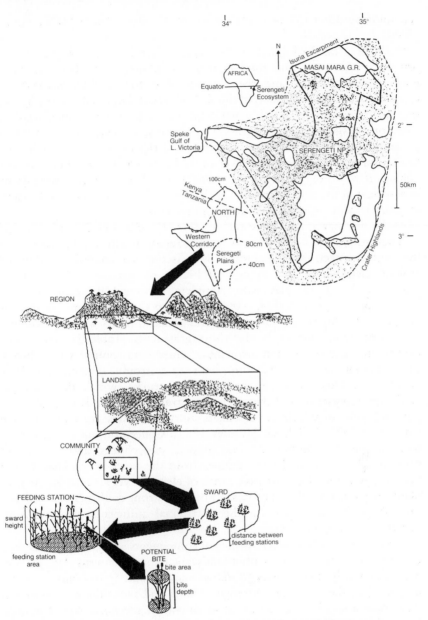

**Fig. 1.** A hierarchical approach to levels at which large ungulates can affect, and be affected by, the distribution of resources in the Serengeti ecosystem (after Laca & Demment 1986; Senft, Coughenour, Bailey, Rittenhouse, Sala & Swift 1987; Demment & Greenwood 1988). Shaded areas in the map are savannas, unshaded areas are open grasslands. The time scale of effects diminishes with spatial scale from an annual basis in the entire region to seconds at the level of bites.

control procedures (Munter, Halvorson & Anderson 1984; McNaughton 1988).

Herbivores can be limited by food quantity, determined by the rate of primary production, and food quality, as influenced by both absolute nutrient requirements and the balance of nutrients (McNaughton & Georgiadis 1986). At the level of the Serengeti ecosystem as a whole (Fig. 1), the most frequent potential mineral limitation on primary productivity, based on foliage analyses and grass requirements, was calcium. Seventy-five percent of all grass samples were potentially deficient in calcium needed for healthy growth. At the same level of scale, potential limitations on the quality of food for grazers due to their mineral requirements were most widespread in the case of sodium, with about 75% of all samples deficient for this element for animal dietary requirements.

At the next level, landscape regions (Fig. 1), in a comparison of forage samples from tall grasses in the north-west Serengeti with short grasses from the south Serengeti (McNaughton 1983), discriminant analysis based on 19 mineral elements in the young leaf blades of actively growing plants separated forage samples with complete accuracy; no samples were misclassified ($\chi^2 = 128$, $P < 0.00001$). Group centroids were $-7$ for tall grasses and 7 for short grasses. Mineral concentrations were consistently and substantially higher in the short grasslands (Table 1), except for chromium, magnesium, and manganese. These data confirm with a broad spectrum of minerals the dichotomy between dystrophic savanna grasslands growing on old, heavily leached basement rocks, which the tall-grass landscape represents, and the eutrophic savanna grasslands that grow on volcanic soils, of which the Serengeti short grasslands are representative (MacVicar 1977; Bell 1982; Huntley 1982). Concentrations of such often limiting animal requirements as copper, phosphorus and zinc were substantially lower in the tall grasses from the north-west. This, to my knowledge, is the most extensive range of elements yet assayed for African plants in a non-agricultural context, providing strong confirmatory evidence for substantial differences in the nutritional sufficiency of forages in different landscape regions.

At lower hierarchical levels, principal components analysis (PC) was used to determine between- and within-landscape mineral patterns (Fig. 1), here demonstrated by two species (*Kyllinga nervosa* and *Sporobolus ioclados*) abundant in the south-east Serengeti Plains and two others (*Pennisetum mezianum* and *Themeda triandra*) abundant in the north-west Serengeti Plains (Fig. 2). The first PC separated species within regions (each point is the mean for a sward). Among the elements loading heavily on PC1 and, therefore, discriminating between landscape regions, was aluminium, which was extremely high in plants from the south-east. Among the elements loading heavily on PC2, therefore separating swards within landscapes, was vanadium.

**Table 1.** Discriminant analysis of elemental content of leaf blades from tall grasses in the north-western Serengeti National Park and short grasses in the southern Serengeti.

| Element | SDF | Concentrations | |
|---------|-----|------|-------|
|         |     | Tall | Short |
| A1      | 3.16  | 133  | 1011 |
| B       | 0.36  | 3.72 | 7.21 |
| Ca      | 0.14  | 4350 | 5458 |
| Cd      | 0.28  | 0.01 | 0.05 |
| Co      | −0.26 | 0.18 | 0.34 |
| Cr      | −0.31 | 0.56 | 0.48 |
| Cu      | 0.38  | 3.54 | 7.30 |
| Fe      | 4.39  | 73.4 | 442  |
| K (%)   | −0.42 | 1.30 | 1.92 |
| Mg      | 0.08  | 2126 | 1957 |
| Mn      | −0.23 | 181  | 73   |
| Mo      | 0.47  | 0.71 | 4.23 |
| Na      | 1.08  | 140  | 671  |
| Ni      | −0.36 | 0.59 | 0.73 |
| P       | 1.15  | 1269 | 6812 |
| Pb      | −0.25 | 3.31 | 6.78 |
| Se      | −0.01 | 0.92 | 1.51 |
| V       | 0.30  | 1.76 | 2.13 |
| Zn      | −0.52 | 16.3 | 28.7 |

SDF, standardized discriminant function coefficients with group centroids of −7 for tall and 7 for short grasses. All concentrations in ppm dry weight, except K as %.

Finally, at the lowest hierarchical level considered here, between individuals (i.e. different ungulate bites) of two species in two swards (Fig. 1) on the south-east Serengeti Plains, there were a wide variety of different patterns, depending upon the mineral (Table 2). For most elements, species differences were more important, suggesting that either the ability of grass species to take up nutrients differed or differences in microsites occupied were important. In some cases, sward was important, suggesting that localized edaphic factors might be influential. And, in addition, some elements were characterized by sward-species interactions, indicating that control of the mineral nutrients in plants is complex. Variation was substantial even in these localized samples, with boron varying by 79%, calcium by 44%, magnesium by 25%, phosphorus by 52%, and sodium by a phenomenal 486%. These data indicate that even within local patches of forage, variation in mineral composition is substantial.

These wet-season data on the mineral contents of Serengeti grasses indicate that grass blades are far from a uniform, homogeneous food source. Instead, (1) there was substantial elemental heterogeneity at all hierarchical levels, from the total region to individual plants within a sward, (2) the broad nutritional gradient in the tropics between eutrophic systems on

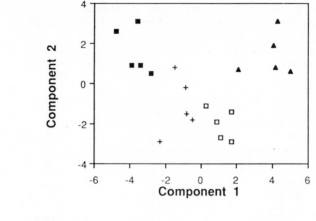

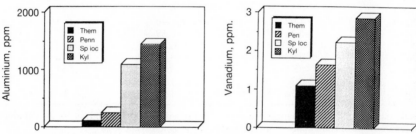

**Fig. 2.** Principal components (PC) analysis of 19 elements in actively growing leaf blades of *Sporobolus ioclados* (unfilled squares) and *Kyllinga nervosa* (triangles) from the south-eastern Serengeti Plains and *Themeda triandra* (filled squares) and *Pennisetum mezianum* (plus signs) from the north-western Serengeti Plains. PC 1 and 2 explained 58% of the variance; first major element loadings are shown for each PC: aluminium for PC1 and vanadium for PC2.

volcanic soils and dystrophic systems on basement rocks is applicable to a very wide range of minerals, and (3) although sodium is widely deficient in forages, as expected, a surprisingly large proportion of samples were sodium-sufficient, an unusual ecological occurrence that might be important to the Serengeti's ability to support large mammal populations (McNaughton 1988).

## Resident animal concentrations and mineral nutrition

The data above indicate that there is substantial mineral heterogeneity in the forages encountered by the nomadic herbivores that dominate such grazing ecosystems as the Serengeti. However, although some animal species undertake long-distance migration, many are residents that do not move over large geographic areas. Those resident species also are very hetero-geneously distributed and certain localities contain substantial multispecific

**Table 2.** Elemental concentrations (ppm except K = %) in the youngest leaf blades of actively growing grasses (*Sporobolus*) in two swards (A & B) in the south-eastern Serengeti Plains.

| Element | Species: | S. ioclados | | S. kentrophyllus | | Significance levels | | |
|---------|----------|-------------|---|------------------|---|---------------------|---|---|
| | Sward: | A | B | A | B | sp | sw | spp × sw |
| Al | | 1340$^c$ | 1093$^{abc}$ | 885$^{ab}$ | 756$^a$ | 0.002 | 0.057 | ns |
| B | | 5.98$^a$ | 10.7$^b$ | 6.05$^a$ | 6.99$^a$ | ns | 0.034 | ns |
| Ca | | 6021$^b$ | 5285$^{ab}$ | 4186$^a$ | 5362$^{ab}$ | 0.039 | ns | 0.028 |
| Cd | | 0.074$^a$ | 0.035$^a$ | 0.059$^a$ | 0.046$^a$ | ns | 0.051 | ns |
| Co | | 0.327$^a$ | 0.381$^a$ | 0.286$^a$ | 0.380$^a$ | ns | ns | ns |
| Cr | | 0.684$^a$ | 0.582$^b$ | 0.435$^a$ | 0.392$^a$ | < 0.001 | 0.014 | ns |
| Cu | | 6.38$^a$ | 6.86$^a$ | 6.72$^a$ | 7.43$^a$ | ns | ns | ns |
| Fe | | 596$^c$ | 485$^{bc}$ | 390$^{ab}$ | 318$^a$ | < 0.001 | 0.017 | ns |
| K | | 1.82$^{ab}$ | 2.23$^b$ | 1.64$^a$ | 1.68$^a$ | 0.003 | 0.035 | 0.067 |
| Mg | | 1771$^a$ | 1898$^{ab}$ | 2215$^b$ | 1995$^{ab}$ | 0.007 | ns | 0.050 |
| Mn | | 87.7$^a$ | 78.6$^a$ | 72.3$^a$ | 84.8$^a$ | ns | ns | 0.023 |
| Mo | | 4.59$^{ab}$ | 3.25$^a$ | 7.25$^b$ | 7.19$^b$ | 0.008 | ns | ns |
| Na | | 164$^a$ | 961$^b$ | 182$^a$ | 302$^a$ | ns | ns | 0.055 |
| Ni | | 0.777$^b$ | 0.699$^{ab}$ | 0.601$^a$ | 0.555$^a$ | 0.034 | ns | ns |
| P | | 6006$^a$ | 6111$^a$ | 7093$^{bc}$ | 9191$^c$ | 0.004 | ns | ns |
| Pb | | 7.99$^b$ | 6.40$^{ab}$ | 6.91$^{ab}$ | 6.34$^a$ | 0.048 | 0.009 | ns |
| Se | | 2.63$^a$ | 1.95$^a$ | 2.79$^a$ | 1.81$^a$ | ns | ns | ns |
| V | | 2.41$^b$ | 2.09$^a$ | 2.27$^{ab}$ | 2.00$^a$ | ns | 0.016 | ns |
| Zn | | 28.2$^a$ | 27.9$^a$ | 35.5$^b$ | 31.8$^{ab}$ | 0.033 | ns | ns |

Significance levels from 2 × 2 ANOVA with d.f. = 1,8; sp = species, sw = sward, spp × sw = interaction. Concentrations with different superscripts are significantly different at least at $P = 0.05$.

resident herds while nearby localities are nearly devoid of animals (McNaughton & Georgiadis 1986).

The traditional explanations for the concentrations of resident animals are (1) that congregation protects the animals from predation, and (2) that different feeding modes of different species facilitate grazing (Sinclair 1985). However, neither of these hypotheses is wholly compelling. If predation drove animal concentration, the animals should move erratically through space, which would reduce their predictability to predators, but residents occupy temporally stable home ranges. If grazing facilitation drove animals to congregate, they should move extensively with localized rainshowers to take advantage of regions of high primary productivity (McNaughton 1985), but, again, the temporal stability of resident home ranges argues against this hypothesis.

An examination of the mineral contents of forages in areas where resident animals concentrate and adjacent, control areas only sparsely occupied by animals revealed significant differences between forages, but not soils, in the two types of localities (McNaughton 1988). In view of both animal requirements and the concentrations of elements in forages, magnesium, phosphorus and sodium appear to be particularly important nutritional factors. Thus, both the capacity of African ecosystems to support large herbivores, and distribution of those herbivores within ecosystems, may be influenced to a substantial degree by the mineral heterogeneity of forages.

## Temporal dynamics

In addition to pronounced spatial heterogeneity, documented here for the first time for the mineral content of forages in a natural ecosystem, African ecosystems are characterized by pronounced temporal heterogeneity at several scales, from diurnal to weekly variation in factors that can influence plant water status (Toft, McNaughton & Georgiadis 1987), to drought cycles coupled to global patterns related to El Nino/Southern Oscillation phenomena (Cane, Zebiak & Dolan 1986). As plant tissues die, either because of internal cues related to plant development state, or because of such external cues as canopy closure or onset of the dry season, mobile elements, such as nitrogen, potassium and phosphorus, will be exported to storage pools (Mengel & Kirkby 1982); in grasses these are primarily in leaf bases and roots. Turgor maintenance by osmolyte accumulation may be a critical adaptation of grasses to arid conditions, prolonging the period during which export can be maintained (Toft et al. 1987).

The mineral sufficiency for ungulates of forage samples in Malawi varied drastically in wet and dry seasons (McDowell 1985). Frequencies of deficiencies for plant-mobile elements increased in the dry season; deficiency frequencies for plant-immobile elements decreased then. During the wet season, only 3% of samples were deficient in potassium; during the dry season the percentage deficient increased to 57%. For phosphorus, deficiency values were 56% in the wet season, 96% in the dry season. And the occurrence of copper deficiency increased from 47% of all samples in the wet season to 91% during the dormant period. Deficiency of less mobile elements, in contrast, was much more likely in the growing season than in the dormant period. For calcium, 81% of samples were deficient in the wet season, only 13% in the dry. Similarly, potential magnesium deficiency dropped from 77% in the wet season to 31% during the dormant period.

These data indicate that large African mammals are confronted with very complex nutritional environments. It is commonly assumed that the wet season is a favourable nutritional season for herbivores, owing particularly to higher nitrogen concentrations and greater digestibilities of forages. There is no doubt that this is true. Nevertheless, widespread mineral deficiencies are also possible during this season. It is, I believe, unwise to ascribe too much importance to a limited array of forage properties until much more is known of the total nutritional environments encountered by herbivores, and their requirements.

## Silicification: grass defence against herbivory

Grasses are remarkably poor in the toxic chemicals that characterize the anti-herbivore defences of dicots (McNaughton 1983). Nevertheless, the presence of cyanogenic glycosides and other potent secondary chemicals in some grass species indicates that such defences are not outside the evolutionary potentials of grasses (Georgiadis & McNaughton 1988). However, the presence of phytoliths, taxonomically diagnostic inclusions of opaline silica, in the earliest plant fossils (Stebbins 1981) and the evolution of hypsodont large mammals over millions of years indicate that physical, rather than chemical, defence has been a predominant evolutionary response of grasses to grazers.

Several aspects of grass morphology, growth form and ecological circumstances seem important for the evolution of predominantly physical rather than chemical defences in the Poaceae. First, grasses grow from intercalary meristems and have parallel-veined leaves so that large herbivores, feeding from above, remove physiologically older portions of leaf blades and do not disrupt the integrative vascular system as they feed. Unless a shoot has converted to a terminal meristem, the growing points remain completely unaffected by defoliation. In fact, opening of the canopy, and the resultant enrichment of red light at the shoot bases compared to conditions under a more closed canopy, activates basal meristems of grasses, a phenomenon called tillering (Deregibus, Sanchez, Casal & Trlica 1985). Second, it has been known for 50 years that one of the major intraspecific evolutionary responses of grasses to grazing is rapid selection for prostrate growth forms less accessible to large herbivores (McNaughton 1984). These small-leaved, often highly rhizomatous or stoloniferous growth forms are capable of generating substantial canopy standing crops within the small volume inaccessible to grazers. Third, many of the traits that protect grasses from grazers by making them physically inaccessible also can confer advantages in the arid to subhumid climates where grasses typically dominate the floras (Coughenour 1985). Grazing may merely amplify trends inherent in grasses, reducing selection by competition for light in desiccating environments and acting on aridity-generated genetic variation through directional selection. Fourth, silica accumulation provides grass tissues with a resistance to abrasion and a tensile strength important to their ability to withstand treading. A grass plant may be walked upon as often as it is bitten, and the combination of heavy bodies with rigid hooves makes ungulates a potentially devastating force to plants with structures that are either fragile or brittle. It is no accident that human playing fields are invariably cloaked in grass. Finally, silica may have other roles in grasses than solely to defend them from herbivores. It may substitute for carbon-based structural support, releasing a significant portion of the energy fixed through photosynthesis to biosynthetic requirements, and to storage pools which are crucial to maintenance and survival in environments inherently stochastic because of unpredictable rainfall and defoliation events (McNaughton, Tarrants, McNaughton & Davis 1985). There also is some evidence that the accumulation of silica may suppress transpiration, thereby conserving water (Jones & Handreck 1967). Thus, although silica accumulation has certainly been a defence against herbivores of considerable selective potency to the latter, it cannot be ascribed a single role, but represents a multi-faceted trait of the Poaceae.

## Conclusion: evolution of grazers and plants of the field layer

It is just that complex of traits in grasses that arose in response to being grazed, as plants and herbivores co-evolved, that provides grasslands with the ability to recover from disturbance so vividly described in the quotation from Ingalls (1904) that opened this essay. Growth from crowns of intercalary meristems capable of regenerating both shoots and roots, production of lateral stems with the capacity to root from nodes, physically resistant tissues laden with silicon, leaf blade expansion to a carbon-exporting state before it becomes physically accessible to large herbivores, all of these traits confer upon grasses a remarkable ability to withstand the vagaries of unpredictable, physically destructive and stringent environments. The nutritional heterogeneity of grasses from the level of geographic regions to single bites belies the idea that grasslands are simple homogeneous resources. Instead, grasses represent a highly heterogeneous food source in both space and time, just as different species of grazers represent heterogeneous forces of natural selection influencing grasses through their diversity of feeding modes.

## Acknowledgements

The U.S. National Science Foundation Ecosystem Studies Program has supported the Serengeti Ecosystem Research Project since 1974. Margaret McNaughton performed ICP analysis. Grass sample collection was assisted by Sean McNaughton, Nicholas Georgiadis, Feetham Banyikwa, Roger Ruess and Barbara Maas.

## References

Barnard, C. & Frankel, O.H. (1964). Grass, grazing animals, and man in historical perspective. In *Grasses and grasslands*: 1–12. (Ed. Barnard, C.). Macmillan, London.
Bell, R.H.V. (1970). The use of the herb layer by grazing ungulates in the Serengeti. In *Animal populations in relation to their food sources*: 111–24. (Ed. Watson, A.). Blackwells, Oxford.

Bell, R.H.V. (1982). The effect of soil nutrient availability on community structure in African ecosystems. In *Ecology of tropical savannas*: 193–216. (Eds Huntley, B.J. & Walker, B.H.). Springer-Verlag, Berlin. (*Ecol. Stud. Anal. Synth.* **42.**)

Cane, M.A., Zebiak, S.E. & Dolan, S.C. (1986). Experimental forecasts of El Nino. *Nature, Lond.* **321**:827–32.

Chacon, E., Stobbs, T.H. & Saldland, R.L. (1976). Estimation of herbage consumption by grazing cattle using measurements of eating behaviour. *J. Br. Grassld Soc.* **31**:81–7.

Chapin, F.S. (1980). The mineral nutrition of wild plants. *A. Rev. Ecol. Syst.* **11**:233–60.

Coughenour, M.B. (1985). Graminoid responses to grazing by large herbivores: adaptations, exaptations and interacting processes. *Ann. Mo. bot. Gdn* **72**:852–63.

Demment, M.W. & Greenwood, G.B. (1988). Forage ingestion: effects of sward characteristics and body size. *J. Anim. Sci.* **66**:2380–92.

Deregibus, V.A., Sanchez, R.A., Casla, J.J. & Trlica, M.J. (1985). Tillering responses to enrichment of red light beneath the canopy in a humid natural grassland. *J. appl. Ecol.* **22**:199–206.

Frenguelli, J. (1930). Particulas de silice organizada en el loess y en los limos pampeanos. Selulas siliceas de Gramineas. *An. Soc. cient. S. Fé* **2**:1–47.

Fryxell, J.M., Greever, J. & Sinclair, A.R.E. (1988). Why are migratory ungulates so abundant? *Am Nat.* **131**:781–98.

Georgiadis, N.J. & McNaughton, S.J. (1988). Interactions between grazers and a cyanogenic grass, *Cynodon plectostachys. Oikos* **51**:343–50.

Grzimek, M. & Grzimek, B. (1960). Census of plains animals in the Serengeti National Park, Tanganyika. *J. Wildl. Mgmt* **24**:27–37.

Higgs, E.S., Fussell, G.E., Nair, K., Rasmussen, W.D., Gray, A.W. & Ordish, G. (1978). History of agriculture. *Encyc. Brit. (Macropedia)* (15) **1**:324–47.

Huntley, B.J. (1982). South African savannas. In *Ecology of tropical savannas*: 101–19. (Eds Huntley, B.J. & Walker, B.H.). Springer-Verlag, Berlin. (*Ecol. Stud. Anal. Synth.* **42.**)

Ingalls, J.J. (1904). *On bluegrass. Kansas description and travel.* M.W. Tennal, Atchison, KS.

Jenny, H. (1930). A study of the influences of climate upon the nitrogen and organic matter content of the soil. *Res. Bull. Mo. agric. Exp. Stn* No. 152:1–62.

Jones, L.H.P. & Handreck, K.A. (1967). Silica in soils, plants, and animals. *Adv. Agron.* **19**:107–49.

Kowalevsky, W. (1873–4). Monographie der Gattung *Anthracotherium* Cuv. und Versuch einer naturlichen Classification der fossilen Hufthier. *Paleontographica* **22**:131–285.

Laca, E.A. & Demment, M.W. (1986). The feeding behavior of a grazing herbivore: harvesting limitations. *Proc. int. Congr. Ecol.* **4**:209.

Leuthold, W. (1977). *African ungulates. A comparative review of their ethology and behavioral ecology.* Springer-Verlag, Berlin. (*Zoophysiology Ecol.* **8**: 1–307.)

MacVicar, C.N. (1977). *Soil classification: a binomial system for South Africa.* Dept. Agr. Tech. Ser., Pretoria.

Maddock, L. (1979). The "migration" and grazing succession. In *Serengeti.*

*Dynamics of an ecosystem*: 104–29. (Eds Sinclair A.R.E. & Norton-Griffiths, M.). University of Chicago Press, Chicago & London.

Martin, W.E. & Matocha, J.E. (1973). Plant analysis as an aid in the fertilization of forage crops. In *Soil testing and plant analysis*: 393–406. (Eds Walsh, L.W. & Beaton, J.D.). Soil Sci. Soc. Amer., Madison WI.

McDowell, L.R. (Ed.) (1985). *Nutrition of grazing ruminants in warm climates.* Academic Press, New York.

McDowell, L.R., Conrad, J.H., Ellis, G.L. & Loosli, J.K. (1983). *Minerals for grazing ruminants in tropical regions.* Dept. Anim. Sci., Center Trop. Agric., U. Florida, Gainesville and U.S.A.I.D.

McNaughton, S.J. (1979). Grassland-herbivore dynamics. In *Serengeti. Dynamics of an ecosystem*: 46–81. (Eds Sinclair, A.R.E. & Norton-Griffiths, M.). University of Chicago Press, Chicago & London.

McNaughton, S.J. (1983). Serengeti grassland ecology: the role of composite environmental factors and contingency in community organization. *Ecol. Monogr.* 53:291–320.

McNaughton, S.J. (1984). Grazing lawns: animals in herds, plant form, and coevolution. *Am. Nat.* 124:863–86.

McNaughton, S.J. (1985). Ecology of a grazing ecosystem: the Serengeti. *Ecol. Monogr.* 55:259–95.

McNaughton, S.J. (1987). Adaptation of herbivores to seasonal changes in nutrient supply. In *The nutrition of herbivores*: 391–408. (Eds Hacker, J.B. & Ternouth, J.H.). Academic Press, Sydney.

McNaughton, S.J. (1988). Mineral nutrition and spatial concentrations of African ungulates. *Nature, Lond.* 334:343–5.

McNaughton, S.J. & Georgiadis, N.J. (1986). Ecology of African grazing and browsing mammals. *A. Rev. Ecol. Syst.* 17:39–65.

McNaughton, S.J., Tarrants, J.L., McNaughton, M.M. & Davis, R.H. (1985). Silica as a defense against herbivory and a growth promoter in African grasses. *Ecology* 66:528–35.

Mengel, K. & Kirkby, E.A. (1982). *Principles of plant nutrition.* International Potash Institute, Bern.

Morison, S.E. (1965). *The Oxford history of the American people.* Oxford University Press, New York.

Munter, R.C., Halvorson, T.L. & Anderson, R.D. (1984). Quality assurance for plant tissue analysis for ICP-AES. *Communs Soil Sci. Pl. Anal.* 15:1285–322.

Sauer, C.O. (1952). *Agricultural origins and dispersals.* American Geographical Society, New York.

Scott, W.B. (1937). *A history of land mammals in the Western Hemisphere.* Macmillan, New York.

Senft, R.L., Coughenour, M.B., Bailey, D.W., Rittenhouse, L.R., Sala, O.E. & Swift, D.W. (1987). Large herbivore foraging and ecological hierarchies. *BioScience* 37:789–99.

Sinclair, A.R.E. (1977). *The African buffalo. A study of resource limitation of populations.* University of Chicago Press, Chicago & London.

Sinclair, A.R.E. (1985). Does interspecific competition or predation shape the African ungulate community? *J. Anim. Ecol.* 54:899–918.

Sinclair, A.R.E. & Norton-Griffiths, M. (1982). Does competition or facilitation regulate migrant ungulate populations in the Serengeti? A test of hypotheses. *Oecologia* 53:364–9.

Stebbins, G.L. (1981). Coevolution of grasses and herbivores. *Ann. Mo. bot. Gdn* 68:75–86.

Talbot, L.M. & Talbot, M.H. (1963). The wildebeest in western Masailand, East Africa. *Wildl. Monogr.* No. 12: 1–88.

Taylor, S.C.S. & Murray, J.I. (1987). Genetic aspects of mammalian survival and growth in relation to body size. In *The nutrition of herbivores*: 487–533. (Eds Hacker, J.B. & Ternouth, J.H.). Academic Press, Sydney.

Teruggi, M.E. (1955). Algunas observaciones microscópicas sobre vidrio volcánico y opalo organogeno en sedimentos pampianos. *Notas Mus. La Plata* (Geol.) 18:17–26.

Toft, N.L., McNaughton, S.J. & Georgiadis, N.J. (1987). Effects of water stress and simulated grazing on leaf elongation and water relations of an East African grass, *Eustachys paspaloides. Aust. J. Pl. Physiol.* 14:211–26.

Vesey-Fitzgerald, D.F. (1960). Grazing succession among East African game animals. *J. Mammal.* 41:161–72.

Watson, R.M. (1967). *The population ecology of the wildebeest* (Connochaetes taurinus taurinus albojubatus *Thomas) in the Serengeti.* Ph.D. thesis: University of Cambridge.

Webb, S.D. (1977). A history of savanna vertebrates in the New World. Part I. North America. *A. Rev. Ecol. Syst.* 8:355–80.

Webb, S.D. (1978). A history of savanna vertebrates in the New World. Part II. South America and the great interchange. *A. Rev. Ecol. Syst.* 9:393–426.

Wells, P.V. (1965). Scarp woodlands, transported soils, and concept of grassland climate in the Great Plains region. *Science, N.Y.* 148:246–9.

Williams, R.E., Allred, B.W., DeNio, R.M. & Paulsen, H.E. (1968). Conservation, development, and use of the world's rangelands. *J. Range Mgmt.* 21:355–60.

Symp. zool. Soc. Lond. (1989) No. 61: 31–50

# Ranging and feeding behaviour of okapi (*Okapia johnstoni*) in the Ituri Forest of Zaire: food limitation in a rain-forest herbivore?

John A. HART and
Terese B. HART

*Wildlife Conservation International*
*New York Zoological Society*
*Bronx, NY 10460, USA*

## Synopsis

Terrestrial vertebrate folivores are uncommon in dense tropical forest. The okapi (*Okapia johnstoni*), a giraffid found only in the forests of north-eastern Zaire, is a strict folivore. Direct observation with the aid of radio telemetry and indirect evidence from browse sign were used to study the movements of free-ranging okapi and the relationship between the okapi's use of space and its food resource. Seasonality had no great effect on either home-range size or animal movements. Okapi used both major forest habitat divisions on the study area: upland and poorly-drained. The okapi is a highly selective feeder so that browse sign was light in all environments. The most concentrated feeding occurred in tree-fall gaps. Light-dependent species found in gaps were preferred browse plants, and even the less favoured shade-tolerant species of the subcanopy were more likely to be browsed when they occurred in tree-fall gaps. Gaps, however, are a scattered resource and account for less than 5% of the forest area. A lactating female did not meet her increasing energy needs by expanding her home-range to include more tree-fall gaps. Instead she increased her foraging time, exploiting more intensively the relatively more dispersed resources of the understorey. It is suggested that the scarcity of high-quality food rather than seasonality or habitat variation is the major constraint affecting okapi feeding behaviour.

## Introduction

The abundant year-round foliage of tropical evergreen forests does not support as large and diverse a community of folivorous mammals as is found subsisting on the tropical savannas. Frugivores rather than herbivores are the most diverse group in most tropical forests (Terborgh 1986).

This predominance of frugivory is found among the ungulate fauna of the

ZOOLOGICAL SYMPOSIUM No. 61
ISBN 0–19–854009–4

Ituri Forest of Zaire despite an indigenous fauna of 15 species. In this community, the diets of the six species of duikers (*Cephalophus* spp.) and the chevrotain *Hyemoschus aquaticus*) are composed primarily of fruits and seeds (Dubost 1984; J. Hart 1986). An additional three ungulate species, the elephant (*Loxodonta africana*) and two wild pigs (*Hylochoerus meinertz-hageni* and *Potamochoerus porcus*), are omnivorous with a large component of fruit in their diets (Alexandre 1978; personal observation). Among the predominantly herbivorous species, the buffalo (*Syncerus caffer*) and bongo (*Tragelaphus euryceros*) are more abundant along the edge of the forest or ecotone while the sitatunga (*Tragelaphus spekii*) and pygmy antelope (*Neotragus batesi*) are restricted to special environments or microsites (Emmons, Gautier-Hion & Dubost 1983). The okapi (*Okapia johnstoni*), a rain forest giraffid, is another folivore. Unlike the other ungulate folivores it is apparently widespread and restricted to the forest interior; however, its habitat use and feeding behaviour have remained little known (Bodmer & Rabb in press).

The giraffe (*Giraffa camelopardalis*), the okapi's nearest relative, is widespread in the savanna woodlands of Africa. The giraffe specializes on a relatively small number of key browse species which provide it with a generally high-quality diet year-round (Pellew 1984). Evidence from at least one study (Pellew 1983b) suggests that reduced availability of these species may limit giraffe populations in some areas.

Rain-forest foliage, in general, is reputed to be of poor quality (Leigh & Smythe 1978; Waterman, Choo, Vedder & Watts 1983). Higher-quality foliage is most abundant in special microsites or habitats. Of these, tree-fall gaps are among the most important. Studies of insect folivory in tropical forests have shown that plants growing in gaps may be more palatable than plants, even of the same species, growing in the understorey (Harrison 1987). Light-dependent tree species growing in disturbed habitats may be more heavily browsed than shade-tolerant species under the canopy (Coley 1983). Similarly, among the ungulates, the pygmy antelope is restricted to gaps and clearings (Feer 1979). Thus tree-fall gaps may be generally important to folivores in the forest understorey.

Dry-season food shortages have been documented for frugivores in a number of tropical forests (Fleming 1979; Gautier-Hion, Duplantier, Emmons, Feer, Heckestweiller, Moungazi, Quris & Sourd 1985; Terborgh 1986) including the Ituri Forest of Zaire (J. Hart 1986). For the case of ungulate folivores, however, the effects of seasonal variation in food supply have been well documented only for savanna species (Maddock 1979; Sinclair, Dublin & Borner 1985).

This paper presents results from the first field study of free-ranging okapi. Data on density, movements, habitat use and feeding in the Ituri Forest of Zaire are discussed in relationship to seasonal and spatial variations in food resources. Observations on changes in foraging behaviour associated with increased energy needs arising from lactation are also used to assess possible constraints in the okapi's use of rain-forest foliage.

## Study area

The vegetation of the Ituri Forest fits the broad category of 'rain forest' as defined by White (1983) for Africa. Most of the study area falls into the subcategory of 'mixed moist semi-evergreen forest'. All the dominant tree species in the Ituri Forest are evergreen but the canopy contains scattered deciduous elements as well.

Bultot (1971) estimated annual rainfall of the Ituri region between 1700 and 1800 mm with a dry season (rainless period) of less than 20 days. During two years, 1986 and 1987, at Epulu, the annual totals were 1719 and 1600 mm of rainfall. The driest season spanned December through February, with three-month totals of 293.2 mm (1986–87) and 159.1 mm (1987–88). The rainy season began in March, with March through May and September through November being the wettest months.

We selected and started to develop our study area in the Ituri Forest in early 1986. The area chosen is located in the Edoro River watershed, northwest of Epulu and 20 km from the single east-west road traversing the forest (Fig. 1). The study area is covered by mature forest with scattered patches of old secondary forest more than 50 years old. It is located beyond the zone of current settlement and is not now used by local hunters. There were no other human activities in the study area although signs of previous elephant poaching remained.

In January 1988, the study area covered 26 km². It was divided by a grid system of paths totalling more than 150 km. Areas used intensively by okapi were divided by quadrats of 250 m on a side. Less frequented areas were divided into 500 m square quadrats. The grid system allowed collared animals to be located rapidly and accurately and served as the basis for mapping animal locations and habitats.

Two major forest types occurred on the study area: upland mixed forest and poorly-drained or swamp forest. Major canopy trees characterizing the former include *Cynometra alexandri* (Caesalpiniaceae), *Fagara macrophylla* (Rutaceae), *Erythrophleum suaveolens* (Caesalpiniaceae), *Brachystegia laurentii* (Caesalpiniaceae), *Cleistanthus michelsonii* (Euphorbiaceae), *Klainedoxa gabonensis* (Irvingiaceae), *Canarium schweinfurthii* (Burseraceae), *Autranella congolensis* (Sapotaceae), and *Alstonia boonei* (Apocynaceae). The swamp forest is generally dominated by the canopy species *Mitragyna stipulosa* (Rubiaceae) but includes, as locally common, *Nauclea xanthoxylon* (Rubiaceae), *Uapaca guineensis* (Euphorbiaceae) and *Neoboutonia manii* (Euphorbiaceae).

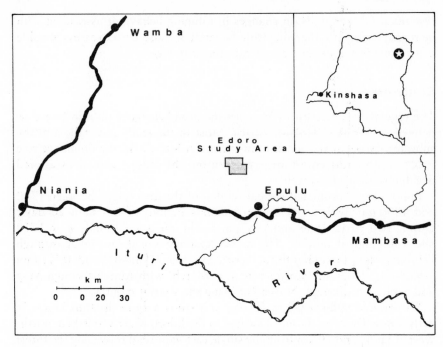

**Fig. 1.** The Edoro Study Area in the Ituri Forest. Major towns and roads are shown. The star on the inset indicates location in Zaire. The sides of the maps run north–south.

Whereas there is a deciduous component to the canopy, the understorey of both upland and swamp forests is entirely evergreen. Woody trees and saplings dominate in both environments although dense stands of herbaceous Marantaceae occur locally in swampy areas.

## Methods

Okapi were captured on the study area in concealed pits. A radio collar carrying a Telonics transmitter (configuration 5b or 6b) was fitted around the neck of each animal while still in the pit. After release the animals were subsequently followed on foot with the aid of portable TR-2 Telonics receivers and hand-held directional antennae.

Data collected on the collared okapi were derived either from remote locations or from following. In remote locations, collared animals were located once or more per day to a particular 250 m quadrat. During following, the animal was approached and its activities observed for several consecutive hours (up to 48 h). Both location and following were carried out at all times of the day and night, with greatest concentration between 06.00 and 18.00 h when okapi were most active.

During following, activity was classed as 'stationary' if the animal was not moving or its movements were confined to a radius of approximately 5 m over a period of 5 min or more. During 'slow movement' the animal's progress was irregular and did not generally exceed 250 m per 20 min. The class 'directional movement' included all movement in a fairly constant direction at a rate greater than 250 m per 20 min.

The information on speed and direction of movement provided by the telemetry equipment was supplemented when possible by direct viewing; such direct observations were successfully made of animals in all three activity classes. The density of the forest understorey, however, often precluded actual sighting without the risk of approaching so close as to disturb the target animal. In these circumstances the information from the radio signal was supplemented by noting the movement of the understorey vegetation or by listening directly to the sounds made by the animal (e.g. an okapi feeding vigorously can be heard from a distance of 15 m or more.)

Observations were made each month from June 1986 through January 1988. Study periods varied in length from two to more than three weeks per month; however, because of the dispersed distribution of the okapi and a dearth of trained personnel, data were not collected on all animals every day. A total of 630 h of direct observation was accumulated over the entire study period.

Movements and locations were mapped and home-range area was calculated for all animals for which observations spanned at least three months and included at least 50 daily locations. All available locations over the entire study period were used to delimit each animal's maximum home-range. Home-ranges were mapped by connecting outermost quadrat locations (excluding the outermost 5% representing isolated outliers) to circumscribe the area used by each animal in a minimum convex polygon (Schoener 1981). Because locations made over short intervals of time may not be independent (Swihart & Slade 1985a, b), a subset of locations separated by at least 8 h, and in most cases by 24 h, was used to assess animals' use of space and habitat associations.

Data on feeding behaviours were gathered from observations of collared animals. These observations allowed us to identify okapi browse and to recognize it up to a month after the feeding event. Browse availability and use were measured on 10 m by 2 m plots in upland forest understorey (12 plots), in swamp forest understorey (12 plots), and in canopy openings caused by tree-fall gaps (20 plots, both upland and swamp forest). Only plants with foliage between 0.5 m and 2 m in height were enumerated as this is the forest layer most frequently browsed by okapi. Fertile specimens of many of the plants had been previously collected and identified as part of an ongoing botanical survey.

The overall availability and productivity of foliage in the okapi browse layer (0.5 m to 2 m) were measured by clipping two replicate 3 m × 2 m plots each in upland understorey and gap, then reclipping the new growth two months later. Four plots were cut in early November during the wet season and four in mid-December during the dry season. Two of the gap plots were surrounded by exclosures to control for the impact of okapi browse. During the original clipping all shoot ends were snipped back approximately 15 cm, regardless of plant form, to simulate okapi browsing. For the second clipping only new growth was taken. After both clippings wet and dry weights of harvested foliage were measured.

## Results

### Home-range

Eight okapi were captured and equipped with radio collars between March 1986 and January 1988 and their movements and activities followed on the study area (Table 1). These consisted of five females and three males. Two females were adult when first captured and both of these, Bahati and Akabi, were reproductively active during the study period. In November 1987, 18 months after her first capture, Bahati had a single calf. This allowed us to compare her movements and use of space during early pregnancy and lactation. Akabi had an engorged udder when first caught in May 1986. Systematic observations were not possible, however, as a path grid was not established in her home-range until two months later at which time her peak lactation was certainly past (Rabb 1978; Bodmer & Rabb 1985).

**Table 1.** Okapi equipped with radio transmitter collars between 1986 and 1988 on the Edoro Study Area, Ituri Forest, Zaire.

| Name | Age at capture | Sex | Date collared Month | Year | Home-range Size (km²) | % Swamp[a] | Status, January 1988 |
|------|---------------|-----|-------|------|---------------|----------|---------------------|
| Paskalina | Subadult | F | 03 | 86 | 1.9 | 22.6 | Emigrated from study area, 12–86 |
| Bahati | Adult | F | 05 | 86 | 3.1 | 12.8 | Alive on study area |
| Akabi | Adult | F | 06 | 86 | 5.1 | 19.8 | Alive on study area |
| Sumbuko | Subadult | F | 02 | 87 | ? | ? | Left study area 2 weeks after capture |
| Edoro | Subadult | F | 03 | 87 | 2.7 | –[b] | Killed by leopard, 11–87 |
| Asoma | Adult | M | 09 | 86 | 9.2 | 8.0 | Alive on study area |
| Semeki | Adult | M | 03 | 87 | 10.5 | –[b] | Alive on study area |
| Mangese | Adult | M | 03 | 87 | 1.6 | 19.2 | Killed by leopard, 06–87 |

[a] Percentage of home-range quadrats with > 50% swamp forest cover.
[b] Substrate type not mapped.

Collared okapi were active (feeding and moving) primarily during daylight hours. Animals occasionally moved about during the first few hours after dark and later as well, especially on moonlit nights. During most of the night, however, animals moved very little and were presumed to be resting.

There was a wide variation in home-range size among the collared okapi followed on the study area (Table 1). Of four females, each tracked for a minimum of five months, home-range size varied between 1.9 km$^2$ and 5.1 km$^2$ (mean = 3.2 ± 1.36). The two largest home-ranges belonged to the two reproductive females. The smallest belonged to a subadult who left the study area after nine months of observation. A second subadult female spent less than a month on the study area before she moved on and her signal was lost. A third subadult female, Edoro, also had a relatively small home-range. She exhibited unusual behaviour, notably, frequent crossing of a large river on the study area. She was killed by a leopard after nine months of observation.

The variation in home-range size was even greater among the collared males (Table 1). The smallest home-range, only 1.6 km$^2$, belonged to an old male whose tracks were first noted on the study area early in 1987. He was collared in March of that year and killed three months later by a leopard. The two other males both had very large home-ranges, exceeding 9 and 10 km$^2$ respectively.

The variation in home-range size did not correspond to any consistent variable in the physical environment. All seven okapi used mature forest on gently rolling terrain. Although their respective areas contained differing proportions of swamp forest, this did not vary with home-range size (Table 1).

Although home-ranges of several collared animals overlapped, animals were rarely seen together. Akabi and Paskalina, two females with extensive home-range overlap, even seemed to avoid each other. The quadrats in the overlap area most used by Akabi were the least frequented by Paskalina.

On the unusual occasions when more than two okapi were seen together, either a mother and calf or an adult male and female were involved. We have two observations of male, female and calf together. The home-ranges of the two wide-ranging collared males bordered or overlapped those of the three adult collared females yet the animals were not often together in the same 250 m quadrat.

There was some turnover on the study area during the first two years of field research as animals were added by birth and immigration or were eliminated by death or emigration. Taking into account this turnover as well as the presence of known but uncollared animals, the density of okapi on the core study area varied between 0.25 and 0.5 animals per km$^2$.

## Seasonality

There was no consistent seasonal variation in the amount of forest area utilized by three okapi for which adequate radio-tracking information was available over at least 1.5 years (Table 2; Fig. 2). When comparable numbers of independent locations per seasons were mapped, the average difference in the number of quadrats utilized in dry and wet seasons by the three okapi did not differ significantly from zero ($t$-test for paired comparisons, $t = 1.22$, 2 $d.f.$, $P > 0.05$).

Similarly, the okapi did not exhibit consistent seasonal differences in the distribution of their locations within their home-range. The frequency distributions of independent locations were neither clumped nor regular during any season for Asoma. Bahati's locations were significantly aggregated during the dry season of 1988, while those of Akabi were clumped during the wet and dry seasons of 1987 (Table 2, Fig. 2). It is unlikely that these differences in use of space were associated with a seasonal change in resource availability as both aggregation and dispersion of locations spanned both the wet and dry seasons.

All the collared okapi used both swamp forest and upland forest. Of the three okapi for which long-term information is available, only Akabi's movements showed a significant association with substrate type (Table 2). During the wet seasons of both 1986 and 1987, she was located in quadrats with a high proportion of swamp forest more frequently than would be expected if her movements were random relative to substrate type. If this reflects a seasonal preference for swamp forest it is curious that it should occur during the wettest season of the year, as other rain-forest ungulates,

**Table 2.** Seasonal changes in home-range use by three okapi.

| Animal | Season[a] | | Total locations | Total quadrats | Location dispersion[b] | Substrate association[b] |
|---|---|---|---|---|---|---|
| Bahati | Wet | 1986 | 65 | 38 | Random | None |
| adult F | Dry | 1986 | 65 | 35 | Random | None |
| | Wet | 1987 | 73 | 37 | Random | None |
| | Dry | 1988 | 70 | 33 | Clumped* | None |
| Akabi | Wet | 1986 | 50 | 31 | Random | Wet* |
| adult F | Dry | 1987 | 47 | 27 | Clumped* | None |
| | Wet | 1987 | 55 | 32 | Clumped* | Wet* |
| | Dry | 1988 | 53 | 38 | Random | None |
| Asoma | Dry | 1987 | 46 | 35 | Random | None |
| adult M | Wet | 1987 | 53 | 44 | Random | None |
| | Dry | 1988 | 51 | 43 | Random | None |

[a] Wet season: March 15–November 30; Dry season: December 1–March 15.
[b] An asterisk (*) indicates a significant deviation ($\chi^2$ goodness of fit; $P < 0.05$) from independent distributions of locations. Independent or random distributions of locations were recorded if quadrat locations followed a poisson distribution (dispersion) or were distributed in proportion to available substrate (wet versus dry).

including buffalo, elephant and pigs, concentrate their activity in swamp forests during the driest months of the year.

While the patterns of okapi locations did not show consistent spatial conformations, all the animals nevertheless favoured and avoided specific areas of their home-ranges (Fig. 2). In some cases the factors contributing to differential use of space are apparent; as, for example, the concentration of Bahati's locations in the south and west of her home-range during the dry season of 1988 while her calf was in the same area. In other cases movement patterns are less easily understood. Akabi shifted the focus of her activity between the north and the south of her home-range while utilizing the central portion relatively less. Asoma, an adult male, appeared to consistently patrol the borders of his area.

The lack of consistent seasonal differences in the okapis' use of their home-ranges was correlated with an apparent availability of suitable browse throughout their areas at all times of the year. Browse renewal followed clipping during both wet and dry seasons (Fig. 3). There were no significant differences in the percentage regrowth on plots clipped during either season in either the subcanopy or tree-fall gaps (2-way ANOVA, $P > 0.5$). Control plots protected from okapi browse did not exhibit consistently greater regrowth (Hart & Hart in prep.).

Differences in foliage production were, however, significant between understorey and gap (2-way ANOVA, $P < 0.05$). The same relationship between habitat types is evident when the total production is considered (Fig. 3). For both the dry and wet season clippings the total dry weight of regrowth foliage produced in tree-fall plots averaged ten times that produced in the subcanopy plots.

## Food availability and selection

Okapi browse damage was very light in all forest habitats. In tree-fall gaps an average of only 14% (range 3–30%) of the stems present had suffered leaf loss from okapi browsing. Plots beneath undisturbed canopy, in both upland and swamp, exhibited even less evidence of browsing, with an average 2% (range 0–3.9%) of the stems of upland forest browsed and 3% (range 0–9.2%) of the stems in swamp forest browsed (Fig. 4). This difference in browse sign between gap and understorey was significant whether plot results were compared by the total number of stems browsed or by the percentage of stems browsed (ANOVA, arcsine transformation of percentages, $P < 0.001$).

The plots examined for browse damage revealed compositional differences that might contribute to the okapi's more intensive use of gaps. Associated with the greater productivity of gaps was their potentially greater food density (Fig. 4). Plots in tree-fall gaps contained more stems in the okapi browse layer than did plots in the subcanopy of either swamp or upland.

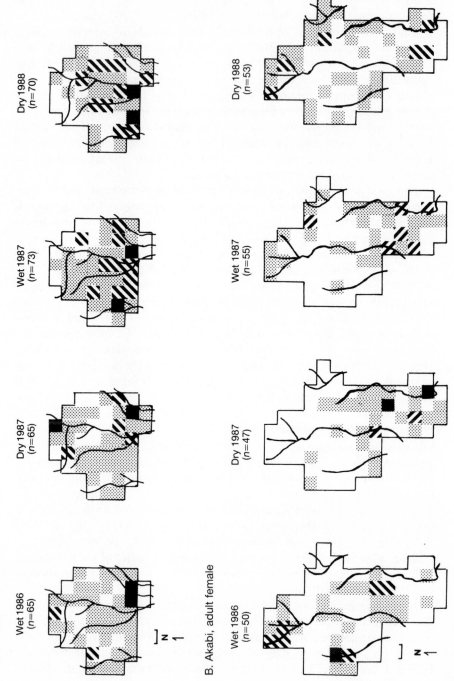

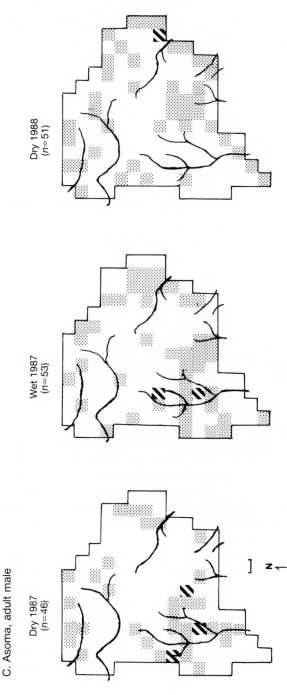

**Fig. 2.** Home-ranges of three radio-collared okapi on the Edoro Study Area. Areas delimited by a solid line circumscribe 250 m quadrat locations for the entire study period with the exception of outliers (5% of total locations). Numbers of locations per quadrat are shown for each of four seasons during the years 1986–88. Blank quadrats had no locations during period; stippled had 1–2 locations; diagonally-barred had 3–4; and black quadrats had more than four locations. Major stream systems on each home-range are shown. The scale bar indicates 250 m.

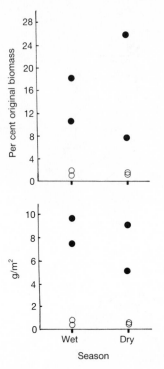

**Fig. 3.** Seasonal and habitat differences in dryweight regrowth on clipped plots in the okapi browse layer. Values for tree-fall gaps are shown by solid symbols; subcanopy values by open symbols.

There were also significantly more species in the gap plots than in subcanopy plots (ANOVA, *a priori* comparisons, $P < 0.001$ in both cases).

Another indication of the greater density of suitable browse in tree-fall gaps was the change in locomotion associated with feeding in gaps. When browsing in the subcanopy, okapi walked steadily, eating as they walked, often moving back and forth over a given area. When feeding in a tree-fall gap, in contrast, okapi moved more slowly, often pausing to feed on several plants within reach from one position.

Comparison of the forage species composition of the three habitat types was complicated by the large component of rare species in all habitats. Nevertheless, a consideration of the most common species was informative (Table 3). Most striking was the similarity in the total contribution of these species to the available forage in all environments. The most common understorey species in the subcanopy, notably *Scaphopetalum dewevrei*, *Drypetes* sp. and *Diospyros bipendensis*, were also among the most common species in gaps and in swamp forest. Even so, these species

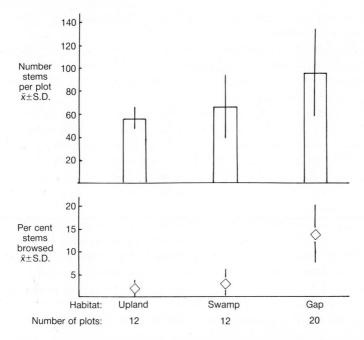

**Fig. 4.** Density of available stems and observed browse damage on 10 m by 2 m plots in upland subcanopy, swamp subcanopy and tree-fall gaps in both upland and swamp. Values shown are mean and standard deviation.

generally accounted for less than one third of available stems. The only exception was *S. dewevrei*, which occasionally accounted for as much as 40% of available stems and then only on subcanopy upland sites. Other species, such as *Pancovia harmsiana* and *Dasylepsis seretii*, though less abundant, nevertheless occurred in all environments. All five of these more common species were shade-tolerant evergreen species that reached maturity in subcanopy layers.

The importance of shade-tolerant species to the composition of tree-fall gaps as well as the understorey of undisturbed forest could be attributed to the fact that gaps were forest understorey that had lost its upper canopy layers and suffered some ground disturbance. The similarity of the understorey of swamp forest with that of upland forest was related to water flow on the study area. Most of the swamp forest was hummocky and although characterized by impeded drainage contained, nonetheless, islets of dry land during most of the year. Although the canopy species of swamp forest were clearly distinct from those of upland forest, the hummocks maintained a shrub flora similar to that of upland forest. Permanently saturated areas, on the other hand, were typically dominated by dense stands of monocots (mainly Marantaceae and Zingiberaceae) which were not browsed by okapi.

In general the shade-tolerant shrubs and tree seedlings of the undisturbed understorey were not favoured okapi browse. This is made evident by the low proportion of available species used by okapi on the subcanopy plots. On the upland plots an average of only 3.4% (range 0–9.5%) and on the swamp plots an average of only 6.1% (range 0–20%) of the species producing foliage between 0.5 m and 2 m had been browsed. Typical subcanopy species were more likely to be browsed in tree-fall gaps than in the shady understorey (Table 3A). This tendency was seen for all the species that were relatively abundant in both gaps and subcanopy and was significant for the two most numerous, *Scaphopetalum dewevrei* and *Drypetes* sp. ($\chi^2$ $P < 0.05$).

Whereas all the most common subcanopy species were also found in gaps, there was an additional group of species that grew only in gaps; these were locally abundant and were preferred okapi browse. This group of light-dependent species was composed of fast-growing tree seedlings, shrubs and lianes (Table 3B). At least 25% of the browsed species in the gap plots fell in this category. Even in gaps, however, generally fewer than one half of the individuals of a given species would be eaten (Table 3).

### Home-range use and feeding during lactation

The area of forest utilized by a lactating female, Bahati, was not markedly different from that used before parturition and during early pregnancy. Seasonal totals of 65 to 73 locations were distributed over an average of 37 quadrats during the non-lactation period (wet season 1986 through wet season 1987). During lactation (dry season 1988) 70 locations were distributed over 33 quadrats (Table 2). The relatively greater aggregation of locations during this period was associated with the presence of a nesting young.

Prior to and during early pregnancy, Bahati was recorded in slow movement only 25% of the time. Most of her time was spent stationary (Fig. 5). After calving and during the first two months of lactation, Bahati spent a significantly greater proportion of her time (45%) in slow movement ($\chi^2 = 10.10$, $P < 0.01$). This was true for morning, noon and evening sampling periods, being most marked during the morning and midday. The increase in slow movement was made at the expense of stationary time; the proportion of fast or directed movement did not change markedly.

Of all activity categories, feeding was most often associated with slow movement. During lactation feeding occurred in both gaps and undisturbed understorey. Slow movement in undisturbed understorey was accompanied by feeding during 66% of 83 h of observation during lactation. Before

**Table 3.** Common plant species browsed by okapi in the Ituri Forest of Zaire.

| Species | Upland (12 plots) | | | Swamp (12 plots) | | | Gap (20 plots) | | |
|---|---|---|---|---|---|---|---|---|---|
| | Plots | Stems per plot (range) | Percent[a] stems browsed | Plots | Stems per plot (range) | Percent stems browsed | Plots | Stems per plot (range) | Percent stems browsed |
| **A. SHADE TOLERANT SPECIES** | | | | | | | | | |
| *Scaphopetalum dewevrei* | 12 | 5–25 | 0 $n=162$ | 8 | 0–18 | 0 $n=61$ | 19 | 0–32 | 5.7 $n=236$ |
| *Drypetes* sp. | 11 | 0–5 | 10.3 $n=39$ | 11 | 0–12 | 16.0 $n=50$ | 17 | 0–13 | 32.3 $n=99$ |
| *Diospyros bipendensis* | 7 | 0–4 | 11.8 $n=17$ | 5 | 0–4 | 16.7 $n=12$ | 12 | 0–6 | 48.3 $n=29$ |
| *Pancovia harmsiana* | 9 | 0–3 | 5.3 $n=19$ | 4 | 0–2 | 0 $n=6$ | 17 | 0–4 | 13.9 $n=36$ |
| *Dasylepsis seretii* | 8 | 0–4 | 0 $n=13$ | 2 | 0–1 | 0 $n=2$ | 9 | 0–6 | 22.2 $n=27$ |
| **B. LIGHT-DEPENDENT SPECIES** | | | | | | | | | |
| *Musanga cecropioides* | 0 | – | – | 0 | – | – | 3 | 0–2 | 33.3 $n=6$ |
| *Macaranga monandra* | 0 | – | – | 5 | 0–4 | 20.0 $n=10$ | 5 | 0–3 | 37.5 $n=8$ |
| *Macaranga spinosa* | 0 | – | – | 0 | – | – | 2 | 0–14 | 53.3 $n=15$ |
| *Manniophyton fulvum* | 2 | 0–1 | 0 $n=2$ | 5 | 0–3 | 11.1 $n=9$ | 8 | 0–9 | 56.3 $n=16$ |
| Unidentified liane | 0 | – | – | 1 | 0–1 | 0 | 8 | 0–3 | 78.6 $n=14$ |

[a] Percentage of total stems ($n$) summed over all plots.

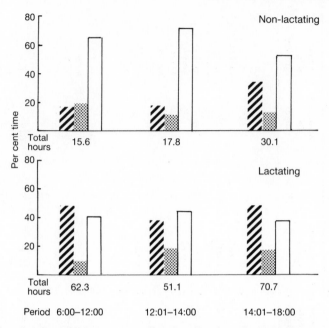

**Fig. 5.** Diurnal activity patterns of an adult female okapi (Bahati) during early pregnancy and peak lactation. Activities, recorded as percentage of total observation time, include 'slow movement' (diagonally-barred histogram), 'directional movement' (stippled histogram) and 'stationary' (open histogram).

lactation, in contrast, only 18% of 16 h of movement through undisturbed subcanopy was accompanied by feeding.

To summarize, it is hypothesized that Bahati met increased energy needs associated with lactation by spending considerably more time browsing in the undisturbed understorey and less time stationary and resting. Pre-gestation and early gestation energy needs may allow for a considerable proportion of necessary browse to be acquired from concentrated sources such as tree-fall gaps and hence allow for longer periods of non-feeding rest time. Bahati did feed in gaps during lactation; however, these resources alone were apparently not adequate. The increased requirements of lactation were met by intensified feeding on the more dispersed resources of the undisturbed understorey.

## Discussion

Not all factors affecting the okapi's use of space are food-related. For example, that the home-ranges of adult males are larger than those of adult females is probably a reflection of social factors rather than of food needs, as males are smaller than females (Landsheere 1957). Likewise the relatively small home-range size of the subadults and the very old male, Mangese, may have been related to the presence of neighbouring 'dominant' okapi. It is also possible that the lactating okapi did not increase home-range size in response to increasing energy needs because of the presence of a nesting young. Perhaps the constraint imposed by distance to the calf was stronger than the need to find more concentrated food supplies.

Size of home-range and, thereby, quantity of available resources may, nevertheless, have a significant effect on the success of individuals. This is indicated by the fact that the two female okapi with the largest home-ranges were the only ones to reproduce successfully during the study period. Also, the three animals with the smallest home-ranges were either killed by leopard or emigrated from the study area.

Seasonality did not appear to be a major factor affecting movements or distribution of okapi. There was no apparent period of food dearth. Leaf flush was at all times evident and certain widespread subcanopy species (e.g. *Pancovia harmsiana*) produced new leaves subsequent to dry-season thinning of the upper canopy. There was no regular seasonal change in either home-range size or movement pattern. Nor was there a strong habitat bias in okapi use of space. Okapi used both major forest types, swamp and upland, throughout the year.

The use of space by okapi indicated that most of the available foliage was of poor quality and that acceptable forage was thinly dispersed without a strong seasonal constraint on availability. This contrasts with the okapi's nearest relative, the giraffe. Studies in the Serengeti woodlands have shown a seasonal element in the giraffe's use of space. Giraffe maintain diet quality by moving into riparian woodlands during the dry season (Pellew 1984). Compared with the okapi's forage plants, the giraffe's food resources are apparent and concentrated. In the relatively species-poor Serengeti woodlands the dominant plant species were also the giraffe's major food species with only 12 species making up more than 85% of the diet (Pellew 1983a, 1984). The okapi, in contrast, is confronted with a diverse forest understorey in which the vast majority of plants are not acceptable forage. The solitary habits of the okapi are consistent with a thinly dispersed food resource.

The food resources of the okapi are not, however, homogeneous. The most intense browsing occurred in small forest openings, notably tree-fall gaps. Such gaps were found in both upland and swamp forests but, overall, accounted for only a small percentage of forest area. An earlier study in the Ituri Forest found that in upland forest, tree-fall disturbances approximately six months to two years old, the age span most attractive to okapi, accounted for less than 5% of total area (T. Hart 1985). Gap size and gap density were also variable from one sample site (of 2.5 ha) to another. It may be that areas of clumped location points found for one female okapi, Akabi, corresponded to areas of higher gap density and hence greater concentration of high-quality food resources.

Notes were kept on the frequency with which okapi visited specific blowdowns. The first browsing forays occurred when regrowth was about six months old. For the next eight months okapi did not visit any one area more than once in three weeks. On successive visits it was unlikely that the same individual plant would be browsed. Instead the animal sought out new seedlings or resprouts which had grown into the okapi browse layer. Thus, although tree-falls represented a concentration of high-quality browse, the resource could only be exploited intermittently and was ultimately transitory.

Even in gaps okapi were selective feeders. Not only did they utilize only a small proportion of the species present but they also ate only the youngest leaves. They never browsed from all the available individuals of a given food species. Despite their large size (200–250 kg, adult), okapi are well modelled for selective feeding. The muzzle is narrow, almost pointed; the lips dexterous and muscular and the tongue exceptionally long and prehensile.

The feeding behaviour of a lactating female supported the supposition that feeding selectivity was a reflection of food quality. As energy needs increased she did not continue to meet food requirements by browsing large gaps ever more intensively. Instead, she adopted a more energetically expensive feeding behaviour. She spent more time walking, exploiting smaller gaps and feeding as she walked, utilizing the less dense food resources available in the subcanopy.

The preliminary results presented in this paper indicate that the main constraints on okapi feeding behaviour and use of space in the Ituri Forest are not seasonal nor habitat-related but, rather, relate to food quality and distribution. High-quality food is nowhere abundant and the patches in which it occurs are small, dispersed and transitory.

## Acknowledgements

Our field research on the okapi is supported by a grant from Wildlife Conservation International, a branch of the New York Zoological Society. The Institut Zairois pour la Conservation de la Nature has sponsored the field work in Zaire and provided important logistical assistance. The Jardin Botanique National de Belgique provided most of the necessary plant identifications. We are deeply indebted to our field assistants, in particular Batido Banangana, Richard Peterson and Sikubwabo Kiyengo. P. Jewell, T. Struhsaker and L. Lyland criticized an earlier draft of this paper.

# References

Alexandre, D.Y. (1978). Le rôle disséminateur des éléphants en Forêt de Taï, Côte d'Ivoire. *Terre Vie* **32**:47–72.

Bodmer, R.E. & Rabb, G.B. (1985). Behavioral development and mother-infant relations in the forest giraffe (*Okapia johnstoni*). In *Zoom op Zoo*:33–53. (Ed. Kruyfhooft, C.). Royal Zoological Society of Antwerp, Antwerp, Belgium.

Bodmer, R.E. & Rabb, G.B. (In press). *Okapia johnstoni*. *Mammalian Sp.*

Bultot, F. (1971). *Atlas climatique du bassin congolais, deuxième partie.* Institut National pour l'Etude Agronomique du Congo (INEAC), Brussels.

Coley, P.D. (1983). Intraspecific variation in herbivory on two tropical tree species. *Ecology* **64** (3):426–33.

Dubost, G. (1984). Comparison of the diets of frugivorous forest ruminants of Gabon. *J. Mammal.* **65** (2):298–316.

Emmons, L.H., Gautier-Hion, A. & Dubost, G. (1983). Community structure of the frugivorous-folivorous forest mammals of Gabon. *J. Zool., Lond.* **199**:209–22.

Feer, F. (1979). Observations écologiques sur le néotrague de Bates (*Neotragus batesi* DeWinton, 1903, artiodactyle, ruminant, bovidé) du nord-est du Gabon. *Terre Vie* **33**:159–237.

Fleming, T.H. (1979). Do tropical frugivores compete for food? *Am. Zool.* **19**:1157–72.

Gautier-Hion, A., Duplantier, J.-M., Emmons, L., Feer, F., Heckestweiller, P., Moungazi, A., Quris, R. & Sourd, C. (1985). Coadaptation entre rythmes de fructification et frugivorie en forêt tropical humide du Gabon: Mythe ou réalité. *Terre Vie* **40**:405–29.

Harrison, S. (1987). Treefall gaps versus forest understory as environments for a defoliating moth on a tropical forest shrub. *Oecologia* **72**:65–8.

Hart, J. (1986). *Comparative dietary ecology of a community of frugivorous forest ungulates in Zaire.* Ph.D. Thesis: Michigan State University.

Hart, T. (1985). *The ecology of a single-species-dominant forest and of a mixed forest in Zaire, Africa.* Ph.D. Thesis: Michigan State University.

Landsheere, J.D. (1957). Observations concernant la capture, l'élévage et les soins de l'okapi. *Zoo, Antwerp* **23**:12–25.

Leigh, E.G., Jr. & Smythe, N. (1978). Leaf production, leaf consumption, and the regulation of folivory on Barro Colorado Island. In *The ecology of arboreal folivores*: 33–50. (Ed. Montgomery, G.G.). Smithsonian Institution Press, Washington, D.C.

Maddock, L. (1979). The 'migration' and grazing succession. In *Serengeti: dynamics of an ecosystem*: 104–29. (Eds Sinclair, A.R.E. & Norton-Griffiths, M.). University of Chicago Press, Chicago & London.

Pellew, R.A. (1983a). The giraffe and its food resource in the Serengeti. I. Composition, biomass and production of available browse. *Afr. J. Ecol.* **21**:241–67.

Pellew, R.A. (1983b). The giraffe and its food resource in the Serengeti. II. Response of the giraffe population to changes in the food supply. *Afr. J. Ecol.* **21**:269–83.

Pellew, R.A. (1984). Food consumption and energy budgets of the giraffe. *J. appl. Ecol.* **21**:141–59.

Rabb, G.B. (1978). Birth, early behavior and clinical data on the okapi. *Acta zool. path. antverp.* No. 71:93–105.

Schoener, T.W. (1981). An empirically based estimate of home range. *Theoret. Pop. Biol.* 20:281–325.

Sinclair, A.R.E., Dublin, H. & Borner, M. (1985). Population regulation of the Serengeti wildebeest: a test of the food hypothesis. *Oecologia* 65:266–8.

Swihart, R.K. & Slade, N.A. (1985a). Influence of sampling interval on estimates of home-range size. *J. Wildl. Mgmt* 49:1019–25.

Swihart, R.K. & Slade, N.A. (1985b). Testing for independence of observations in animal movements. *Ecology* 66 (4):1176–84.

Terborgh, J. (1986). Keystone plant resources in the tropical forest. In *Conservation biology: the science of scarcity and diversity*: 330–44. (Ed. Soulé, M.). Sinauer Publishers, Sunderland, MA.

Waterman, P.G., Choo, G.M., Vedder, A.L. & Watts, D. (1983). Digestibility, digestion-inhibitors and nutrients of herbaceous foliage and green stems from an African montane flora and comparison with other tropical flora. *Oecologia* 60:244–9.

White, F. (1983). *Vegetation map of Africa south of the Sahara*. (2nd edn). UNESCO/AETFAT/UNSO, Paris.

*Symp. zool. Soc. Lond.* (1989) No. 61: 51–71

# Buffalo and their food resources: the exploitation of Kariba lakeshore pastures

R.D. TAYLOR

*Department of National Parks and Wild Life Management P. Bag 2003, Kariba, Zimbabwe*

## Synopsis

Aspects of the spatial and social organization of buffalo and their feeding behaviour were studied at Kariba, Zimbabwe, where lakeshore pastures provide an abundant high quality food source during the hot dry season. Buffalo distributed themselves seasonally according to the availability of food with lakeshore pasture being the most preferred habitat. Adjacent food and water resources contributed to reduced movement and home range size. Compared to other buffalo populations, the Kariba buffalo grazed longer and ruminated less during the dry season. Within the daily cycle, grazing peaks were interspersed with rest periods, either in shade or wallowing, during the middle of the day, with 50% or more of grazing time occurring at night. Most buffalo were rarely more than 3 km away from water so that drinking bouts were short and frequent. The number and size of buffalo herds and male groups were greatest on the lakeshore pastures during the hot dry season when densities reached 40 buffalo $km^{-2}$. Bachelor male groups preferred the lakeshore all year round, with group size doubling during the dry season. Seasonal change in herd size was different to other buffalo populations and was related more to food distribution and abundance than to breeding. The non-random spacing of buffalo along the lakeshore resulted in a separation between breeding herds and male groups, which reduced competition for both space and food. These behavioural responses to a new environment maximized the advantages of a seasonally abundant food resource adjacent to a permanent water supply.

## Introduction

Spatial and social organization reflects the response of a population to its environment, and so will vary with change in that environment (Ewer 1968; Leuthold 1977). In 1958 flooding of the middle Zambezi valley above Kariba gorge reduced the range of many large herbivores, depriving them of important dry-season habitats (P.J. Jarman 1972a,b). By 1963 the filling of

ZOOLOGICAL SYMPOSIUM No. 61
ISBN 0–19–854009–4

Lake Kariba was complete. Since then an important ecological development has been the colonization of the lakeshore by the swamp grass *Panicum repens* L. (Magadza 1970; Bowmaker 1973). The resultant pastures have become a major food source for a number of large herbivores, in particular the African buffalo, *Syncerus caffer* (Sparrman) (Begg 1973; Taylor 1979).

Successful exploitation of this new food resource has necessitated a number of behavioural adjustments to a changed environment. Such adjustments are usually reflected in the seasonal distribution of buffalo and the associated occupation of different habitats at different times of the year, often in response to fluctuations in the availability of food or water or both. Home range size, seasonal splitting and spacing of buffalo herds are likewise related (Sinclair 1977; Monfort 1979, 1980) whilst climate has an important influence on daily movement, feeding behaviour and other related activities (Leuthold 1977). The present study was undertaken to measure adjustments in the spatial and social organization of buffalo and their feeding behaviour in relation to the Kariba environment, in particular the lakeshore pastures.

## The study area

Buffalo were studied in Matusadona National Park, Zimbabwe, on the southern shores of Lake Kariba below the Zambezi escarpment (Fig. 1). A single hot wet rainy season extends from November to April, a cool dry season from May to July and a hot dry season from August to October. Annual rainfall varies between 400 and 800 mm and mean minimum and maximum temperatures are 17 and 35°C respectively.

A perennial pasture of *Panicum repens* along the lakeshore has been maintained by an annual fluctuation in lake level. During lake level recession, pasture is increasingly exposed, so that from July until the onset of the rains in November green food is available to grazing herbivores during the hot dry season. This increase in available green pasture is inversely proportional to the decline in biomass and quality of savanna grasslands away from the lakeshore (Taylor 1979, 1985). Between the lakeshore and the escarpment the valley vegetation is dry deciduous *Colophospermum mopane* woodland and *Combretum* − *Terminalia* spp. tree and shrub savanna. In addition to elephant, *Loxodonta africana* (Blumenbach), and other large mammals typical of the Zambezi valley, the area supports a buffalo population upwards of 4000 animals (10 buffalo $km^{-2}$) which has been increasing in response to the food supply for the past 20 years (Taylor 1985).

## Methods

Over four years, systematic flight lines 2 km apart were flown monthly

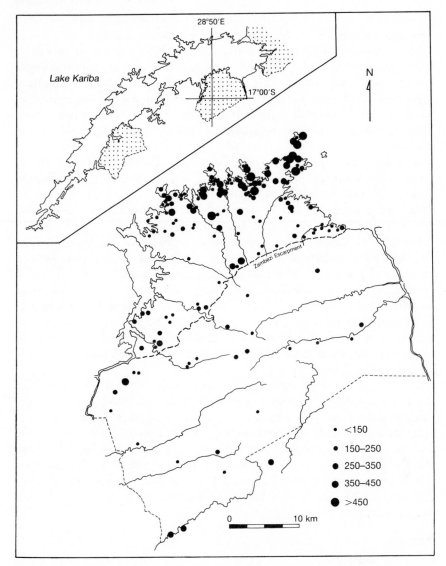

**Fig. 1.** The study area location on the southern shores of Lake Kariba, Zimbabwe, showing the distribution and size of buffalo herds in Matusadona National Park. Stippling indicates protected wildlife areas.

across the valley and along the lakeshore of Matusadona National Park to locate buffalo breeding herds and bachelor male groups. Their positions were plotted to within 500 m on 1:100 000 maps and herds were photographed following visual estimates of their size. Herds were counted subsequently from the photographic prints (Sinclair 1973) and where photography was not possible, bias in visual estimates was corrected (Norton-Griffiths 1974). Male groups were counted visually.

For each survey the presence or absence of each herd and group in one or more vegetation types was recorded on a 0.25 km$^2$ grid superimposed over a vegetation map of the area. Chi-squared $(\chi^2)$ values, association coeffecients and their standard errors were calculated using the method of Sinclair (1977), adapted from that of Cole (1949). Mean monthly and seasonal values were used in analysing buffalo distribution and habitat selection.

Following the capture of two buffalo herds (Coetsee & Taylor 1978), three young adult females (age 4–6 years) in each herd were immobilized with drugs and marked with brightly coloured neck collars. Subsequent aerial and ground monitoring of marked and other herds provided data on seasonal, weekly and daily movement patterns.

Feeding behaviour was measured using groups of known resident males and herds of up to 200 buffalo. Animals were located on foot and observed at 30–50 m using 8 × 30 binoculars. Continuous diurnal activity times for grazing, ruminating, drinking, wallowing and time spent in shade were recorded over a period of 2–5 days each month. Nocturnal measurements were limited to full moon periods during the hot dry season. Since herd behaviour is closely synchronized (Sinclair 1977), the activity of 50% or more of the herd was recorded. Additional information on weekly and daily movement was also obtained.

For each of the three seasons the proportion of each half-hour period of the day buffalo spent grazing and in shade or wallowing, or both, was calculated from the measurements made above. Ambient shade temperatures were recorded hourly during periods of observation in the hot dry season and mean monthly temperatures were obtained from Park records. Using the aerial survey data, the distance between individual herds, and their distance from permanent and seasonal water supplies, were measured. Error estimates are 95% confidence limits on either side of the mean values unless otherwise stated.

## Results

### Seasonal distribution

During the hot wet season buffalo herds were dispersed throughout the valley with no selection or avoidance of the lakeshore $(\chi^2 = 0.14$; n.s.) or the valley $(\chi^2 = 0.05$; n.s.). Although selection of the lakeshore was significant in the cool dry season $(\chi^2 = 8$; $p < 0.01)$, the valley was neither selected or avoided $(\chi^2 = 1.33$; n.s.). During the hot dry season there was a major concentration of buffalo herds on the lakeshore for which selection was highly significant $(\chi^2 = 120.3$; $p < 0.001)$. There was also a significant avoidance of the valley at the same time $(\chi^2 = 12.03$; $p < 0.001)$.

## Habitat selection

The association of buffalo herds with different vegetation types is shown in Fig. 2. The highest association was with lakeshore pasture in September during the hot dry season ($C = 0.480$; $p < 0.001$). At the same time there was a negative association with all other habitat types. After lakeshore pasture, the next best habitat was woodland savanna with which there was a positive association during the cool dry season. The least preferred habitat was *C. mopane* woodland.

Over the period 1976–80, association with lakeshore pasture was highest during each hot dry season when lake levels were receding and an increasing area of pasture was being exposed (Fig. 3). Following the onset of the annual rains, usually in mid-November, the level of association declined as buffalo herds dispersed into the valley away from the lakeshore. There was a negative association in the cool dry season of each year when peak lake levels were reached and buffalo were denied large areas of pasture because it was inundated.

Buffalo males showed a positive and significant selection for, and association with, lakeshore pasture throughout the year ($p < 0.001$, Table 1). There was a significant avoidance of the valley ($p < 0.001$).

## Movement patterns

Sightings of collared animals were infrequent, only 14 observations of one marked herd and 13 of the other being made over a five-month period

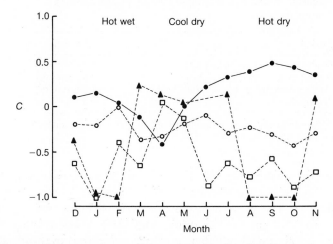

**Fig. 2.** The monthly association coefficients (C) of buffalo breeding herds with different vegetation types in Matusadona National Park. Closed circles (●), lakeshore pasture; closed triangles (▲), woodland savanna; open circles (○), tree and shrub savanna; open squares (□), *C. mopane* woodland. The solid line with closed circles (●) indicates the best available habitat.

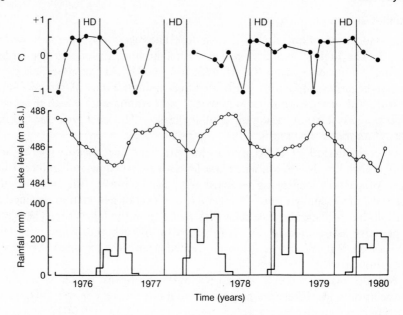

**Fig. 3.** The seasonal change in association (C) of buffalo breeding herds with lakeshore pasture and the relationship to lake level fluctuation and rainfall over the period 1976–1980. HD indicates the hot dry season.

**Table 1.** Seasonal association coefficients (C) of buffalo bachelor males with lakeshore pasture in Matusadona National Park.

| Season | $C^a$ ± S.E. |
|---|---|
| April–May (cool dry, lake high) | 0.79 ± 0.066 |
| June–July (cool dry, lake falling) | 0.81 ± 0.065 |
| August–November (hot dry, lake low) | 0.64 ± 0.061 |
| December–March (hot wet, lake rising) | 0.53 ± 0.049 |

[a] All values of C are significant at the 0.1% level of probability. S.E. = standard error.

during the dry season (Fig. 4). The two herds overlapped slightly and remained adjacent to the lakeshore, apart from one observation at the escarpment base. The greatest distance between sightings was 8.5 km for one herd and 11.7 km for the other, suggesting a home range size of 60–110 km². The mean number of herds observed over an area of approximately 400 km² was 5.3. Assuming little or no overlap between herds (Sinclair 1977; Prins 1987) a similar home range size of 60–80 km² is indicated.

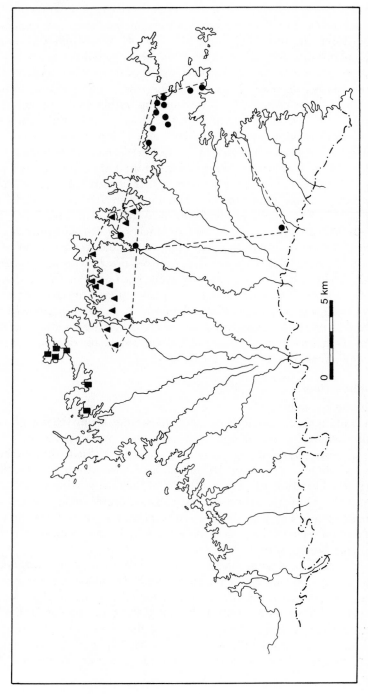

**Fig. 4.** The dry-season movement of buffalo herds in Matusadona National Park. Distances between like symbols indicate weekly movement of the two marked herds, represented by closed circles (●) and triangles (▲), and the approximate home ranges are indicated by the broken lines.

Over weekly intervals the furthest distance between sightings of individual herds observed from the air varied from 1 to 5 km, indicating that each herd had moved over an area of up to 20 km² (Fig. 4). When followed on the ground, the distance covered by each of two herds amounted to 6 km over six days in September and 11 km over 10 days in October respectively. These two herds behaved in a similar fashion by remaining in one locality for four and 10 days respectively, before moving, within 48 h, distances of 5 and 8 km to new localities. The total area covered by each herd during the six- or 10-day period was not more than 8 and 17 km² respectively.

The mean daily distance covered by marked herds, measured from the air, was 1.3 ± 0.4 km. For herds observed on the ground, the mean distance was 1.1 km. Overall, minimum daily movement varied from 1.2–2.2 km (95% C.L.) which is lower than the minimum figure of 2.8 km measured by Sinclair (1977). The minimum distances measured underestimate the real distance travelled and should be doubled (Sinclair 1977). This provides an estimate of 3.4 km moved during 24 h; but the distance travelled can be as great as 6–7 km per day when herds move from one locality to another.

Buffalo males that lived permanently away from breeding herds moved considerably less and occupied very much smaller areas. One group of up to 13 males occupied an area no larger than 50 ha. The area occupied by a second group of 16–18 animals did not exceed 2 km² whilst a third group of two to six males occupied an area of 2–3 km². The daily distance covered by any one animal did not exceed 2 km.

### Feeding behaviour

### Grazing and ruminating

For much of the year, buffalo males spent most time during the day grazing lakeshore pasture, but during peak lake levels there was a marked switch to grazing savanna grassland (Fig. 5). During the hot dry season the lakeshore was grazed almost exclusively. During 66 h of observation in September and October, buffalo herds spent only 1% of the time grazing savanna. In November the proportion of time grazing pasture decreased as time grazing savanna increased. These grazing patterns support the seasonal distribution of buffalo and their habitat associations (Figs 2, 3) and confirm that they selected lakeshore pasture for feeding.

Both seasonally and within the 24-h cycle some 76% of time was taken up with grazing and ruminating (Table 2), which is similar to the values of 65% and 85% measured by Grimsdell & Field (1976) and Sinclair (1977) respectively. Although time spent in diurnal grazing was similar (Table 2), ruminating declined during the dry season so that males spent on average 43% of time grazing and 31% ruminating. Similarly, buffalo herds spent 47% of time grazing and 41% ruminating during the day in the hot dry season.

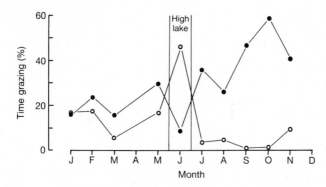

**Fig. 5.** The seasonal variation in the proportion of time spent grazing lakeshore pasture (closed circles, ●) and savanna grassland (open circles, ○), by buffalo bachelor males in Matusadona National Park. The period of high lake level is also shown.

Between May and November consistently lower ruminating times closely followed the grazing pattern (Fig. 6), which was also significantly correlated with digestibility and forage intake ($p < 0.01$). From August onwards, abundant high-quality food allowed both longer grazing times and greater intake for a similar or lower ruminating time (Taylor 1985). Dry-season increases in grazing time elsewhere (M.V. Jarman & Jarman 1973;

**Table 2.** Grazing and ruminating times for buffalo bachelor males in Matusadona National Park

| Season | Diurnal | | Nocturnal | | 24-hour mean | |
|---|---|---|---|---|---|---|
| | Hours | % | Hours | % | Hours | % |
| Grazing | | | | | | |
| Hot wet | 3.9 ± 0.35 | 32.5 | | | | |
| Cool dry | 6.2 ± 0.89 | 51.6 | | | | |
| Hot dry | 5.4 ± 0.45 | 45.0 | 6.9 ± 1.05 | 57.5 | 12.3 ± 0.77 | 51.3 |
| Mean | 5.1 ± 0.98 | 42.5 | | | | |
| Ruminating | | | | | | |
| Hot wet | 6.5 ± 1.91 | 54.2 | | | | |
| Cool dry | 3.0 ± 1.91 | 25.0 | | | | |
| Hot dry | 2.2 ± 1.30 | 18.3 | 3.9 (3.1 − 4.7)[a] | 32.5 | 6.1 ± 1.63 | 25.4 |
| Mean | 3.7 ± 1.52 | 30.8 | | | | |
| Combined | | | | | | |
| Hot wet | 10.4 ± 1.19 | 86.7 | | | | |
| Cool dry | 9.2 ± 1.66 | 76.7 | | | | |
| Hot dry | 7.6 ± 0.93 | 62.5 | 10.8 ± 1.52 | 90.0 | 18.4 ± 1.20 | 76.7 |
| Mean | 8.8 ± 1.21 | 74.2 | | | | |

[a] indicates range

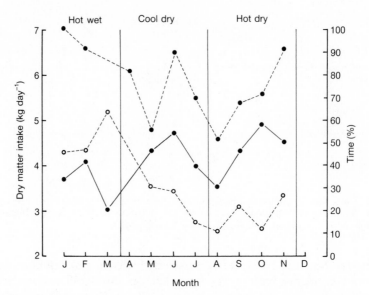

**Fig. 6.** The seasonal relationship between the proportions of time spent grazing (solid line with closed circles, ●) and ruminating (broken line with open circles, ○) and dry matter forage intake (broken line with closed circles, ●) for buffalo in Matusadona National Park.

Grimsdell & Field 1976) are associated with declining quality of herbage; they are also accompanied by a substantial increase in rumination.

Compared with data for the Serengeti (Sinclair 1974), buffalo males on lakeshore pastures grazed longer ($p < 0.001$; $\chi^2 = 78.96$) and ruminated less ($p < 0.001$; $\chi^2 = 56.07$) over a 24-h period during the hot dry season when food was plentiful and of good quality. Although time spent in diurnal grazing over all seasons was not significantly different between males from the two areas, buffalo herds at Kariba grazed more during the day ($p < 0.001$; $\chi^2 = 28.0$) in the dry season than did buffalo herds in the Queen Elizabeth National Park, Uganda (Grimsdell & Field 1976). In this instance there was no significant difference in rumination time.

### Activity cycles

For buffalo males throughout the year there were two grazing peaks evident during the day, early to mid-morning and mid- to late afternoon, interspersed with a rest period from about 12.00 to 14.00 h (Fig. 7). In the wet season the two grazing peaks were least obvious. Short grazing bouts alternated with longer rest periods at fairly regular intervals. Grazing bouts increased in duration into the cool dry season, with a reduction in the period spent in shade at midday. Although herds grazed longer in the hot dry season, both social groups displayed two grazing peaks at similar times during the day. In addition to the midday rest period, both herds and males also spent part of the early morning in shade, so that two rest periods emerged in contrast to the single peak of the previous season (Fig. 7). At night males showed two grazing peaks, from 20.00 to 22.00 h and from 02.00 h onwards. Nocturnal grazing time was slightly longer than diurnal (Table 2).

## Shade and wallowing

Although the seasonal variation in time spent grazing was inversely correlated with that spent in shade or wallowing ($p < 0.01$) this relationship was not well correlated with seasonal variation in ambient shade temperature. Only during the cool dry season was there a positive relationship between time in shade and ambient temperature ($p < 0.05$). During the hot dry season time spent in shade decreased as grazing increased, despite maximum temperatures being reached. Within the daily cycle, however, rest periods in shade or wallowing occurred during hours of peak daily temperatures (Fig. 7), usually between 13.00 and 17.00 h. On average, 40% of time was spent on these activities, of which 27% was in shade and 13% wallowing on the lakeshore.

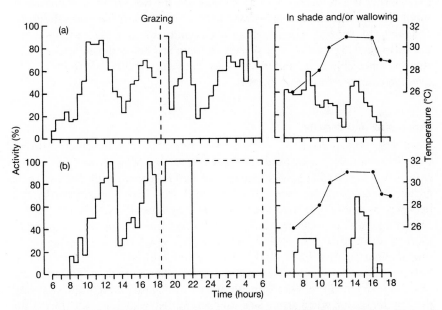

**Fig. 7.** Daily activity cycles during the hot dry season for buffalo in Matusadona National Park. The proportion of each half-hour spent grazing and in shade and/or wallowing is shown for bachelor males in (a) and for breeding herds in (b). The ambient shade temperature is also indicated. Herd activity between 22.00 and 06.00 h was not recorded.

## Water resources and drinking activity

Permanent water supplies for buffalo were provided by the lake and by seasonal rivers and pans which hold water in pools. The greatest distance between the lakeshore and the escarpment is about 15 km. Above-average rainfall during the study period ensured adequate water supplies inland for much of the year. During the wet season buffalo herds were observed a greater distance away from permanent water ($\bar{x} = 2.8 \pm 1.0$ km) than during the dry season when the mean distance from permanent water was $1.3 \pm 0.6$ km. The converse held for seasonal water supplies. Buffalo herds were closer to these ($\bar{x} = 1.5 \pm 0.2$ km) in the wet season but further away in the dry season ($\bar{x} = 2.4 \pm 0.5$ km). Overall, 60% of herds were less than 1.5 km away from permanent water and over 70% between 1 and 3 km away from seasonal water.

For buffalo males, over 50% of drinking bouts occurred between 10.00 and 14.00 h with a peak from 11.00 to 12.00 h. Drinking occurred throughout the day from before 06.00 until 20.00 h and was also observed on two occasions in the early morning hours between 02.00 and 03.00 h. During the wet season, an individual might drink two or three times from the lake's edge, but usually once in the morning and once in the afternoon. During the cooler months, drinking during the day usually occurred only once, either in late morning or at midday. As the hot season advanced, the number of drinks increased to two or three in the morning and afternoon. Drinking bouts for individual animals were, on average, of 1–3 min duration. During the hot dry season in October, herds drank once during the daytime, either around midday or between 19.00 and 16.00 h. The average drinking time was 20 min.

## Group size and density

Herd size ranged from 31 to 669 with over 50% of herds comprising 100–300 animals; mean herd size was $232 \pm 26$. Male group size varied from 1 to 14 individuals, with one or two buffalo constituting 64% of male groups. Mean group size was $2.9 \pm 0.5$. Herd size varied both seasonally and between lakeshore pasture and other vegetation types (Fig. 8). In the hot dry season a greater number of herds on the lakeshore were significantly larger than those inland ($p < 0.001$), and conversely a lesser number of herds inland were smaller ($p < 0.05$), than at other times. The proportion of buffalo males inland amounted to $8 \pm 1.8\%$ so that more than 90% of males observed occupied the lakeshore. Numbers and density were highest during the hot dry season when group size almost doubled from $2.4 \pm 0.6$ during the rains to $4.6 \pm 1.1$.

The density of buffalo on the lakeshore, herds and males combined, ranged from 1 buffalo km$^{-2}$ at the end of the wet season to almost 40

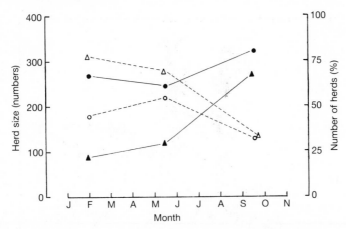

**Fig. 8.** The seasonal change in the mean size and numbers of buffalo breeding herds observed on lakeshore pasture and in other habitat types in Matusadona National Park. Closed circles (●) and triangles (▲) with solid lines indicate size and numbers on pasture, and open circles (○) and triangles (△) with broken lines size and numbers in other habitats.

buffalo km⁻² at the height of the dry season. With the start of the rains density declined rapidly (Table 3).

### Herd spacing

The distances between neighbouring herds were measured; 26% of these measurements showed herds to be between 3 and 4 km apart; the mean distance between herds was $5.7 \pm 0.8$ km. Both the Poisson Index of Dispersion (Southwood 1978) and the Kolmogorov-Smirnov test (Siegel 1956) indicated that the spacing distribution of herds was significantly different from random (Patterson 1965; Sinclair 1977). The distribution of buffalo males shows more clearly this spacing effect (Fig. 9). Males occupied residential areas, which did not appear to be defended in the sense of a territory (Sinclair 1977), and within which they were very sedentary. Herds were

**Table 3.** Seasonal density of buffalo (no. km⁻²) on lakeshore pastures in Matusadona National Park

| Season | Breeding herds | Bachelor males | Combined density |
|---|---|---|---|
| Hot dry – hot wet | 13.7 | 2.4 | 16.1 |
| Hot wet – cool dry | 1.1 | 1.6 | 2.7 |
| Cool dry – hot dry | 9.6 | 1.5 | 11.1 |
| Hot dry | 35.4 | 2.2 | 37.6 |

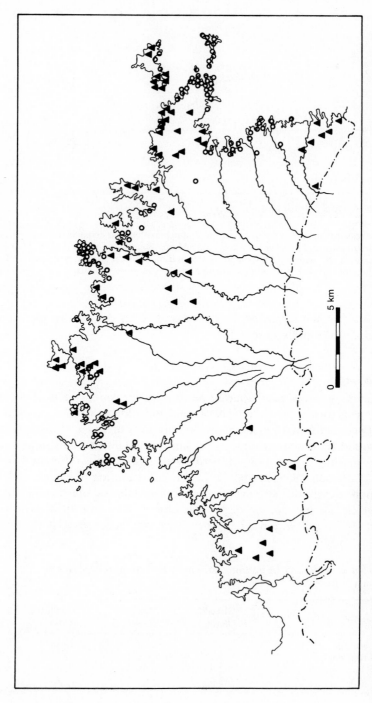

**Fig. 9.** The distribution of buffalo breeding herds (closed triangles, ▲) and bachelor male groups (open circles, ○) showing the spacing effect along the lakeshore between the two social groups. Data collected over one season.

uncommon in these areas, particularly where there were concentrations of
male groups. Thus herds and bachelor males were fairly evenly spaced out
along the lakeshore, with herds moving inland according to the seasonal
availability of food. Most male groups remained along the lakeshore.

## Discussion

Buffalo in the flooded Kariba basin distribute themselves seasonally
according to the available food resources. Wet-season dispersals and dry-
season concentrations are related to the availability of food and water,
which are both highly seasonal in semi-arid environments (P.J. Jarman
1972b; Western 1975). Whereas food is generally a diminishing resource
until the onset of the rains (Sinclair 1975), the lakeshore pastures provide an
increasing dry-season food supply (Taylor 1985). Although one of the most
widespread bovids in Africa, buffalo are water-dependent grazers (Western
1975) preferring green food (Hofmann 1973) and are not found in areas
with rainfall below 250 mm (Stewart & Stewart 1963). Thus adjacent
grazing and water resources at Kariba have reduced considerably two of the
most important environmental constraints to buffalo productivity.

The optimum or best available habitat is that with which buffalo are
associated most during the dry season and for which there has been a
positive selection (Sinclair 1977). Since water and food occurred together at
Kariba, water requirements could explain the high association with
lakeshore pasture. However, buffalo were always close to water (see p. 67)
so that it can be assumed habitat choice was not dependent upon this
resource (Prins 1987). The open treeless pastures provided no shade even
though time spent grazing pasture increased during the dry season; nor did
this habitat provide protection from predators. Clearly, habitat choice was
determined primarily by the availability of food.

Movement and home range size in many African ungulates tend to be
highly flexible and can be adapted to a range of environmental conditions
(Leuthold 1977). Reported home range sizes for buffalo in East Africa vary
from 10 km$^2$ in the Queen Elizabeth National Park (Eltringham &
Woodford 1973) to an average of 222 km$^2$ in the Serengeti (Sinclair 1977).
In Kafue National Park, Zambia, home range size averaged over 700 km$^2$
and in Zimbabwe, around 250 km$^2$ (Conybeare 1980; Mloszewski 1983).
The estimates of about 80 km$^2$ for buffalo in Matusadona National Park,
although not conclusive, are very much smaller.

A feature of decreasing home range size is the strong correlation between
density of animals within the home range and rainfall and primary
production (Sinclair 1977; Leuthold 1977). The highly productive lakeshore
pastures enable seasonally high densities of buffalo to be supported. This
may well be facilitating a decreasing home range size together with a

corresponding increase in density within the home range. Whilst home range may contract owing to dry-season concentrations on rivers (P.J. Jarman 1972a), within the range the distance covered in searching for food is probably greater (Conybeare 1980). The opposite occurs at Kariba, where the localized availability of dry-season green food, together with water, enables animals to move less and to concentrate in areas of limited size.

Weekly and daily movement, although similar in pattern and function to that of other buffalo populations (Leuthold 1972; Grimsdell & Field 1976), were less extensive, with lower mean distances covered than elsewhere. Sinclair (1977), Conybeare (1980), Mloszewski (1983), and Stark (1986) all report distances of 5–8 km covered daily by buffalo herds. By comparison the 3 km measured at Kariba indicates that buffalo are much more sedentary, remaining either on or close to the lakeshore for much of the year. Movement away from the lakeshore also involves relatively short distances.

Seasonal changes in feeding behaviour and related activity patterns are governed primarily by the quantity and quality of food (Hancock 1953; Leuthold 1977). This can be related to expected grazing and ruminating times during the climatic seasons experienced under African conditions and compared with observed data for buffalo in East Africa (Sinclair 1974; Grimsdell & Field 1976; Sinclair 1977) and for buffalo at Kariba (Table 4). Although the interpretation is subjective, the observed data are similar to those expected and suggest that Kariba buffalo were not having to adjust their grazing–rumination pattern to accommodate food of poor quality. Although wet-season results for buffalo males are somewhat anomalous

**Table 4.** Expected and observed grazing and ruminating times for buffalo in relation to food quantity and quality during the climatic seasons experienced under African conditions.

|  |  | Hot wet | Cool dry | Hot dry |
|---|---|---|---|---|
| Food quantity |  | High | Intermediate | Low    High[a] |
| Food quality |  | Good | Mixed | Poor   Good[a] |
| Grazing | Expected | Intermediate | Long | Long |
|  | Observed[b] | Short | Long | Intermediate |
|  | Observed[c] | Short | Long | Long |
| Rumination | Expected | Short | Intermed./ Long | Intermediate |
|  | Observed[b] | Short | Intermediate | Long |
|  | Observed[c] | Long | Intermediate | Short |

[a] Kariba lakeshore pastures
[b] East African buffalo
[c] Kariba buffalo

with unexpectedly long rumination periods, grazing periods are similar to those observed elsewhere. Abundant and high-quality food probably allows buffalo to maximize nutrient intake over a relatively short space of time and thereafter ruminate at will.

The distribution of daytime activity for Kariba buffalo is similar to that seen elsewhere (Sinclair 1974, 1977; Lewis 1977; Mloszewski 1983; Stark 1986) with the mobile activities of grazing and moving concentrated in the early morning and late afternoon. In their alternation of feeding and ruminating phases (Balch 1955) the buffalo are comparable to cattle, with the buffalo displaying relatively longer alternating bouts of feeding and ruminating. A third, although minor, grazing peak in the cool dry season is probably associated with the restricted availability of good-quality food due to high lake levels. Although cattle prefer to graze in daylight in temperate zones (Hancock 1953), under tropical conditions up to a third of grazing time may be at night (Schottler, Efi & Williams 1975) and in wild ruminants, grazing is usually interspersed with other activities throughout the 24-h period (Leuthold 1977). For buffalo at Kariba, grazing at night occupied half or more of their feeding time, which is similar to that reported by Grimsdell & Field (1976), Sinclair (1977) and Mloszewski (1983).

Observations of wild ungulates under a semi-domesticated regime showed that their activity patterns varied mainly in relation to the heat load acting upon them (Lewis 1977) and that ambient temperature accounted for over 80% of the variation in the daily activity of buffalo. At Kariba, although time spent in shade by buffalo was not correlated with the seasonal variation in ambient temperature, rest periods, either in shade or wallowing, nevertheless occurred during hours of peak daily temperatures. Of the four species investigated by Lewis (1977) buffalo showed the greatest susceptibility to heat stress, which was attributed largely to coat charac-teristics. Near the lake, heat stress is alleviated by the prevailing on-shore breezes (Anon. 1981) and the opportunities for wallowing. All social classes were observed wallowing on the lakeshore, although Sinclair (1977) found that in the Serengeti only bachelor males wallowed. This activity involves very little extra energy expenditure, since both food resources and wallowing areas are adjacent to each other.

Whereas limited water supplies are restrictive in other semi-arid environments during the dry season, buffalo on the lakeshore meet their water requirements with little difficulty. As well as the lake itself, there is also widespread surface water during the wet season; shortage of water cannot therefore be a limiting factor. This is especially important since buffalo need to drink daily (Mloszewski 1983). In Amboseli, Western (1975) found that 99.5% of the water-dependent large herbivore biomass was contained within a radius of 15 km from water. This is the greatest distance buffalo below the Zambezi escarpment in Matusadona National

Park have to travel to reach lake water. Most buffalo were rarely more than 3 km away from either seasonal or permanent water.

Although drinking patterns are similar to those of buffalo elsewhere (Weir & Davison 1965; Sinclair 1977; Mloszewski 1983; Stark 1986), for both herds and individual bachelor males the time taken to drink was half that recorded by Mloszewski (1983). Whereas most buffalo drink only once during the day, they are able to do so more frequently at Kariba, thus reducing time spent per drinking bout. Undoubtedly the high moisture content of the food itself (Taylor 1985) also contributes to minimizing water stress (Vesey-Fitzgerald 1960).

The size of buffalo herds is related to the habitat types in which they occur and the dispersion of food in those habitats (P.J. Jarman 1974; Leuthold 1977). Large herds (up to 1500) are found in the open grasslands of the Serengeti (Sinclair 1977) in contrast to very much smaller herds (20–50) in the montane and lowland forest areas of Africa (Sidney 1965). In the southern savannas herd sizes lie between these extremes (Mloszewski 1983). Buffalo herds of 450 animals or more were consistently observed on the lakeshore, for which two reasons are suggested. First, a continuous and evenly dispersed sward of pasture is optimal for both the feeding style and typical group size of buffalo (P.J. Jarman 1974). Second, herd size is a function of food distribution and abundance, which is also a seasonal phenomenon.

Sinclair (1977) attributed a regular seasonal splitting of herds in Serengeti buffalo to change in the food supply. Abundant wet-season food resources permitted large aggregations of buffalo which had to split in the dry season because their preferred habitats fragmented into areas too small for a large herd to occupy as one entity. At Kariba, the optimum habitat in the dry season is the open expanse of lakeshore pasture on which buffalo can concentrate and maintain large herds. During the wet season the widespread availability of both food and water allows dispersal and fragmentation of herds. Although Prins (1987) found no clear relationship between seasonality and fusion–fission patterns in buffalo, the evidence from Kariba suggests that such social organization in buffalo is related to the seasonality of food density and its dispersion. But the nature of its occurrence must vary between different environments.

Seasonal variation in male group size is largely a function of breeding behaviour (Grimsdell 1969; Taylor 1985) and Sinclair (1977) suggested that breeding induced the formation of large herds in the wet season. Although breeding also occurs during the wet season at Kariba, at which time bachelor males rejoin herds (Taylor 1985), the proportion doing so is very small ($\pm$ 5%) and statistically insignificant. Consequently herd size would appear unrelated to season of breeding.

An important organizational feature of Kariba buffalo is the manner in which males separate themselves out from areas occupied by herds. Herds occupy the large expanses of lakeshore pasture whereas the numerous bays, inlets and drainage lines, often too small to support even small breeding herds, provide a year-round grassland habitat which male groups successfully exploit. Competition between herds and male groups is reduced because males select areas unsuitable for herds (Sinclair 1977).

An apparently unique feature of the Kariba environment is the atypical dry-season production of lakeshore *Panicum repens* pastures. The implications for grazing herbivores are considerable and, clearly, buffalo at Kariba have adjusted to a changed environment in a number of ways. Their response in terms of spatial, temporal and social organization and feeding behaviour has been to maximize the advantages of a seasonally abundant food resource adjacent to a permanent water supply.

## Acknowledgements

I thank the late K.J. Fynn and G.G. Hall for piloting the aircraft during the aerial surveys and D. Mapungu and F. Mushori for assistance in the field. D.H.M. Cumming kindly commented on the manuscript. I am most grateful to the Zoological Society of London and the British Council in Zimbabwe for supporting my attendance at this Symposium. The work is published with the approval of the Director, Department of National Parks and Wild Life Management, Zimbabwe.

## References

Anon. (1981). *Climate handbook of Zimbabwe*. Issued by the Department of Meteorological Services. Government Printer, Harare.

Balch, C.C. (1955). Sleep in ruminants. *Nature, Lond.* 175:940–1.

Begg, G.W. (1973). The biological consequences of discharge above and below Kariba Dam. In *Proceedings of the 11th congress on large dams, Madrid, 11–25 June, 1973* 1:421–30. International Commission on Large Dams, Paris.

Bowmaker, A.P. (1973). Hydrophyte dynamics in Mwenda Bay, Lake Kariba. *Kariba Stud.* 3:42–59.

Coetsee, A.M. & Taylor, R.D. (1978). Hand capture of buffalo calves for research purposes. *S. Afr. J. Wildl. Res.* 8:173.

Cole, L.C. (1949). The measurement of interspecific association. *Ecology* 30:411–24.

Conybeare, A. (1980). Buffalo numbers, home range and daily movement in the Sengwa Wildlife Research Area, Zimbabwe. *S. Afr. J. Wildl. Res.* 10:89–93.

Eltringham, S.K. & Woodford, M.H. (1973). The numbers and distribution of buffalo in the Ruwenzori National Park, Uganda. *E. Afr. Wildl. J.* 11:151–64.

Ewer, R.F. (1968). *Ethology of mammals*. Logos Press, London.

Grimsdell, J.J.R. (1969). *Ecology of the buffalo*, Syncerus caffer *in western Uganda*. Ph.D. Thesis: Cambridge University.

Grimsdell, J.J.R. & Field, C.R. (1976). Grazing patterns of buffaloes in the Rwenzori National Park, Uganda. *E. Afr. Wildl. J.* 14:339–44.

Hancock, J. (1953). Grazing behaviour of cattle. *Anim. Breed. Abstr.* 21:1–13.

Hofmann, R.R. (1973). *The ruminant stomach. Stomach structure and feeding habits of East African game ruminants*. East African Literature Bureau, Nairobi. (*E. Afr. Monogr. Biol.* 2:1–354.)

Jarman, M.V. & Jarman, P.J. (1973). Daily activity of impala. *E. Afr. Wildl. J.* 11:75–92.

Jarman, P.J. (1972a). Seasonal distribution of large mammal populations in the unflooded Middle Zambezi Valley. *J. appl. Ecol.* 9:283–99.

Jarman, P.J. (1972b). The use of drinking sites, wallows and salt licks by herbivores in the flooded Middle Zambezi Valley. *E. Afr. Wildl. J.* 10:193–209.

Jarman, P.J. (1974). The social organisation of antelope in relation to their ecology. *Behaviour* 48:215–67.

Leuthold, W. (1972). Home range, movements and food of a buffalo herd in Tsavo National Park. *E. Afr. Wildl. J.* 10:237–43.

Leuthold, W. (1977). *African ungulates. A comparative review of their ethology and behavioral ecology*. Springer-Verlag, Berlin. (*Zoophysiology Ecol.* 8).

Lewis, J.G. (1977). Game domestication for animal production in Kenya: activity patterns of eland, oryx, buffalo and zebu cattle. *J. agric. Sci., Camb.* 89:551–63.

Magadza, C.H.D. (1970). A preliminary survey of the vegetation of the shore of Lake Kariba. *Kirkia* 7:253–67.

Mloszewski, M.J. (1983). *The behaviour and ecology of the African buffalo*. Cambridge University Press, Cambridge, London etc.

Monfort, N. (1979). Étude des populations de buffles, Syncerus caffer (Sparrman) du Parc National de l'Akagera (Rwanda). 1. Répartition spatiale et distribution. *Z. Säugetierk.* 44:111–27.

Monfort, N. (1980). Étude des populations de buffles Syncerus caffer (Sparrman) du Parc National de l'Akagera (Rwanda). 2. Organisation et adaptation au milieu. *Z. Säugetierk.* 45:173–88.

Norton-Griffiths, M. (1974). Reducing counting bias in aerial censuses by photography. *E. Afr. Wildl. J.* 12:245–8.

Patterson, I.J. (1965). Timing and spacing of broods in the black-headed gull *Larus ridibundus*. *Ibis* 107:433–59.

Prins, H.T.T. (1987). *The buffalo of Manyara. The individual in the context of herd life in a seasonal environment of East Africa*. Ph.D. Thesis: University of Groningen.

Schottler, J.H., Efi, P. & Williams, W.T. (1975). Behaviour of beef cattle in equatorial lowlands. *Aust. J. exp. Agric. Anim. Husb.* 15:725–30.

Sidney, J. (1965). The past and present distribution of some African ungulates. *Trans. zool. Soc. Lond.* 30:1–397.

Siegel, S. (1956). *Non-parametric statistics for the behavioral sciences*. McGraw-Hill, New York & London.

Sinclair, A.R.E. (1973). Population increases of buffalo and wildebeest in the Serengeti. *E. Afr. Wildl. J.* 11:93–107.

Sinclair, A.R.E. (1974). The natural regulation of buffalo populations in East Africa. IV. The food supply as a regulating factor, and competition. *E. Afr. Wildl. J.* **12**:291–311.

Sinclair, A.R.E. (1975). The resource limitation of trophic levels in tropical grassland ecosystems. *J. Anim. Ecol.* **44**:497–520.

Sinclair, A.R.E. (1977). *The African buffalo. A study of resource limitation of populations*. The University of Chicago Press, Chicago & London.

Southwood, T.R.E. (1978). *Ecological methods, with particular reference to the study of insect populations*. (2nd edn). English Language Book Society, Cambridge.

Stark, M.A. (1986). Daily movement, grazing activity and diet of savanna buffalo, *Syncerus caffer brachyceros*, in Benoue National Park, Cameroon. *Afr. J. Ecol.* **24**:255–62.

Stewart, D.R.M. & Stewart, J. (1963). The distribution of some large mammals in Kenya. *Jl E. Afr. nat. Hist. Soc.* **24**:1–52.

Taylor, R.D. (1979). Biological conservation in Matusadona National Park, Kariba. *Proc. Trans. Rhod. scient. Ass.* **59**:30–40.

Taylor, R.D. (1985). *The response of buffalo*, Syncerus caffer *(Sparrman) to the Kariba lakeshore grassland* (Panicum repens L.) *in Matusadona National Park*. D. Phil. Thesis: University of Zimbabwe.

Vesey-Fitzgerald, D.F. (1960). Grazing succession among East African game animals. *J. Mammal.* **41**:161–72.

Weir, J. & Davison, E. (1965). Daily occurrence of African game animals at water holes during dry weather. *Zool. Afr.* **1**:353–68.

Western, D. (1975). Water availability and its influence on the structure and dynamics of a savannah large mammal community. *E. Afr. Wildl. J.* **13**:265–86.

*Symp. zool. Soc. Lond.* (1989) No. 61: 73–87

# Strategies for water economy amongst cattle pastoralists and in wild ruminants

P.A. JEWELL and
M.J. NICHOLSON

*Research Group in Mammalian Ecology and Reproduction
University of Cambridge
Physiological Laboratory
Downing Street
Cambridge CB2 3EG*

*International Livestock Centre for Africa
P.O. Box 568a
Addis Ababa, Ethiopia*

## Synopsis

The nature of primary production in the semi-arid zone and the assessment of carrying capacity are fundamental to an understanding of how the exploitation of this zone can be sustained. The absolute availability of water or nutrients (whichever is the limiting factor) determines the annual biomass of vegetation but its quality depends on the relative availability of these two factors. In the most arid regions the biomass of annual plant production is very low; the protein content is, however, high. An ability to exploit the evanescent production of the wet season is clearly highly desirable, and only mobile pastoralists can do this, their cattle showing good performance and achieving useful growth rates.

The manner in which pastoralists have coped with seasonal fluctuations of food and water for their cattle closely resembles the migrations of many wild species of grazing herbivores. Wildebeest, for example, are dependent on surface water but their annual migrations, as in the Serengeti, allow them to maintain a high animal biomass. The white-eared kob in the Sudan show a similar adaptation.

If excessive pressures were not placed on pastoralists they themselves could effect greater efficiency in the way they use the range. Recent work at the International Livestock Centre for Africa has measured the effects of the infrequent watering of cattle that is practised under traditional management by the Borana tribe in Ethiopia and Kenya. Restricting water to every third day had little effect on the productivity of their cattle. Conception and calving rates were unaffected, although birthweights and weaning weights were somewhat reduced. Walking and night enclosing had little effect on productivity despite a reduced dry matter intake due to the loss of

ZOOLOGICAL SYMPOSIUM No. 61
ISBN 0–19–854009–4

grazing time. The results contradict several assumptions concerning pastoralist husbandry practice: watering of Zebu cattle once every third day can be continued indefinitely in *Bos indicus* and the adverse effects of such a practice are small. It is of interest that a species of ruminant like the oryx (*Oryx beisa*) that is highly adapted to arid conditions voluntarily adopts a strategy of drinking once every three or four days.

## Introduction

This paper will examine the parallels between the ranging strategies of wild ruminants and the movements of pastoralists with their livestock. Cattle will provide the focus of interest, and, in addition to a review of the problems of adaptation to a limited supply of water, new data derived from field work in Ethiopia will be presented. Throughout extensive regions of Africa the indigenous ruminants and local populations of domestic cattle have to face the privations imposed by aridity. Shortage of water may be chronic or seasonal, or spatial with regard to drinking water, and associated with the annual cycle is a shortage of protein (Taylor 1968; Breman & de Wit 1983). Calcium is also important for grazing ruminants, as Kreulen (1975) has emphasized; moreover, some African soils are notorious for their low mineral status and the grazers may be confronted with other mineral deficiencies (Bell 1982). Extensive movement and seasonal migration offer a solution to these problems (Sinclair 1983).

### Migration

The seasonal migrations of certain species of African grazing ruminants are spectacular: wildebeest (*Connochaetes taurinus*) in the Serengeti in Tanzania (Maddock 1979), white-eared kob (*Kobus kob leucotis*) in the Sudan (Fryxell & Sinclair 1988) and (at least formerly) springbok (*Antidorcas marsupialis*) in South Africa (Cronwright-Schreiner 1925).

Pastoralists depend on the same resources as the herds of wild ungulates and a close parallel can be seen in their ranging strategies: a clear example is seen in the Ngorongoro Conservation Area in Tanzania, used by both Maasai pastoralists and several species of wild ungulates. Not surprisingly, conflict develops over the use of resources (Homewood, Rodgers & Arhem 1987; Jewell in press).

The nutritional advantages of extensive seasonal movement are clearly seen in the Sahelian zone south of the Sahara. In the most arid northern part of this region the biomass of annual plant production is less than 1t/ha but the protein content is, relatively, high. In the less arid regions of the southern Sahel and savanna border the annual plant biomass may be 2–5 t/ha but levels of protein are seasonally very low. An ability to exploit the evanescent production of the wet season in the north is clearly highly desirable, and only mobile pastoralists can do this, their cattle showing good performance and achieving good growth rates (Breman & de Wit 1983).

Elsewhere in Africa the true pastoralist systems are characterized by transhumance (the seasonal movement of livestock to places of different climate or terrain) and may become quite elaborate in their organization. A variant of this system is that of the Borana of southern Ethiopia (which will be analysed in more detail later) in which there is separation of dry free-ranging herds (*fora*), managed by young men, and the village-based herds of mainly lactating cows and calves (*warra*) managed by the family. The *fora* herds may move up to 100 km away from the permanent camps (ILCA 1986).

### Water resources

In the dry season water is at a premium. Large mammals exhibit a wide range of physiological adaptations that enhance the conservation of water. Prominent amongst these are kidneys that can produce a concentrated urine, dry faeces, a coat of high reflectance, and behavioural adpatations such as feeding at night (Schmidt-Nielsen 1964; Taylor 1968; King 1979). An important adaptation to shortage of water and food of low quality is a reduction of metabolic rate. This was revealed by studies on cattle (Western & Finch 1986) and goats (Brosh, Shkolnik & Choshniak 1986).

One of the most successful of the dry-country ruminants is the oryx (*Oryx beisa*). Stanley-Price (1985) has shown that this species may drink only once in every three or four days. It has been generally supposed that to water domestic cattle as infrequently as once in every three days is life-threatening but research supporting this conclusion has mainly been conducted on European breeds (Balch, Balch, Johnson & Turner 1953; Leitch & Thompson 1944). Nevertheless, using zebu *steers*, Payne (1965) concluded that while giving water every three days had a significant effect on growth, the effect was masked by the much greater influence of seasonal changes in pasture quality and quantity. There are, however, pastoral people who adopt the practice of infrequent watering. The cattle of the Borana are generally watered on a three-day cycle throughout the dry season. It is believed that in the past the Maasai and other pastoralist tribes used also to practise infrequent watering, but ceased to do so largely as a result of an improvement in water supplies.

## Experimental trials and observations

As far as was known, no data were available on the effects of chronic water restriction on adult *breeding* cattle. Trials were therefore established on a ranch in southern Ethiopia (Nicholson 1987a) in order to examine whether giving water to cattle every second or third day constituted a major constraint on productivity. The aim was to assess the effects of water restriction on the lactational performance of cows and on the growth of calves before and after weaning. Since walking is normally associated with infrequent watering, a walking component was added to the trials together with the added constraint of enclosing the cattle at night. The major results from this four-year study relating to productivity have been published (Nicholson 1987a,b) and will only be summarized here, but new data on the behavioural adaptations of the cattle will be presented.

## The Borana system

The Borana water all their livestock infrequently during the dry season primarily because sources of water are scarce and it is not easy to obtain water from them. Most drinking water is lifted by hand from deep wells: these occur in well-systems which are unevenly situated throughout the southern rangelands of Ethiopia. The frequency of watering varies according to species: horses are watered daily or on alternate days, sheep and goats every three to five days, and camels as infrequently as once every ten days (Nicholson 1985). The two principal advantages of infrequent watering are that it reduces water consumption, thereby alleviating the work load in the wells, and allows exploitation of distant grazing resources. Cattle watered daily could have access to a maximum of 150 km$^2$ of grazing around a well assuming a maximum daily walk of 7 km. Watered every third day they could range 20 km or more from water and could exploit the grazing in an area of more than 1300 km$^2$ (Fig. 1). By dividing their herds into milking and non-milking units, the Borana ensure greater exploitation of grass because the non-milkers do not have to return to the village every night.

## Productivity under water restriction

Infrequent watering of ruminants restricts not only the times of drinking but also the total quantity of water drunk. This is largely owing to the constraint imposed by the volume of the rumen (Nicholson in press): a lactating cow watered once in three days is unable to imbibe at one brief visit to the well three times the volume of water that can be drunk by a cow watered daily. The result is that cows watered every third day drink about 30% less water (Fig. 2) than those watered daily despite voluminous intakes of up to 104 l (Nicholson 1985). Superimposed on this restricted intake is a continuous cycle of dehydration and rehydration. Despite this, signs of distress to the animals, either in the herds of pastoralists or under experimental conditions, were never observed.

Since overall water turnover declined also by 30%, one might have

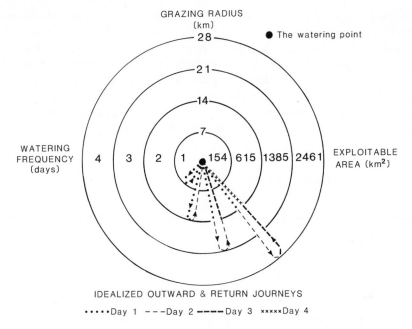

**Fig. 1.** Diagram to illustrate the grazing available for cattle to exploit depending upon watering frequency.

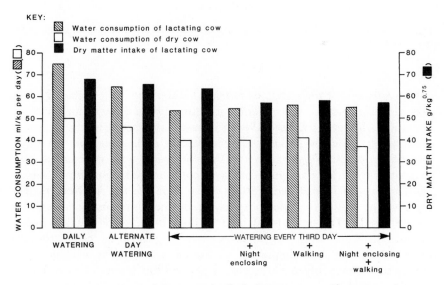

**Fig. 2.** Water consumption and dry matter intake by Boran cows on Abernosa ranch.

expected a marked loss of productivity. In fact, over the four years during which productivity parameters were measured (milk production, calf growth, weaning weights, age at maturity of first calving, calving weights and percentages), the costs to productivity were very slight. The largest differences tended to be weaning weights which water restriction reduced by as much as 10% (Fig. 3). With the onset of the rains, compensatory growth took place and no permanent stunting or growth retardation occurred. The differences were muscular rather than skeletal in origin.

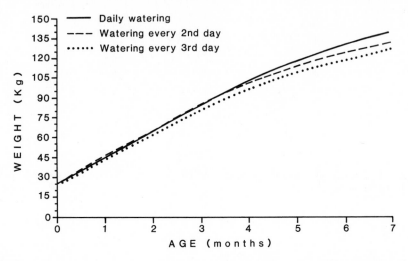

**Fig. 3.** Calf growth curves on Abernosa ranch (1983 and 1984 calves combined; total number of calves 115).

Measurement of the effects of restricted watering on the lactating cows was a crucial part of the experiment. Using live weight as an index of condition, the results for the 1984–85 season are shown in Fig. 4. There were about 20 cows suckling their calves in each group on the three watering regimes. Weight gain from May onwards was confounded with pregnancy while June and July data were confounded by parturition which caused a weight loss of about 50 kg per animal (see Fig. 4). Nevertheless there was no significant difference between the mean weights for each group in any month. The animals adapted to their lower water intakes by an increase in water economy manifested by higher urine concentration, marked adiuresis and alterations in diurnal activity. The evidence was strong that metabolic rate was also depressed but it was not feasible to measure it under the conditions of the trial.

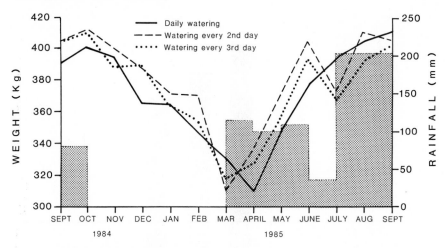

**Fig. 4.** Change in weight of lactating Boran cows, month by month 1984–85; about 20 cows in each treatment. Stippled histograms show rainfall.

### Behavioural adaptations of cattle on infrequent watering

As might have been expected, the differences between cattle on three different watering regimes (daily, every two days and every three days) were larger in the dry season than during the wet season. During the rains all the groups behaved similarly owing to the high moisture content of the grass, the availability of surface water and the frequent cloud cover which reduced solar radiation and hence evaporation losses.

The effect of a longer period between watering was to reduce appetite according to the degree of dehydration. For example, when food intake was measured in the three groups, a significant negative linear trend on watering interval was observed in feed intake, with cows eating 68.0, 65.6 and 61.8 g per $kg^{0.75}$ liveweight per day (S.E. 1.58; $n=12$ crossover design) when watered daily, every other day and every third day respectively (Fig. 2). When the cattle watered once every three days were studied in more detail, it was found that a significant linear trend also existed according to the hours since watering. On the first day, cows ate 5.7 kg of dry matter per day, on the second 4.9 kg and only 4.3 kg on the third day (Fig. 5). Dehydration gave rise to inappetence. However, digestibility was unchanged between groups.

The pattern of grazing and ruminating changed with time in the group of cows watered every third day. Grazing was intense on the day of watering (42% of the day) falling to 16% on the third day. Conversely, rumination increased from 20% on the first day to 34% on the third day (Fig. 6).

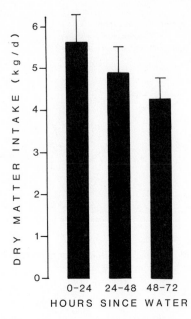

**Fig. 5.** Dry matter intake by lactating Boran cows that were watered once every three days.

## Effects of walking and night enclosure on cattle watered every third day

Four groups of cows were recorded over a two-year period and each group was given water on a three-day cycle. The control group were given water every third day; the second group walked 40 km every third day, 20 km before watering, and 20 km immediately afterwards. The third group were enclosed at night and the fourth group were both enclosed and walked. Walking continued over an eight-month dry season (during which the cows that were walked covered over 3000 km) but was discontinued two to three

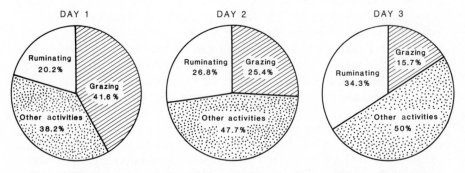

**Fig. 6.** The activity patterns displayed by lactating Boran cows watered once every three days.

months before parturition and the onset of the rains. During the dry season all the cattle had access to drinking water for 10 min every third day.

A set of trials carried out in 1985–87 were the first in which a regime of watering every third day was combined with night enclosing and walking. Moreover, in these trials, body condition, milk yield, calf growth and the fertility of cows were measured, whereas in all previous work on walking and night enclosing, the liveweight change in steers had been the sole parameter used to reflect differences in productivity.

Scoring body condition is a useful way of assessing environmental effects on animals. In this work, the scoring system was based on a visual, subjective appraisal of the bodily appearance of the cows, using a nine-point system (Nicholson & Butterworth 1986). It was shown that such condition-scoring is repeatable and reproducible and was highly correlated with body weight and heart-girth (Nicholson 1987c; Nicholson & Sayers 1987).

Walking greatly reduced the total time spent grazing, whereas night enclosure had less effect. When the cattle walked the 40 km on the day of watering, 8 h of grazing time was lost. Over the three-day cycle, however, the cattle compensated somewhat, and their total grazing time was 25% below the control cattle which were not subjected to either walking or enclosing (Fig. 7).

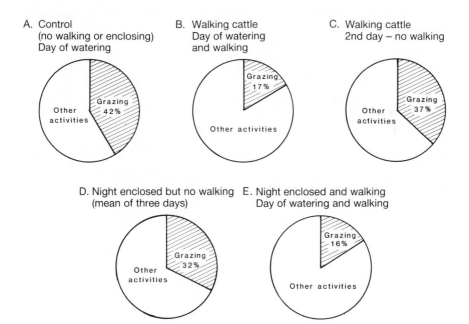

**Fig. 7.** Percentage of time spent grazing on selected days by Boran cattle being watered once every three days.

In contrast, the effect of night enclosure superimposed on watering once every three days was to reduce grazing time by only 5% against the controls. The combination of night enclosure and walking had an additive effect so that these cattle showed a 40% reduction in grazing time.

Simultaneous measurements of food intake were made which showed that the reduction in dry matter intake nowhere near approached the observed reduction in grazing time. Walking and night enclosing each depressed food intake in lactating cows by 5% compared with the control cows subjected only to watering once every three days. The combination of walking with enclosure resulted in a similar additive effect which reduced dry matter intake by 10% against the controls (see Fig. 2).

Despite these marked effects on feeding behaviour, and the energetic requirements of walking, the overall effects of walking and enclosing on productivity as seen in the condition scores (Fig. 8) and determined by adult weight loss, calf and weaner growth, calving percentage and birth weight, were negligible when compared with the controls (see Nicholson 1987b).

The findings showed that the cows made alterations in their feeding behaviour by an intensification of eating during those hours that grazing was possible. Under many conditions, particularly in wild ruminants during migration, walking and eating are not mutually exclusive and may even be synergistic as new pastures are reached. In contrast, livestock are often trekked along predetermined routes where the vegetation is frequently overgrazed.

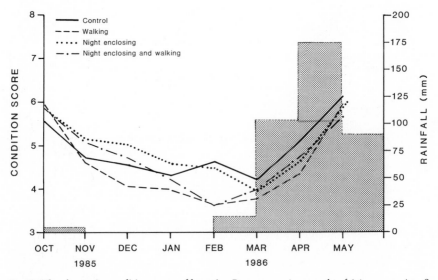

**Fig. 8.** The change in condition score of lactating Boran cows (on a scale of 1 (very poor) to 9 (very good) ) through the dry season. Stippled histograms show rainfall.

It was of interest to note that walking speed was greater before drinking (5 kph) than after drinking (4 kph). A number of factors could explain this, including the motivation of thirst when dehydrated or the extra weight or fatigue following rehydration.

The patience with which thirsty cattle would await their turn to drink in the wells or at the trough was remarkable. Even when severely dehydrated (as measured by plasma osmolality), the animals would wait hours until it was their turn to drink. Once at the trough, rehydration would be rapid with intakes as high as 35 l/min and total consumption of between 80 and 100 l for lactating cows at the peak of their lactation in the early dry season. The extent of dehydration was comparable with data from other workers on camels and goats (Macfarlane & Howard 1972; Schmidt-Nielson 1964; Shkolnik, Borut, Choshniak & Maltz 1974): these species are noted for their ability to conserve water. Cattle, on the other hand, have a poor ability to conserve water and they have achieved their adaptation by developing a tolerance to dehydration.

## Discussion

Large herbivorous mammals are able to cover great distances to meet their needs for fresh grazing and for water, and this they can do with energetic efficiency (Alexander & Maloiy, this symposium). Their movement may be attenuated by the need to meet specific nutritional requirements, particularly for protein and trace elements (McNaughton, this symposium). The exact nature of this annual quest, and what nutrients the animals seek, has yet to be investigated but the phenomenon of migration is conspicuous enough. One of the best-known is that of wildebeest in the Serengeti and some of the more modern data about them were analysed by Pennycuick (1975). She showed that annual differences in the pattern of the migration were correlated with differences in rainfall and that rainfall affects particularly the timing of the movement of the wildebeest on and off the plains. An equally spectacular migration is that of tiang (*Damaliscus korrigum*) in the Sudan where the herds make annual migrations between areas that are 400 km apart. In the dry season (January to April) the tiang are in well-watered areas in the Sudd but when the wet season starts they move to the south-east, covering an average distance of 10 km a day between April and May. It is across these very migration routes that the proposed Jonglei canal is being constructed (J. Kingdon pers comm., and Anon. 1983:35). To the east of the range used by tiang is the Borana National Park in Sudan where seasonal migrations are exhibited by a population of 800 000 white-eared kob (*Kobus kob leucotis*). Their dry-season and wet-season ranges are 150–200 km apart, and they apparently move much further in the wet season than would be necessary to reach fresh green grazing: perhaps they

seek some critical nutritional resources (Fryxell & Sinclair 1988). The kob
are dependent on drinking water daily (Schoen 1971) and are never far from
water. In contrast, the fringe-eared oryx (*Oryx beisa*), a species that is
physiologically adapted to arid areas, may hold territory in places far from
water and trek up to 20 km only once in every three to five days to drink
(Wacher 1986). These strategies adopted by natural populations of
ruminants have parallels in the traditional methods of livestock husbandry
used by pastoralists. This lends support to the assertion that nomadic
pastoralism is the most successful way to exploit the resources of arid
regions.

Free-ranging cattle prefer to drink at least once each day (Arnold &
Dudzinski 1978) and it may be supposed that new adaptations have been
called for where pastoral herders have imposed less frequent watering. The
experimental work reviewed here was undertaken in order to reveal the
nature of these adaptations. The major finding in the study was that when
water was provided at intervals of two or of three days, overall water
consumption was depressed by about 8% and 30%, respectively, below that
of cattle watered once a day. It was therefore accurate to describe the
provision of water at two- or three-day intervals as water restriction. The
magnitude of this reduction in water intake was related more to the interval
between drinking than to the category of cattle so restricted, since total
water consumption was equally depressed in bulls, steers, dry cows and
lactating cows.

A significant linear reduction in food intake occurred as a result of
infrequent watering, but the magnitude of the reduction was small in
comparison to the reduction in water consumption. Appetite fluctuated
according to the stage in the watering cycle, and dehydration led to marked
inappetence. No loss of productivity resulted from watering the cattle every
two days, either on Abernosa ranch in Ethiopia or on Galana ranch in
Kenya. A three-day interval between watering had a small but significant
negative effect which was larger in the hotter climate of Galana ranch,
where growing steers were between 10 and 23 kg smaller at the end of a 17-
month period of water restriction.

The ability of cattle to adapt to a substantially diminished water budget
was clearly apparent, since they demonstrated an ability to operate
efficiently with a 30% reduction in water turnover. This raises the question
as to why animals are inherently profligate with their resources and yet
display considerable thrift when the need arises. The same question applies
to the use of energy, since it was clear that the observed cost of walking and
the theoretical cost were quite different, and it is suggested that animals
cope with such adverse circumstances by an ability to lower their basal
metabolic rate.

Despite a substantial loss of grazing time as a result of both night enclosing and walking, the additional effect of these practices on breeding cattle was small. However, when the feed is not as freely available as it was at Abernosa, these effects are likely to be detrimental. This latter point refers to most of sub-Sahelian Africa under traditional management.

The experiments that we have described had the objective of measuring the ways in which infrequent watering affects productivity in cattle. It was found that the effect was not really adverse and some light has been thrown on the efficiency of the Borana system. Infrequent watering permits more extensive use of grazings and raises carrying capacity (Fig. 1). In this way it reduces the threat of overgrazing and also avoids the danger of severe overgrazing that develops round bore-holes (Arntzen & Veenendaal 1986). Recently some anthropologists have questioned the assertion that pastoralists cause overgrazing and have argued that such assertions have persisted from colonial attitudes and reflect political antagonisms towards pastoralists (Homewood & Rodgers 1987; Collett 1987). These re-assessments are timely and fit our contention that there is no more efficient or convivial (in the sense used by Illich 1973) way of exploiting huge tracts of semi-arid land than nomadic pastoralism. More ecological assessment is urgently needed, however, particularly of changes in plant communities, changes in ground cover, and the status of woody plants. We do not doubt that an increase in the density of bore-holes may lead to overgrazing (Sinclair & Fryxell 1985), or that the loss of woody plants reflects well-established degradation associated with over-abundant domestic stock (IUCN 1986). For the rangelands of some pastoralists the conclusion that these lands are badly overgrazed is inescapable (Carr 1977). The condition in which pastoralists find themselves, however, is generated by many factors outside their control. Amartya Sen (1981) instances the fact that in the recent droughts in Africa nomadic pastoralists suffered far higher mortality than other sections of the human population and he identifies three factors that made them the destitutes: first, the direct loss of their animals and their food resource, second, the fall in value of the animals and loss of exchange entitlement, and third, the loss of dry-season grazing through the progressive incursion of agricultural settlers onto their land. The adaptations of pastoral people and their domestic stock to semi-arid rangelands, which we see as basically environmentally sound, need to be recognized and preserved if the resources of these rangelands are not to be wasted.

# References

Anon. (1983). *Development studies in the Jonglei canal area. Final report. 5. Wildlife studies.* (National Council for Development of the Jonglei Canal Area: Government of the Democratic Republic of the Sudan, Ministry of Finance and Economic Planning.) Mefit-Babtie Srl, Glasgow, Rome & Khartoum.

Arnold, G.W. & Dudzinski, M.L. (1978). *Ethology of free-ranging domestic animals*. Elsevier, Amsterdam.

Arntzen, J.W. & Veenendaal, E.M. (1986). *A profile of environment and development in Botswana*. Institute for Environmental Studies, Free University, Amsterdam.

Balch, C.C., Balch, D.A., Johnson, W.W. & Turner, J. (1953). Factors affecting the utilization of food by dairy cows. 7. The effect of limited water intake on the digestibility and rate of passage of hay. *Br. J. Ntr.* 7:212–24.

Bell, R.H.V. (1982). The effect of soil nutrient availability on community structure in African ecosystems. In *Ecology of tropical savannas*: 193–216. (Eds Huntley, B.J. & Walker, B.H.). Springer-Verlag, Berlin (*Ecol. Stud. Anal. Synth.* 42.)

Breman, H. & de Wit, C.T. (1983). Rangeland productivity and exploitation in the Sahel. *Science, N.Y.* 221:1341–7.

Brosh, A., Shkolnik, A. & Choshniak, I. (1986). Metabolic effects of infrequent drinking and low-quality feed on Bedouin goats. *Ecology* 67:1086–90.

Carr, C.J. (1977). Pastoralism in crisis. The Dasanetch and their Ethiopian lands. *Univ. Chicago, Dept. Geography Research Paper* No. 180:319.

Collett, D. (1987). Pastoralists and wildlife: image and reality in Kenya Maasailand. In *Conservation in Africa, people, policies and practice*: 129–48. (Eds Anderson, D. & Grove, R.). Cambridge University Press, Cambridge.

Cronwright-Schreiner, S.C. (1925). *The migratory springbucks of South Africa*. Fisher-Unwin, London.

Fryxell, J.M. & Sinclair, A.R.E. (1988). Seasonal migration by white-eared kob in relation to resources. *Afr. J. Ecol.* 26:17–31.

Homewood, K. & Rodgers, W.A. (1987). Pastoralism, conservation and the overgrazing controversy. In *Conservation in Africa, people, policies and practice*: 111–28. (Eds Anderson, D. & Grove, R.). Cambridge University Press, Cambridge.

Homewood, K., Rodgers, W.A. & Arhem, K. (1987). Ecology of pastoralism in Ngorongoro Conservation Area, Tanzania. *J. agric. Sci., Camb.* 108:47–72.

ILCA (1986). *The productivity and potential of the southern rangelands of Ethiopia*. Joint ILCA/RDP Ethiopian Pastoral Systems Study Programme, ILCA, Addis Ababa.

Illich, I. (1973). *Tools for conviviality*. Calder & Boyars, Fontana, London.

IUCN (1986). *The Sahel report*. IUCN, Gland Switzerland.

Jewell, P.A. (In press). Conflicts between pastoralism, cattle rearing and wildlife. In *African wildlife: research and management*. (Eds Kayanja, F.I.B. & Edroma, E.L.). ICSU Press, Paris.

King, J.M. (1979). Game domestication for animal production in Kenya: field studies of the body-water turnover of game and livestock. *J. agric. Sci., Camb.* 93:71–9.

Kreulen, D. (1975). Wildebeest habitat selection on the Serengeti Plains, Tanzania, in relation to calcium and lactation: a preliminary report. *E. Afr. Wildl. J.* 13:297–304.

Leitch, I. & Thompson, J.S. (1944). The water economy of farm animals. *Nutr. Abstr. Rev.* 14:197–223.

Macfarlane, W.V. & Howard, B. (1972). Comparative water and energy economy of wild and domestic mammals. *Symp. zool. Soc. Lond.* No. 31:261–96.

Maddock, L. (1979). The 'migration' and grazing succession. In *Serengeti: dynamics of an ecosystem*:104–29. (Eds Sinclair, A.R.E. & Norton-Griffiths, M.). University of Chicago Press, Chicago.

Nicholson, M.J. (1985). The water requirements of livestock in Africa. *Outl. Agric.* 14(4):156–64.

Nicholson, M.J. (1987a). The effect of drinking frequency on some aspects of the productivity of zebu cattle. *J. agric. Sci., Camb.* 108:111–20.

Nicholson, M.J. (1987b). Effects of night enclosure and extensive walking on the productivity of zebu cattle. *J. agric. Sci., Camb.* 109:445–52.

Nicholson, M.J. (1987c). *The effects of infrequent drinking intervals on the productivity of Boran cattle.* Ph.D. Thesis: University of Cambridge.

Nicholson, M.J. (In press). Depression of dry matter and water intake in Boran cattle owing to physiological, volumetric and temporal limitations. *Anim. Prod.*

Nicholson, M.J. & Butterworth, M.H. (1986). *Cattle condition scoring manual.* International Livestock Centre for Africa, Addis Ababa.

Nicholson, M.J. & Sayers, A.R. (1987). Relationships between body weight, condition score and heart girth changes in Borana cattle. *Trop. anim. Hlth Prodn* 19(2):115–20.

Payne, W.J.A. (1965). Specific problems of semi-arid environments. *Qualitas Pl. Mater. veg.* 12:269–94.

Pennycuick, L. (1975). Movements of the migratory wildebeest population in the Serengeti area between 1960 and 1973. *E. Afr. Wildl. J.* 13:65–87.

Schmidt-Nielsen, K. (1964). *Desert animals: physiological problems of heat and water.* Clarendon Press, Oxford.

Schoen, A. (1971). The effect of heat stress and water deprivation on the environmental physiology of the bushbuck, the reedbuck, and the Uganda kob. *E. Afr. agric. For. J.* 37:1–7.

Sen, A. (1981). *Poverty and famines.* Clarendon Press, Oxford.

Shkolnik, S., Borut, A., Choshniak, I. & Maltz, E. (1974). Water economy and drinking regime in the Bedouin goat. In *Ecological research on development of arid zones with winter precipitation*: 79–80. Publication spéciale, Département de Publications Scientifiques, Organisation de la Recherche Agronomique, Bet-Dagan, Israel.

Sinclair, A.R.E. (1983). The function of distance movements in vertebrates. In *The ecology of animal movement*: 240–58. (Eds Swingland, I.R. & Greenwood, P.J.). Clarendon Press, Oxford.

Sinclair, A.R.E. & Fryxell, J.M. (1985). The Sahel of Africa: ecology of a disaster. *Can. J. Zool.* 63:987–94.

Stanley-Price, M.R. (1985). Game domestication for animal production in Kenya: the nutritional ecology of oryx, zebu cattle and sheep under free-range conditions. *J. agric. Sci., Camb.* 104:375–82.

Taylor, C.R. (1968). The minimum water requirements of some East African bovids. *Symp. zool. Soc. Lond.* No. 21:195–206.

Wacher, T.J. (1986). *The ecology and social organization of fringe-eared oryx on the Galana Ranch, Kenya.* D. Phil. Thesis: University of Oxford.

Western, D. & Finch, V. (1986). Cattle and pastoralism; survival and production in arid lands. *Hum. Ecol.* 14:77–9.

Symp. zool. Soc. Lond. (1989) No. 61: 89–110

# The ecology of female behaviour and male mating success in the Grevy's zebra

Joshua R. GINSBERG[1]

Department of Biology
Princeton University
Princeton, NJ 08544, USA

## Synopsis

In the Grevy's zebra (*Equus grevyi*) reproductive condition determines the priority a female places on food and water. The interaction between the requirements of individual females and ecological conditions produces a range of behavioural responses. Females with young foals are shown to be highly predictable in their patterns of movement and association: they are always found near permanent water and form close associations with several other females with whom they are in reproductive synchrony. Other females are less predictable: their associations are fluid and their movements are determined by the availability and abundance of food. The mating success of breeding (territorial) males is determined, in part, by the resulting changes in female distribution and association. Reproductive success among territorial males is correlated with the ecological variables affecting the distribution of females: males with access to water attract females with young foals; males with access to abundant food attract other classes of females. Male reproductive success does not correlate with other measures of male quality such as body size or territory area.

## Introduction

Early studies of the social organization and mating behaviour of African mammals, particularly ungulates, frequently concluded that the behaviour and social relationships of females were determined, or at the very least, strongly influenced by the reproductive and social behaviour of a dominant male of the species. A male, portrayed as the dominant individual in a social group, determined female behaviour by direct control of the group. The historical emphasis on male behaviour is noted by Moss (1982) and can be

---

[1] Present address: Dept. of Zoology, University of Oxford, South Parks Road, Oxford OX1 3PS.

ZOOLOGICAL SYMPOSIUM No. 61
ISBN 0–19–854009–4

seen in a variety of studies of African mammals: e.g. hartebeest (Backhaus 1959); impala (Schenkel 1966); baboons (Hall & Devore 1965); and plains zebra (Klingel 1965, but see 1969).

By the early 1970s, the great diversity in social behaviour shown by the African ungulates had been extensively examined (see Geist & Walther 1974). Following studies of birds (Crook 1965) and primates (Crook & Gartlan 1966), several authors (Jarman 1974; Estes 1974; Geist 1974) attempted to quantify how differences in evolution, ecology and resource availability affect social organization. These studies focused on large scale, interspecific differences in ecology and morphology and the correlation of these factors with patterns of resource dispersion. It became apparent that patterns of social behaviour are formed not by the direct control of some individuals by others, but through the control of essential resources.

Recent theories of animal social organization have sought to identify the factors that limit reproduction in each sex and to predict how these limitations constrain breeding behaviour. In the great majority of species, females reproduce with greater or lesser success according to their ability to acquire critical resources and subsequently convert these resources into offspring; males, according to their ability to gain access to oestrous females. Given this inequality, female behaviour can be predicted by studying the spatial and temporal distribution of critical ecological resources. Male breeding behaviour will be determined by the predictability in the resulting distribution of females and the male's ability to monopolize either female groups or, in most ungulates, the resources females require for reproduction (Emlen & Oring 1977; Bradbury & Vehrencamp 1977; papers in Rubenstein & Wrangham 1986).

Interspecific comparisons provided hypotheses concerning the adaptive value of a particular social organization or mating system and its ecological correlates. By their very nature, however, these comparisons were limited in their ability to provide answers as to how subtle, or not so subtle, differences in ecology produce the differences observed among species (Clutton-Brock & Harvey 1984). Such mechanisms are best studied in the field in intraspecific studies. For instance, in an intraspecific comparison, Clutton-Brock, Guinness & Albon (1982) showed that the social organization and ecology of red deer (*Cervus elaphus*) can best be understood by treating either sex as the equivalent of a separate species.

An animal's social organization, however, may be strongly influenced by both the ecological and historical variables (e.g. family inheritance of rank or home range; phylogenetic inertia). There may be strong interactions between existing sociality and a female's response to environmental change. For instance, in the red deer, near synchrony of breeding means that most females encounter the same ecological conditions (food and water availability) while in the same state of reproduction. In the plains zebra, females are less closely synchronized in their breeding but live in tightly bonded groups (Klingel 1969); hence, although individual females in a group may differ in their resource requirements, the overriding benefits of group membership appear to obscure individual female response to changing ecological conditions. Many of the questions which remain unanswered in the study of mating systems, therefore, relate to the interaction between ecology, sociality and patterns of female distribution (Wrangham & Rubenstein 1986).

To minimize these potential confounding effects, the direct effect of ecological variation on female behaviour might therefore best be studied in a species which meets the following criteria: females are not highly synchronized in their breeding; social bonds are easily formed and broken; and ecological conditions vary immensely throughout the year. The Grevy's zebra (*Equus grevyi*) is such a species. Its distribution is limited to the semi-desert of northern Kenya and southern Ethiopia, an area of highly variable and unpredictable environmental fluctuations. Food and water both show large variation in their distribution and abundance (Barkham & Rainy 1976; Ginsberg 1987). While breeding males are territorial (Klingel 1974), females do not appear to show great fidelity to particular sites or fidelity in their associations with other individuals (Klingel 1974; Rubenstein 1986; Ginsberg 1987). Patterns of movement and associations among females were so variable that Klingel (1974) concluded that they were essentially random.

Klingel, however, noted a single exception to this apparent lack of structure in female associations: 'in one case, two mares, each with a newborn foal, were kept together (by a territorial male) for at least 30 days in a certain territory' (Klingel 1974). In a short-term study of a different population of Grevy's zebra, Rubenstein (1986; unpublished data) found that associations among many females were far less random. Females with young foals nearly always formed groups, and these groups remained together for at least several weeks.

This stability in the associations of females in a species in which associations among individuals are usually so fluid is striking. Intraspecific differences in patterns of social organization have been observed in several ungulate species (Duncan 1975; Gosling 1975; Jarman & Jarman 1979; Rubenstein 1981). Such variation appears to result from large temporal and spatial variation in the availability of resources. In these studies, response to changes in the distribution of resources appears to be consistent among all females. In the Grevy's zebra, however, different females faced with identical patterns of resource distribution appear to exhibit different patterns of behaviour. Variation in female behaviour appears to be related to both changes in female reproductive state and differences in prevailing ecological conditions.

This paper will examine some of the ecological variables which might influence these changes in the pattern of movement and association of Grevy's zebra mares. It will investigate how external conditions (availability of resources) and internal reproductive state (hence, need for resources) interact to affect the predictability of association and movement in females. Male reproductive success will be discussed in the light of the observed patterns of female dispersion.

## Methods and background information

### Study animals

The Grevy's zebra is the largest wild equid, weighing approximately 400 kg. As in all equids, sexual dimorphism is slight, with males weighing approximately 10% more than females (King 1965). Males also have large upper canines; such enlargement of the canines is lacking in the females. An earlier study of the Grevy's zebra (Klingel 1974) discovered that breeding males are territorial. Males hold their territories for an extremely long time, up to seven years (J.R. Ginsberg, unpublished data). The territories are among the largest of any known ungulate (Owen-Smith 1977), ranging in size from 2 to 12 km$^2$ (Klingel 1974; Ginsberg 1987).

Breeding is not highly seasonal in the Grevy's zebra (Klingel 1974). Gestation lasts 409 days, although variation (387–428 days) has been noted in captive and semi-captive populations (Iaderosa 1983). Females exhibit behavioural oestrus (sexual receptivity) six to 15 days after giving birth (9.3 ± 1.08, $n=8$, this study). Unless impregnated, females will continue to cycle every 27 days (26.8 ± 0.28, $n=9$; this study, calculated from number of days between last day of first oestrus observed and first day of second oestrus observed).

Females in different reproductive conditions have different resource requirements. In the congeneric horse, *E. caballus*, reproductive females increase their nutritional requirements over those of non-reproductive females by the following amounts: in late pregnancy, females need 80% more energy and 60% more protein; in peak lactation, defined as the period in which the female is providing the maximum amount of milk to its foal (Oftedal 1984), the female needs more than twice as much energy (120%), 95% more protein and at least 50% more water (NRC 1978; Pollock 1980; Oftedal, Hintz & Schryver 1983).

In this discussion of female behaviour and ecology, these differences in energetic requirements will be used to define five classes of reproductive females. Females in the last trimester of pregnancy will be referred to as in late pregnancy. From birth until three months, the age when foals are first observed drinking, the foal is entirely dependent on the mare for water. This period will be defined as early lactation and encompasses the entire period of peak lactation. From the end of early lactation until the time a mare begins to wean her foal (six months), the energetic and water requirements of lactation are fairly constant. This period describes mid-lactation. Late lactation encompasses the period from the onset of weaning until independence of the foal. The class of non-reproductive females includes nulliparous females, or any female who, for a number of reasons (death of her foal, spontaneous abortion, temporary infertility), is neither in late pregnancy nor in association with a foal.

## Study area

This research was conducted just to the north and east of Mt. Kenya in the region around the town of Archer's Post, Kenya (37° 30' East, and 0° 40' North). In particular, the study area focused on the south-eastern 100 km$^2$ of the Buffalo Springs Reserve. The region is classified as a mosaic of semi-desert bushland and woodland, generally conforming to Pratt, Greenway & Gwynne's (1966) vegetation classification Zone V, wooded and bush-covered grasslands. The area, while climatologically homogeneous (dry and hot), is vegetationally diverse. This diversity is apparent even on a small scale (Barkham & Rainy 1976).

Average rainfall for the Archer's Post Station, calculated over the 33-year period 1947–80, is 365 mm (Government of Kenya 1971, 1984). The range in rainfall is from 120 mm to 875 mm per annum. Rain usually occurs in two seasons, March/April and October/November. However, the predictability of both the timing of rainfall and the quantity of rain that will fall at any one time is low (Barkham & Rainy 1976; Ginsberg 1987).

Grass is the dominant food of the Grevy's zebra, as it is of all equids. The abundance of the grass was monitored consistently at 54 locations in the study area. These locations were chosen to represent all the vegetational types in the study area and the contiguous patch of a particular habitat represented by a transect location is referred to as a transect area. The 54 locations were sampled at three-month intervals. The estimation of the biomass of the grass was made using a pin-frame intercept method (McNaughton 1979; C.L. Jensen unpublished); for this study area, the regression relating biomass to pin intercept was $Y = 25.4 X + 9.4$, $r^2 = 0.89$ where $Y$ = biomass, and $X$ = mean number hits/pin.

## Definition of seasons

Because of the temporal unpredictability of rainfall, criteria related to the distribution of surface water have been used to define seasons. The seasonal cycle begins with the season defined as the Rains: water is essentially ubiquitous, available in puddles, streams, water holes and permanent springs. In seasons which follow the end of the Rains the environment becomes progressively drier. The cycle can be interrupted at almost any season by another period of rain.

At the end of the Rains, the distribution of water changes quickly. Over a period of approximately two months after the last significant rain, ephemeral water disappears. This season is called the Early Dry and is defined as the period when the rain has stopped and ephemeral water sources are drying up. The total disappearance of ephemeral water marks the beginning of the Dry season. During the Dry season, water holes and intermittent streams dry up over the course of perhaps two months. The beginning of the Late Dry is defined as the time when only permanent water is available in springs and rivers. A fifth season, Drought, is defined as the failure of rains for over a year. This season persists until it rains. This paper will focus on patterns of female behaviour in the Early Dry, Dry and Late Dry seasons.

### Measuring behaviour and ecology of the zebra

Behavioural measurements were taken on 1084 individual Grevy's zebra identified by their stripe patterns. Zebra stripes are individually distinct (Klingel 1974) and identifications were made with an accuracy of at least 98% (Ginsberg 1987).

Information was collected daily on the distribution and identity of the zebra in the study area. The entire study area was surveyed for zebra every two weeks. Point-sample data, collected for 9795 sightings (5061 males; 4734 females) provide: (1) information on the distribution of the zebra; (2) independent point-sample data on the behaviour of individual zebra (interval between samples $\geq$ 12 h); (3) biweekly assessment of body condition (Pollock 1980) and (4) records of associations of individual zebra.

Measurements of body size were made using photometric analysis. A 500 mm lens was calibrated for distances less than 30 m at 1 m intervals. Using the lens as a range-finder (error of $\pm$ 0.6 m at 30 m) photographs were made of individual zebra at distances of less than 30 m. All photographs were made at right angles to the individual. Distance to the animal was noted at time of exposure. Each individual was photographed a maximum of three times on different days. From photographs enlarged to exactly $\times$ 2, the distance from the withers to the belly (linear girth) was measured. This measurement was converted into an actual linear girth using the geometric proportionality: girth on film/focal length of lens = actual girth/distance to zebra. For any particular individual, estimates of linear girth from one photograph to the next had an error of approximately 8% (Ginsberg 1987).

Observations were made on individuals and groups of zebra. Three types of data were collected: focal, sequential scan, and *ad lib.* (see Altmann 1974 for definitions). In all three categories, data were collected from a Landrover. Observations were often made with binoculars (10 $\times$ 40) and a

20×–60× spotting scope. The average distance to the zebra was 50–80 m. Focal data were collected on an NEC 8201A lap computer. Sequential scan samples were made on groups of up to 20 individuals. During focal sample periods, *ad lib.* data were not collected. At all other times, data were collected on the following behaviours: fights, sexual behaviour, drinking times and behaviour, nursing behaviour and territorial behaviour (vocal and dung marking). For all behaviours, the time, location and individuals involved in the event were recorded.

## Results

### Distribution of females around water in relation to their use of water

Point-sample data collected on daily surveys were used in this analysis. Sample sizes are shown in Fig. 1 in which the mean distance females were observed from permanent water is plotted for each reproductive class in each season. Also plotted is the mean distance to permanent water of 100 randomly selected locations in the study area representing the expected distance to permanent water if females are randomly distributed. If females are found closer to water than would be expected at random, they are taken to be water-bound. If females are found, on average, at a distance from permanent water greater than would be expected at random, they are considered to be less water-dependent. Because sources of water become

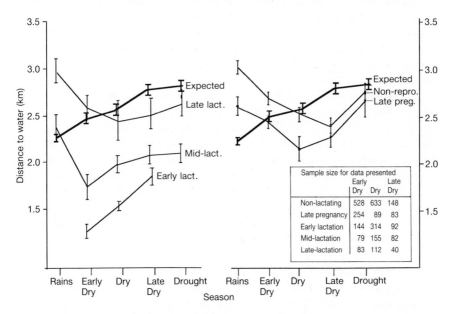

**Fig. 1.** Distance to water for five classes of females across five seasons. See text for discussion.

fewer as conditions grow drier, mean distance to water increases across season. If the movement of a particular class of females depends on the distribution of water, their mean distance to water should parallel this increase.

Both season (2-way ANOVA, d.f. = 3, $F$ = 2.66, $p < 0.05$) and reproductive condition (d.f. = 4, $F = 4.92, p < 0.01$) have significant effects on female ranging patterns in relationship to water distribution; there is a strong interaction between the two variables ($F = 5.58, p < 0.01$). Females in late lactation, non-reproductive females, and females in late pregnancy follow a similar pattern across seasons. In the Rains, all three categories are independent of permanent water. When water becomes less widely distributed in the Early Dry season, their mean distance from water declines. Females in late pregnancy become water-bound during the Early Dry season, while non-reproductive females and those in late lactation remain less dependent on permanent water until the Dry or Late Dry season. In the Drought the pattern reverses: mean distance to water increases for both non-reproductive females and those in late pregnancy.

Females in mid-lactation show a pattern similar to that of the three classes discussed above; however, in all seasons, they are significantly closer to water than any class except those females in early lactation. In the three seasons for which data are available, females in early lactation are found closer to water than all other classes of females.

Females in early lactation react differently from other classes of females to increasingly dry conditions. They are water-bound in all seasons. As seasons become drier, these females are found at significantly greater distances from water. This suggests that while females in early lactation may be sedentary around water in the Early Dry season, they cannot pursue this strategy in the Dry or Late Dry season.

In addition to limiting their movements to areas closer to water, lactating females make more frequent trips to water and spend more time drinking on those trips. Data are pooled for the Early Dry, Dry and Late Dry seasons. On days in which females were continuously observed for at least 8 h, females in early lactation visited water on 33 of 36 days (92%), females in mid-lactation visited on 12 of 17 days (71%), and females which were not lactating made visits on 16 of 28 days (57%) (three way contingency test, $G = 11.05, p < 0.01$). When visiting water holes, females in early and mid-lactation drank for $172 \pm 22$ s whereas non-lactating females drank for $104 \pm 28$ s (d.f. = 42, $t = 2.61, p < 0.05$).

### Distribution of females in relation to food resources

To examine whether female distribution is sensitive to differences in the food supply densities, a test was made of the null hypothesis that female habitat preference was independent of grass biomass. In each season, transect areas were classified by the biomass of grass. The frequency with which each class of females was found in areas of different biomass categories was compared to an expected distribution equal to the frequency with which these areas were surveyed. Frequencies were compared using a 2 × 5 contingency table and a $\chi^2$ statistic.

As the seasons become progressively drier, the abundance of grass declines consistently (Fig. 2). In the Early Dry season, the density of grass was greater than 100 g/m² on 35% of the areas surveyed and greater than 50 g/m² on 68% of those areas. By the Dry season, these percentages had declined to 10% with more than 100 g/m² and 40% with more than 50 g/m². By the Late Dry season, only 5% of the areas surveyed had more than 100 g/m², and only 35% of them had more than 50 g/m². If the movements of females were not influenced by preference for grass at a particular density, or by the movements of the observer, their distribution would be expected to reflect the pattern shown in Fig. 2. It does not.

The data presented in Fig. 3 show the percentage by which actual use of transect areas deviated from expected use in each season and for each class of females. Daily point samples were used for this analysis and sample sizes are the same as in Fig. 1. During all seasons all reproductive classes differ significantly from expectation ($p < 0.05$) in their patterns of distribution. In the Early Dry season, all females showed greater than expected use of the areas in which grass abundance was high, at 100–199 g/m² and

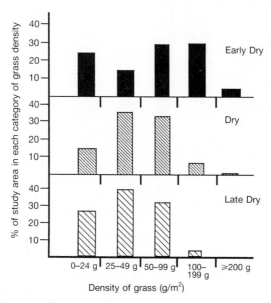

**Fig. 2.** Density of grass (g/m²) in the study area in each of three seasons.

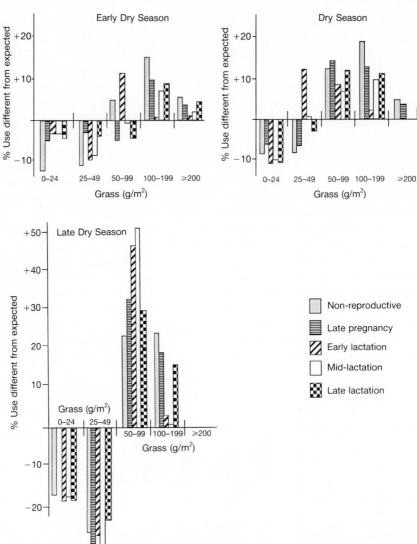

**Fig. 3.** Female habitat preference as measured by use of areas with different densities of grass. The expected use would reflect the availability of grass and would show a pattern similar to that in Fig. 2. The histograms express the differences between this expected use and actual use.

$\geq 200$ g/m$^2$; the pattern was least noticeable among females in early lactation. These females showed a preference for areas of medium abundance (50–99 g/m$^2$). Non-reproductive females particularly avoided areas of low biomass.

In the Dry season, similar differences among reproductive classes were observed (Fig. 3). Females in early lactation differed significantly from all other classes. No significant differences were observed between females in late pregnancy (LP) and non-reproductive (NR) females ($\chi^2 = 3.5$, $p > 0.05$) or between either of these classes and females in late lactation (LL) (LP/LL $\chi^2 = 2.9$, $p > 0.05$; LL/NR $\chi^2 = 3.3$, $p > 0.05$). Females in mid-lactation and late pregnancy differed significantly, but the differences are small ($\chi^2 = 9.9$, $p < 0.04$). Females in early and mid-lactation appear to spend more time in areas of low grass biomass.

During the Late Dry season, all females followed a similar pattern of distribution in avoiding the areas where grass was least abundant and heavily using those where it was relatively plentiful. Females in early lactation did not differ from those in mid-lactation in their distribution; nor were significant differences observed among females in the late lactation, late pregnancy or non-reproductive categories. However, females in early and mid-lactation differed in their habitat use from all other females, avoiding the areas with the most abundant grass ($100-199$ g/m$^2$); other females showed a strong preference for these areas.

### Patterns of home range use

Home range has been estimated for each individual as the area of a circle with its radius equal to the mean distance from the geometric centre of all survey sightings for that individual in any particular season. Individuals are treated as independent points in each season; however, the same individuals do not occur in all seasons. Only individuals which were sighted on more than four days in a season are included in the analysis.

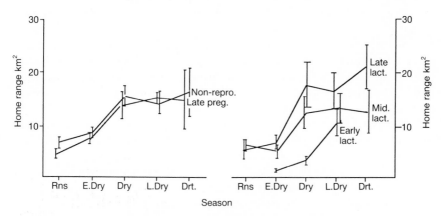

**Fig. 4.** Home range by reproductive class and seasons. Home range was calculated as a circle with the radius equal to the mean distance from the geometric centre of all sightings.

Females of all classes except those in early lactation show similar patterns in their changes in home range use across seasons (Fig. 4). As resources become scarce, and grass and water become disjunct, females increase their home range size; however, this increase does not continue past the Dry season. Home range use is also influenced by reproductive condition. Within each season, females with increased water demands (early/mid-lactation) have smaller home range areas than other females (2-Way ANOVA, Season d.f. $= 4$, $F = 13.0$, $p < 0.01$; reproductive condition d.f. $= 4$, $p < 0.01$, interaction $F = 1.6$, $p > 0.05$).

### Consistency of female associations

To measure consistency of associations, data on group composition were examined in the Early Dry, Dry, and Late Dry seasons. A group was sighted on day 1, then monitored on successive days to determine changes in group composition. The group was monitored until one member of the original group remained. The measure, therefore, represents the time for a focal female to experience a 100% turnover in group membership. If several sub-groups were sighted on successive days, each sub-group was considered to be an independent sample. The percentage of the original group is plotted as a function of the number of days since first observation. No significant differences in consistency of association were observed among females in mid-lactation, late lactation, and late pregnancy. These groups were pooled for this analysis.

In all females, associations were more consistent when ecological conditions were more mesic (Fig. 5). In the Early Dry season, for females other than those in early lactation, the maximum time before a 100% change in group membership was approximately eight days. However, groups of females in mid-lactation and late pregnancy showed less fluidity in the first few days after sighting than did groups of non-reproductive females. In the Dry season, all females except those in early lactation showed a similar pattern of group stability. Groups decayed more slowly than in the Late Dry season, yet total turnover occurred in under six days. During the Late Dry season, all groups changed composition entirely within four days.

In the Dry and Early seasons, females in early lactation showed patterns of group fidelity quite different from those of other females. In the Early Dry season, on the first day after initial observation, approximately 60% of the original group was still together. This core group remained consistently in association for up to two months, after which the group broke up. These core groups tended to be composed of females with young foals of similar age. The pattern in the Dry season was similar to that of the Early Dry, except that there was greater variation in group consistency. Some groups

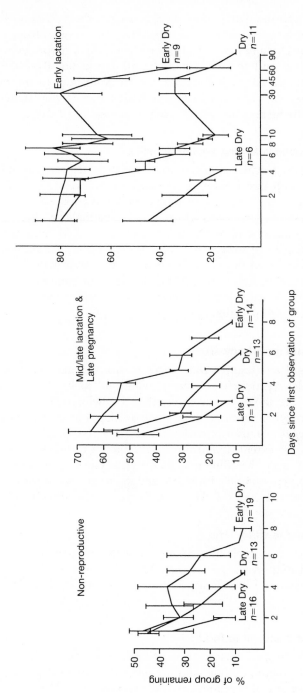

**Fig. 5.** Group stability over time. A group sighted on day 1 is monitored at 1-day intervals. The composition of the original group is compared with that of the group sighted on successive days. For each successive day, the percentage of the original group remaining is plotted. Group sizes on day 1 are $\geq 3$, $\leq 15$. Mean values ± s.e. $n$ = number of females in each analysis. Note that for females in early lactation, scale is $\text{Log}_2$. Y axis (percentage of group remaining) of similar scale in all comparisons.

turned over in approximately one week, while others continued in association for up to three months. The rise in the graph (Fig. 5) from 10 to 30 days in both the Dry and Early Dry season reflects this variation. Groups which are decaying tend to pull down the 10-day mean. By 30 days, unstable groups dissolved (and therefore do not contribute to the mean values): any group still intact at 30 days had a higher consistency of association than the mean value for all groups sighted at 10 days.

### Relative mating success among territorial males

A territorial male that achieved access to an oestrous female was able to mate successfully with that female. In all observations greater than 2 h of oestrous females and territorial males, at least one copulation was observed. The number of oestrous females to which a territorial male had access was therefore used as a measure of male reproductive success. To minimize bias from unequal intensity of observation of different males, the number of oestrous females observed in association with a territorial male has been divided by the number of sightings of that male. Oestrous females have been classified into those in early lactation and those in cycling oestrus either with older foals or with no attendant young (cycling).

### Mating success and male body size

In dimorphic ungulates, relative differences in size explain much of the variation in male reproductive success (Clutton-Brock *et al.* 1982). Territorial males are larger, on average, than bachelor males ($t = 5.2$, $p < 0.001$). The two distributions show some overlap, but are significantly different (Kolmogorov-Smirnov two-sample test $p < 0.01$). This suggests size may be, to some extent, important to attaining a territory. As in red deer, so in Grevy's zebra breeding males vary greatly in size. However, no correlation is seen between male body size and the frequency of association with oestrous females (EL: $r^2 = 0.06$, $p > 0.7$; cycling $r^2 = 0.09$, $p > 0.25$).

### Mating success and territory size

Territory size does not explain variation in male reproductive success. Despite its large variation (range 2.1–11.6 km$^2$) there was no relationship between territory size and association with oestrous females, either in early lactation ($r = 0.10$, $p > 0.13$) or cycling ($r^2 = 0.09$, $p > 0.25$).

### Mating success and access to food

In the following analyses the association of territorial males with oestrous females (in early lactation, cycling, and the sum of the two categories) will be compared to ranking of vegetation abundance as measured by the richest patch of grass in a male's territory (maximum g/m²). For each season, absolute measures were ranked from 1 (best) to N (worst) where N equals the number of territorial males present in that season. If vegetation measures were not significantly different from one another (compared with a Student's t-test), an average rank was assigned (e.g. if ranks 3 and 4 were not different, each was assigned a rank of 3.5). Each season was considered separately, provided that there were sufficient data on male associations with oestrous females. In the Rains and the Drought, insufficient numbers of females in early lactation were present to conduct analyses. A summary analysis, in which all pairs of rank and association are considered, is presented in Table 1. By comparing ranks across season, problems of varying absolute abundance of vegetation are minimized.

In all three seasons examined, and in the summary of all seasons, all correlations between the rank of a male's territory in terms of the maximum biomass of grass found on it, and his association with oestrous females in early lactation, were positive, but insignificant. In the summary of all seasons, and in each season except for the Early Dry, a male's association with oestrous females that were cycling was positively and significantly correlated with the rank of his territory. In those seasons when both early lactation and cycling females were present, if a male's association with all oestrous females was compared to the rank of his territory, no significant correlations were seen in any season.

**Table 1.** Male mating success in each season as a function of territorial quality. In all comparisons, the richest patch of grass in each male's territory was measured (g/m²) and ranked from 1 to N. Mating success (access to oestrous females) was compared to this rank. In the Dry and Late Dry seasons, and across the entire study, males with the higher ranked territories had greater mating success with cycling females. Mating success with females in the early stages of lactation did not correlate with the ranking of the territory in any season. If females are treated as a single group, male mating success did not correlate with this measure of territory quality. All tests Kendall's T, significant results boxed for emphasis.

| | Season | | | |
|---|---|---|---|---|
| Type of female | Early Dry n=13 | Dry n=13 | Late Dry n=14 | Entire study n=67 |
| Early lactation | $T = 0.267$ $p > 0.05$ | $T = 0.08$ $p > 0.05$ | $T = 0.107$ $p > 0.05$ | $T = 0.108$ $p > 0.05$ |
| Cycling | $T = 0.192$ $p > 0.05$ | $T = 0.579$ $p < 0.01$ | $T = 0.458$ $p < 0.05$ | $T = 0.237$ $p < 0.05$ |
| All females | $T = 0.31$ $p > 0.05$ | $T = 0.29$ $p > 0.05$ | $T = 0.327$ $p > 0.05$ | $T = 0.069$ $p > 0.05$ |

n, number of territorial males

## Mating success and access to water

Of the 16 territorial males observed over the course of the study, nine had access to water. Table 2 presents data on the relationship between access to water and association with oestrous females. Access to water had no significant influence on the level of association with cycling females in oestrus. However, in the Early Dry season, the Dry season, and across all seasons, males with access to water were seen more frequently in association with oestrous females in early lactation than were males without access to water. If analysis is restricted to males that had territories with access to water, relative abundance of vegetation did not appear to be important in attracting early lactating females in oestrus. In the two seasons in which a male's association with early lactating females in oestrus was significantly correlated with access to water, it did not correlate with maximum grass abundance (Early Dry, $T = 0.519$, $N = 7$, $p > 0.05$; Dry, $T = 0.17$, $N = 8$ $p > 0.05$).

**Table 2.** Male mating success (association with oestrous females) in each season as a function of access to permanent water. In the Early Dry and Dry seasons, and across the entire study, males with access to permanent water had greater mating success with females in early lactation. Mating success with females in later stages of lactation or with no dependent foals (cycling) did not correlate with access to permanent water in any season. If females are treated as a single group, male mating success did not correlate with this measure of territory quality. All tests Mann Whitney $U$ tests, significant results boxed for emphasis.

| Type of female | Season | | | | |
|---|---|---|---|---|---|
| | Early Dry | Dry | Late Dry | Entire study | |
| Early lactation | $U = 38.5$ $p < 0.01$ | $U = 39.0$ $p < 0.01$ | $U = 33$ $p > 0.05$ | $U = 330$ $p < 0.01$ | $n_1 = 19$ $n_2 = 22$ |
| Cycling | $U = 22.5$ $p > 0.05$ | $U = 33$ $p > 0.05$ | $U = 30$ $p > 0.05$ | $U = 625.5$ $p > 0.05$ | $n_1 = 31$ $n_2 = 37$ |
| All females | $U = 28.5$ $p > 0.05$ $n_1 = 6$ $n_2 = 7$ | $U = 25$ $p > 0.05$ $n_1 = 6$ $n_2 = 7$ | $U = 29.5$ $p > 0.05$ $n_1 = 6$ $n_2 = 8$ | $U = 244$ $p > 0.05$ | $n_1 = 19$ $n_2 = 22$ |

$n_1$, access to water; $n_2$, no water.

Areas without access to water had a higher average maximum density of grass (Mann-Whitney $U = 509$, $n_1 = 44$, $n_2 = 38$, $p < 0.01$). Access to water and high abundance of forage are not mutually exclusive; males with territories which had high maximum biomass of grass and access to water could attract both classes of females.

## Discussion

In the Grevy's zebra, differences in reproductive condition strongly affect patterns of female behaviour. In many of the variables examined, there is a strong interaction between the behaviour and ecology common to a particular reproductive condition and the season in which those variables were examined. The clearest differences are seen when comparing females in early lactation to all other classes of females.

In the Early Dry and Dry seasons, females in early lactation differed significantly from other classes of females in their patterns of movement and association, and in the priority they placed on different resources (grass and water). During these relatively mesic seasons, females with young foals more closely resemble individuals of a different species – the plains zebra, *Equus burchelli* – than they do the archetypal pattern for Grevy's zebra described by Klingel (1974). They are found near water, form close associations with a small number of individuals with whom they are in reproductive synchrony, and remain on the territory of a single male (Ginsberg 1987).

Many differences seen in the behaviour and ecology of females in early lactation may be explained by their increased demands for water during lactation (NRC 1978; Pollock 1980; Oftedal *et al.* 1983). Although all Grevy's zebra require free-standing water in their diet, lactating females are more dependent on water. They require larger quantities of water at shorter intervals. They restrict their movement to areas near permanent water and, as a result, their home ranges are smaller. By limiting the area of their home ranges, females in early lactation appear to sacrifice feeding opportunities to achieve access to water.

In a comparative study of 14 species of domestic and wild ungulates in a semi-desert in southern Kenya, Western (1975) found that water availability, not grass abundance, imposed the most severe restrictions on the distribution of water-dependent species. Among these 14 species, those that needed less water were found at a greater distance from water than those that needed to drink more often. This increased grazing radius offered a greater variety and probably greater quality of habitats to species with low water requirements. The data presented for female Grevy's zebra, in which the distance to water is negatively correlated with water use, parallels this pattern.

Among female Grevy's zebra in early and mid-lactation, however, mean distance to water increased as the land grew drier. This would not appear to agree with Western's (1975) results. Patterns in the distribution of various species of hoofed stock belonging to nomadic pastoralists suggest an explanation. As ephemeral water disappears, stock become concentrated around permanent water holes. This increased density rapidly reduces the abundance of grass in these areas. Pastoralists are then forced to move their animals farther from water to find food (Coughenour, Ellis, Swift, Coppock, Galvin, McCabe & Hart 1985; Robinson 1985). As the dry season progresses, food abundance becomes correlated with the distance travelled from water.

In this study too, the density of grass showed a positive correlation with distance from permanent water during the Dry and Late Dry seasons (Ginsberg 1987). Hence, during the more mesic seasons, lactating females may be able to find enough food and remain near water. In more xeric conditions, these females must abandon sites near water if they are to have access to even marginally adequate food resources.

Changes in area of home range may also be explained by the disjunction of food and water in the drier seasons. In the reproductive classes that are not lactating, or among females in late lactation, home range area increases sharply with the disappearance of ephemeral water (beginning of the Dry season—see Fig. 4). This increase in home range area coincides with a seemingly contradictory decrease in the mean distance to permanent water. This suggests that in the Rains and Early Dry season the movements of these females may not have been influenced by the distribution of permanent water. Ephemeral water was abundant and the females were moderately sedentary. With the disappearance of ephemeral water, even non-lactating females or those in late lactation became, to a greater or lesser extent, dependent on permanent water. Because food and water were no longer to be found together, their home range areas increased as they began moving between water and grazing areas further away from water.

Patterns of association were also seen to vary in response to changes in both reproductive condition and season. Particularly striking were the changes in the consistency of association among females in early lactation. In the Early Dry and Dry seasons associations last up to three months; in the Late Dry, individual associations persist for a maximum of eight days.

These data support theories of female grouping which suggest that associations among Grevy's zebra are not imposed by the male (Klingel 1974), but are the result of ecological conditions which allow females to form bonded groups (Rubenstein 1986; Wrangham 1980). Further support for this argument can be found in studies of equids adapted to more mesic conditions. In horses, a female's reproductive condition influences both patterns of movement and her associations among females of her own harem (Rubenstein 1986; Berger 1986). Similar patterns of association among females within harems have been seen in the plains zebra, *Equus burchelli* (J.R. Ginsberg, unpublished data).

Associations among female Grevy's zebra are fairly unpredictable even among those which show the most durable associations when compared to many other ungulates. Grevy's zebra females, however, are spatially predictable in two ways: those with young foals are found near water; those without young foals are found in areas of abundant food. Given that the location of females is predictable, it has been argued that males will do best to hold those resources to which females are attracted (Emlen & Oring 1977; Bradbury & Vehrencamp 1977). Alternative strategies, all of which involve following females, require temporal predictability of female association, oestrus, movements or some combination of all three variables (Gosling 1986).

In the Grevy's zebra, as in many ungulate species, territoriality does confer the potential for reproductive activity: 91% of all copulations observed were made by a territorial male while on his territory (Ginsberg 1987). However, amongst territorial males, no correlation could be found between male body size and reproductive success. Nor did absolute territorial area determine a male's access to oestrous females. Reproductive success is correlated only with the quality of the resources which males hold on their territories.

Differences in male mating success, however, only become apparent if differences in female ecological requirements are first recognized. Males with patches of relatively dense forage on their territories differentially attracted and mated with all females except those in early lactation. Males with access to permanent water had a greater reproductive success with females in early lactation, but not with other classes of females. When females are grouped into one category, regardless of their reproductive condition, neither access to water nor abundance of food was a good predictor of male reproductive success.

The variation in the predictability of female association and behaviour which results from the unpredictable environmental conditions in which the Grevy's zebra is found is large: associations are fluid, patterns of movement seasonally variable. This variability is, in large measure, due to seasonal changes in the relative abundance of food and water and their interaction with the resource requirements of females in different states of reproduction. In the Grevy's zebra, male reproductive success appears to be correlated with precisely the same ecological variables as affect the spatial distribution and patterns of association of females.

## Acknowledgements

The author wishes to thank the Office of the President of the Republic of Kenya, the Wildlife Conservation and Management Department, Kenya, and the Department of Zoology, University of Nairobi, for sponsoring this research. Special thanks to D.I. Rubenstein for years of support and advice. The work was funded by the National Geographic Society (Grants #'s 3022–85; 2638–83), the New York Zoological Society and Princeton

University. This manuscript was written while receiving support from the N.S.F. in the form of a N.A.T.O. Postdoctoral Fellowship.

## References

Altmann, J. (1974). Observational study of behavior: sampling methods. *Behaviour* **49**:227–67.

Backhaus, D. (1959). Experimentelle Untersuchungen über die Sehschärfe und das Farbsehen einiger Huftiere. *Z. Tierpsychol.* **16**:445–67.

Barkham, J.P. & Rainy, M.E. (1976). The vegetation of the Samburu-Isiolo Game Reserve. *E. Afr. Wildl. J.* **14**:297–329.

Berger, J. (1986). *Wild horses of the Great Basin. Social competition and population size.* Chicago University Press, Chicago & London.

Bradbury, J.W. & Vehrencamp, S.L. (1977). Social organization and foraging in emballonurid bats. 3. Mating systems. *Behav. Ecol. Sociobiol.* **2**:1–17.

Clutton-Brock, T.H., Guinness, F.E. & Albon, S.D. (1982). *Red deer: behavior and ecology of two sexes.* Chicago University Press, Chicago: Edinburgh University Press, Edinburgh.

Clutton-Brock, T.H. & Harvey, P.H. (1984). Comparative approaches to investigating adaptation. In *Behavioural ecology: an evolutionary approach* (2nd edn): 7–29. (Eds Krebs, J.R. & Davies, N.B.). Blackwell Scientific, Oxford.

Coughenour, M.B., Ellis, J.E., Swift, D.M., Coppock, D.L., Galvin, K., McCabe, J.T. & Hart, T.C. (1985). Energy extraction and use in a nomadic pastoral ecosystem. *Science, N.Y.* **230**:619–25.

Crook, J.H. (1965). The adaptive significance of avian social organizations. *Symp. zool. Lond.* No. **14**:181–218.

Crook, J.H. & Gartlan, J.S. (1966). Evolution of primate societies. *Nature, Lond.* **210**:1200–3.

Duncan, P. (1975). *Topi and their food supply.* PhD thesis: University of Nairobi.

Emlen, S.T. & Oring, L.W. (1977). Ecology, sexual selection, and the evolution of mating systems. *Sience, N.Y.* **197**:215–23.

Estes, R. (1974). Social organization of the African Bovidae. In *The behavior of ungulates and its relation to management*: 166–205. (Eds Geist, V. & Walther, F.R.). I.U.C.N., Morges, Switzerland. (*IUCN Publs* (N.S.) No. 24.)

Geist, V. (1974). On the relationship of social evolution and ecology in ungulates. *Am. Zool.* **14**:205–20.

Geist, V. & Walther, F.R. (Eds). (1974). *The behavior of ungulates and its relation to management.* I.U.C.N., Morges, Switzerland. (*IUCN Publs* (N.S.) No. 24).

Ginsberg, J.R. (1987). *Social behavior and mating strategies of an arid adapted equid: the Grevy's zebra.* PhD thesis: Princeton University.

Gosling, L.M. (1975). *The ecological significance of male behaviour in Coke's hartebeest, Alcelaphus buselaphus cokei, Günther.* PhD thesis: University of Nairobi.

Gosling, L.M. (1986). The evolution of mating strategies in male antelopes. In *Ecological aspects of social evolution*: 244–81. (Eds Rubenstein, D.I. & Wrangham, R.W.). Princeton University Press, Princeton.

Government of Kenya (1971). *Summary rainfall in Kenya*. Kenya Meteorological Department, Nairobi.

Government of Kenya (1984). *Climatological statistics for Kenya*. Kenya Meteorological Department, Nairobi.

Hall, K.R.L. & DeVore, I. (1965). Baboon social behaviour. In *Primate behaviour. Field studies of monkeys and apes*: 53–110. (Ed. DeVore, I.). Holt, Rinehart & Winston, New York.

Iaderosa, J. (1983). Gestation period in Grevy's zebra: managerial and evolutionary considerations. *Proc. reg. Conf. AAZPA* 1983:622–6.

Jarman, P.J. (1974). The social organisation of antelope in relation to their ecology. *Behaviour* 48:215–67.

Jarman, P.J. & Jarman, M.V. (1979). The dynamics of ungulate social organization. In *Serengeti: dynamics of an ecosystem*: 185–220. (Eds Sinclair, A.R.E. & Norton-Griffiths, M.). Chicago University Press: Chicago & London.

King, J.M. (1965). A field guide to the reproduction of the Grant's zebra and Grevy's zebra. *E. Afr. Wildl. J.* 3:99–117.

Klingel, H. (1965). Notes on the biology of the plains zebra *Equus quagga boehmi* Matschie. *E. Afr. Wildl. J.* 3:86–8.

Klingel, H. (1969). The social organization and population ecology of the plains zebra (*Equus quagga*). *Zool Afr.* 4:249–63.

Klingel, H. (1974). Soziale Organisation und Verhalten des Grevy-Zebras (*Equus grevyi*). *Z. Tierpsychol.* 36:37–70. [English summary.]

McNaughton, S.J. (1979). Grassland–herbivore dynamics. In *Serengeti. Dynamics of an ecosystem*: 46–81. (Eds Sinclair, A.R.E. & Norton-Griffiths, M.). University of Chicago Press, Chicago & London.

Moss, C. (1982). *Portraits in the wild.* (2nd edn). Chicago University Press, Chicago.

NRC (1978). *Nutritional requirements of horses*. National Academy Press, Washington.

Oftedal, O.T. (1984). Milk composition, milk yield and energy output at peak lactation: a comparative review. *Symp. zool. Soc. Lond.* No. 51: 33–85.

Oftedal, O.T., Hintz, H.F. & Schryver, H.F. (1983). Lactation in the horse: milk composition and intake by foals. *J. Nutr.* 113:2196–206.

Owen-Smith, N. (1977). On territoriality in ungulates and an evolutionary model. *Q. Rev. Biol.* 52:1–38.

Pollock, J. (1980). *Behavioural ecology and body condition changes in New Forest ponies*. The Farm Livestock Committee: RSPCA Scientific Publications, The Causeway, Horsham, U.K.

Pratt, D.J., Greenway, P.J. & Gwynne, M.D. (1966). A classification of East African rangeland, with an appendix on terminology. *J. appl. Ecol.* 3:369–82.

Robinson, P.W. (1985). *Gabra nomadic pastoralism in nineteenth and twentieth century northern Kenya: strategies for survival*. PhD thesis: Northwestern University, Evanston, Il.

Rubenstein, D.I. (1981). Behavioral ecology of island feral horses. *Equine Vet. J.* 13:27–34.

Rubenstein, D.I. (1986). Ecology and sociality in horses and zebras. In *Ecological aspects of social evolution*:282–302. (Eds Rubenstein, D.I. & Wrangham, R.W.). Princeton University Press, Princeton.

Rubenstein, D.I. & Wrangham, R.W. (Eds). (1986). *Ecological aspects of social evolution*. Princeton University Press, Princeton.

Schenkel, R. (1966). On sociology and behaviour in impala (*Aepyceros melampus* Lichtenstein). *E. Afr. Wildl. J.* 4:99–114.

Western, D. (1975). Water availability and its influence on the structure and dynamics of a savannah large mammal community. *E. Afr. Wildl. J.* 13:265–86.

Wrangham, R.W. (1980). An ecological model of female-bonded primate groups. *Behaviour* 75:262–300.

Wrangham, R.W. & Rubenstein, D.I. (1986). Social evolution in birds and mammals. In *Ecological aspects of social evolution*: 452–70. (Eds Rubenstein, D.I. & Wrangham, R.W.). Princeton University Press, Princeton.

Symp. zool. Soc. Lond. (1989) No. 61:111–125

# Elephant mate searching: group dynamics and vocal and olfactory communication

JOYCE H. POOLE[1,2] and
CYNTHIA J. MOSS[2]

[1]*Biology Department*
*Princeton University*
*Princeton, NJ 08544, USA*

[2]*Amboseli Elephant Project*
*African Wildlife Foundation*
*P.O. Box 48177*
*Nairobi, Kenya*

## Synopsis

Male and female African elephants, *Loxodonta africana*, live in two different social systems and range over large areas, often at very low densities. The availability of both oestrous females and musth males at any particular time is extremely low. The results of our study suggest that elephants use a variety of searching and attracting strategies to overcome the potential difficulty of locating these few and often distant mates. Females search for musth males by moving in large aggregations and possibly by listening for their calls and following their urine trails. Oestrous females attract males to them by conspicuous postures and behaviours, by particular vocalizations and by olfactory signalling with urine. Musth males search for females by associating with big cow/calf groups, by covering large distances, by listening for the females' loud, very low-frequency calls and by monitoring female urine and dung. Males attract females by calling and by leaving trails of strong-smelling urine.

## Introduction

In a social and ecological system in which males and females range separately over wide areas the problem of finding mates may require special solutions. Most large African mammals live in social systems in which males and females either live together in semi-permanent groups (e.g. plains zebra, *Equus quagga*: Klingel 1965; gelada baboon, *Theropithecus gelada*: Dunbar & Dunbar 1975; gorilla, *Gorilla gorilla*: Schaller 1963; yellow baboon, *Papio cynocephalus*: Washburn & DeVore 1961; Cape buffalo, *Syncerus caffer*: Grimsdell 1969; lion, *Panthera leo*: Schaller 1972; spotted hyena, *Crocuta crocuta*: Kruuk 1972) or are found in predictable locations (e.g. impala, *Aepyceros melampus*: Jarman 1979; Grevy's zebra, *Equus grevyi*: Klingel 1974; Uganda kob, *Adenota kob*: Buechner 1961; white rhinoceros, *Ceratotherium simum*: Owen-Smith 1975), and thus mates are not difficult to find. In all of these mating systems the proportion of time allocated to searching for mates is relatively small.

African elephants, *Loxodonta africana*, on the other hand, may spend a large proportion of their reproductive effort in the search for mates. Elephants live in a fluid and dynamic system of spatial use and grouping patterns in which males and females live in separate but overlapping spheres (Douglas-Hamilton 1972; Moss & Poole 1983). Neither sex is territorial. Females live in matriarchal family units (Buss 1961) and adult males live independently, alternating between associating with females and with other males (Moss & Poole 1983). In most areas where they have been studied, females live in predictable dry-season home ranges but migrate over large areas during the wet season (Leuthold & Sale 1973; Leuthold 1977; Western & Lindsay 1984). The size of group in which an individual finds itself may change from day to day and season to season, with small scattered groups typical in the dry season and large aggregations typical in the wet season (Western & Lindsay 1984). Moving singly or in groups of up to several thousand, individual elephants may travel as far as 50 km in a few days (Leuthold 1977) and may range over 3700 km$^2$ (Leuthold 1977). In some areas they live at densities as low as 0.024 km$^{-2}$ (Poché 1974). Finding a mate under these conditions can be problematical for both sexes.

Elephants do not exhibit a pronounced breeding season (Perry 1953; Buss & Smith 1966; Poole 1987). Oestrous females may be observed in any month of the year although the frequency of oestrus is higher during and following the wet season (Poole 1987; Moss 1988). A female elephant is in oestrus for three to six days (Moss 1983) and may conceive only once every three to nine years (Laws, Parker & Johnstone 1975). Given the short receptive period and high investment in each offspring (Lee & Moss 1986) it is crucial for a female to find a high-quality mate (Moss 1983; Poole in press). Large males in the rutting period of musth (Poole & Moss 1981; Hall-Martin & van der Walt 1984; Poole 1987) are a scarce resource, yet females are almost always guarded and mated by a musth male (Poole in press). At the same time, for an individual male receptive females are extremely rare not only in space and time but as a result of intense male-male competition. Once an oestrous female is found and guarded by a high-ranking male she is no longer available to lower-ranking males (Poole 1989).

In this paper we examine some of the possible ways in which musth males and oestrous females find each other.

## Methods

### Study animals and study site

A long-term study of social and reproductive behaviour of the elephant population in Amboseli National Park, Kenya, was begun in 1972 and continues. The Park covers an area of 390 km$^2$ and consists of semi-arid wooded, bushed and open grasslands interspersed with a series of permanent swamps. The Park provides a dry-season concentration area for migratory herbivores including elephants. In the wet season the elephants may leave the Park and move onto privately owned ranchland, where water availability is highly seasonal, thereby increasing their range to approximately 3500 km$^2$ (Western & Lindsay 1984). This larger area can be considered the Amboseli ecosystem (Western 1975). In this ecosystem live 670 elephants including 50 matriarchal families. The number of adults has remained constant over the last 10 years at approximately 160 males and 225 females.

### Sampling methods

At each sighting of a single elephant or group the following data were recorded on computer coding sheets: date, time, location, habitat, group size, presence or absence of musth male(s) or oestrous female(s), and identification of family units and individual males. The results on group size and association patterns presented in this paper were drawn from subsets of the long-term dataset. Rates of behaviour were derived from 3-h focal samples (Altmann 1974) collected in 1985–86 on adult males during either their musth or non-musth periods.

The majority of vocalizations made by African elephants contain frequencies below the range of human hearing (Poole, Payne, Langbauer & Moss 1988). Audio recordings of elephant vocalizations were made on a Nagra IVSJ recorder equipped with two Sennheiser MKH 110 microphones capable of recording these very low frequencies. Manufacturer's curves were used to determine the frequency and amplitude of all components, and field calibrations were made by means of a B&K model 14230 acoustic calibrator.

### Definitions

A family unit consists of a single adult female or two or more adult females, their immature offspring and other calves. The members show a high frequency of association over time, act in a co-ordinated manner, exhibit affiliative behaviour toward one another and are known to be related or are putatively related.

A group of elephants is defined as any number of elephants of any age or sex moving together in a co-ordinated manner with no single member or subgroup at a distance greater than the diameter of the main body of the group at its widest point.

An aggregation is a group of elephants consisting of two or more family units with or without adult males.

A cow/calf group can consist of a fragment of a family unit, a family unit or an aggregation of family units and may or may not contain associating males. A bull group consists of only males.

An oestrous female can be identified by particular behaviours in the presence of males (see Moss 1983) as well as by specific vocalizations (Poole *et al.* 1988).

Musth is a male rutting period which can be identified by physical and behavioural characteristics such as urine dribbling, swollen and secreting temporal glands, aggressive and sexual behaviour and specific vocalizations (Poole 1987). A musth male is a male elephant exhibiting these character-istics. The mean age of first musth is $29 \pm 3$ years old (Poole 1987).

The low-frequency vocalizations (fundamental frequencies range from 10–30 Hz) made by elephants in a variety of social contexts are referred to as rumbles. In this paper we discuss several different rumbles: post-copulatory or oestrous, musth and female chorus (see Poole *et al.* 1988).

Testing consists of one elephant using the tip of its trunk to inspect the genitals, urine or dung of another individual. In the case of males testing females or their urine or dung, the male often exhibits a flehmen-like response (Rasmussen, Schmidt, Henneous & Groves 1982), placing his trunk tip against his vomeronasal organ.

## Results

### Group dynamics

### Cow/calf groups

In the dry season females are typically found in many small dispersed groups while during and following the rains females tend to aggregate into large groups. For example, in 1985 in the dry months of August, September and October the median cow/calf group size class was 1–9 ($N = 206$; interquartile range from 1–9 to 10–24) while during the wet months of February, March and April the median group size class was 10–24 ($N = 175$, interquartile range from 10–24 to 40–54). There were significantly more large groups during the wet months than during the dry months (Kolmogorov-Smirnov $\chi^2 = 38.0$, *d.f.* $= 2$, $P < 0.001$; Fig. 1). Groups numbering 55 or more individuals accounted for 22.2% of the sightings in February–April, but for only 1.0% of the sightings in August–October.

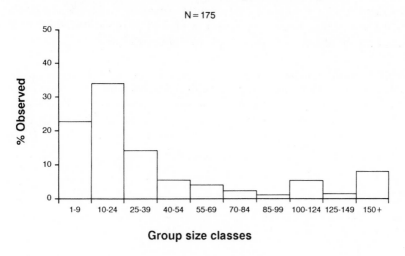

**Wet months 1985 - February, March, April**

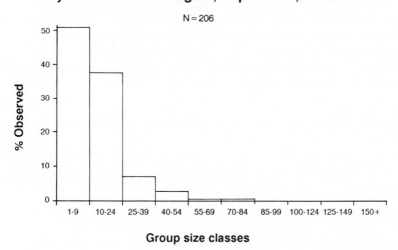

**Dry months 1985 - August, September, October**

**Fig. 1.** Histograms comparing the frequency distribution of cow/calf group sizes during three wet months and three dry months in 1985 (Kolmogorov-Smirnov $\chi^2 = 38.0$, *d.f.* $= 2$, $P < 0.001$).

## Bull groups

Adult males are found singly, in small all-male groups or in association with females. Among the breeding-aged males (Poole in press) the grouping and association pattern is greatly influenced by the phenomenon of musth.

During musth, males are typically found in close association with females (median = 62.5% of sightings) or alone (median = 32.5% of sightings). When not in musth they are found in all-bull groups (median = 65% of sightings) or alone (median = 24% of sightings; Table 1).

**Table 1.** Association patterns (percent observations alone, with females, with males) of the six oldest males during musth and non-musth.

| Male | During musth | | | | During non-musth | | | |
|------|-------------|-----------|-----------|-----|-------------|-----------|-----------|-----|
|      | Alone %     | w/Cows %  | w/Bulls % | N   | Alone %     | w/Cows %  | w/Bulls % | N   |
| 7    | 34          | 59        | 7         | 56  | 22          | 10        | 68        | 59  |
| 10   | 31          | 64        | 5         | 75  | 35          | 3         | 62        | 69  |
| 13   | 16          | 77        | 7         | 143 | 8           | 5         | 87        | 40  |
| 22   | 36          | 61        | 3         | 112 | 48          | 5         | 47        | 114 |
| 99   | 35          | 61        | 4         | 72  | 26          | 12        | 62        | 136 |
| 126  | 23          | 73        | 4         | 75  | 16          | 4         | 80        | 51  |

### Occurrence of oestrous females

The estimated number of oestrous females per year is presented in Table 2. The estimates are derived from 129 births recorded between December 1984 and April 1987. Sixty-eight of these mothers were seen in oestrus 22 months before, which suggests that we observed approximately half of all females who were in oestrus each year. In an average year there were an estimated 75 females in oestrus.

**Table 2.** The number of observed and estimated oestrous females for the years 1980–86.

| No. of oestrous females | 1980 | 1981 | 1982 | 1983 | 1984 | 1985 | 1986 |
|-------------------------|------|------|------|------|------|------|------|
| Observed                | 43   | 87   | 17   | 81   | 27   | 64   | 20   |
| Estimated               | 86   | 174  | 34   | 162  | 54   | 128  | 40   |

Although there was a peak in the number of females in oestrus during the first eight months of the year, females were observed in oestrus throughout the year. Table 3 shows the mean frequency of oestrous females per month over a period of 10 years. The number of females in oestrus per month ranged from a low of 0 during the driest months to a high of 24 after the rains in June 1981.

Oestrous females tended to be located in large groups. For example, in 1985 cow/calf groups containing an oestrous female ($N = 92$) were significantly larger than groups without an oestrous female ($N = 685$;

**Table 3.** Mean number of observed and estimated oestrous females per month (1976–86).

| No. of oestrous females | J | F | M | A | M | J | J | A | S | O | N | D |
|---|---|---|---|---|---|---|---|---|---|---|---|---|
| **Observed** | | | | | | | | | | | | |
| Mean | 3.2 | 3.5 | 5.5 | 4.7 | 3.1 | 5.1 | 5.1 | 2.5 | 2.3 | 1.7 | 1.1 | 1.6 |
| s.d. | 2.6 | 3.1 | 4.7 | 5.6 | 3.5 | 6.8 | 5.3 | 3.6 | 2.8 | 2.3 | 1.1 | 1.8 |
| Estimated | 6 | 7 | 11 | 10 | 6 | 10 | 10 | 3 | 5 | 3 | 2 | 3 |

Kolmogorov-Smirnov $\chi^2 = 31.3$, $d.f. = 2$, $P < 0.001$; Fig. 2). The median size-class of groups containing an oestrous female was 25–39 individuals (interquartile range from 10–24 to 55–69) while the median size-class of groups without an oestrous female was 10–24 (interquartile range from 1–9 to 10–24). Groups containing oestrous females were larger than groups without oestrous females during both the wet and the dry seasons.

### Occurrence of musth males

The median number of musth males per month was three (range 0–13). Musth males tended to associate with large cow/calf groups. Again in 1985, groups containing a musth male were significantly larger than groups without an associating musth male (Kolmogorov-Smirnov $\chi^2 = 76.7$, $d.f. = 2$, $P < 0.001$; Fig. 3). The median size-class of groups containing a musth male ($N = 152$) was 25–39 (interquartile range from 10–24 to 70–84) while the median size-class of groups without an associating musth male ($N = 625$) was 10–24, interquartile range from 1–9 to 10–24). Groups containing a musth male were large whether an oestrous female was present ($N = 80$) or not ($N = 72$) and there was no difference in the frequency distribution of group size-classes between these two groups (Kolmogorov-Smirnov $\chi^2 = 2.1$, $d.f. = 2$, $P = 0.35$).

### Communication

#### Vocal

In 1986 32 tape recordings of 20 min each were made in the presence of cow/calf groups ranging in size from 12 to 150 elephants. The median percentage of rumbles heard by the human auditors was 27% of those actually recorded and present on the spectrograms (interquartile range = 21–46%). By counting the number of vocalizations exhibited on the spectrograms we were able to compare the number of vocalizations made by different-sized cow/calf groups. The rate of calling in large groups was higher than the rate in smaller groups (Spearman rank correlation $R_s = 0.48$, $N = 32$, $P < 0.01$; range = 0.2–9.2 calls per minute).

Although both musth and non-musth males were observed listening, sexually active musth males listened more often than did non-musth males

## Cow/calf groups with an oestrous female

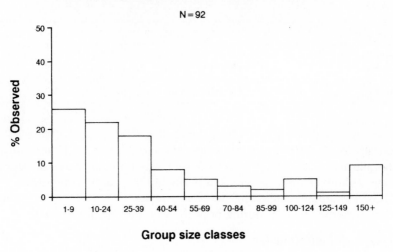

## Cow/calf groups without an oestrous female

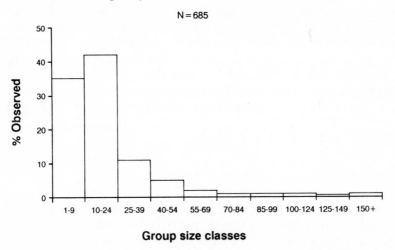

**Fig. 2.** Histograms comparing the frequency distribution of the size of cow/calf groups with and without an oestrous female present (Kolmogorov-Smirnov $\chi^2 = 31.3$, $d.f. = 2$, $P < 0.001$).

(Mann-Whitney $U = 4$, $N_1 = 6$, $N_2 = 8$, $P = 0.004$; median rate per 3 h: of musth males = 2.5, interquartile range 2.0–3.75; of non-musth males = 0.75, interquartile range 0–1). Of the 165 times that musth males were seen listening during 62 3-h focal samples, the listening male was alone 124 times, with other males 34 times and with females only seven times.

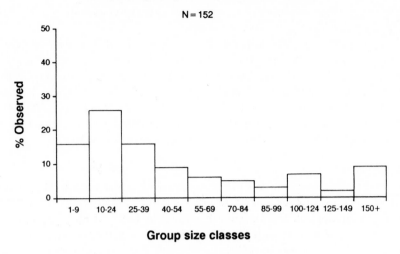

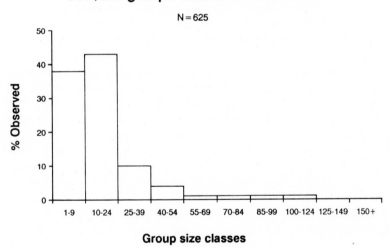

**Fig. 3.** Histograms comparing the frequency distribution of the size of cow/calf groups with and without an associating musth male (Kolmogorov-Smirnov $\chi^2 = 76.7$, $d.f. = 2$, $P < 0.001$).

Occasionally musth rumbling was associated with listening behaviour by the rumbling male. Of the 200 musth rumbles made by males when they were with females during focal samples collected in 1985, 11% were answered with a loud female chorus by nearby females. Since many of the rumbles made by elephants were inaudible to humans, some answers may have gone undetected. Preliminary playback experiments using musth rumbles resulted in females urinating and defecating and approaching the speaker (J. H. Poole unpublished data).

Copulations are often followed by a series of loud, very low-frequency post-copulatory calls by the mated female. During 57 of the 69 matings observed during 1985 and 1986 we were close enough to the mating pair to determine whether the female called. In 45 cases the mated female rumbled, in 12 she did not ($\chi^2 = 19.1$, $d.f. = 1$, $N = 57$, $P < 0.001$). Preliminary playback experiments using a post-copulatory rumble attracted males from up to 400 m away (J. H. Poole unpublished data).

**Olfactory**

Males moved from cow/calf group to cow/calf group testing the vulva of each female in turn. Females often responded to the approach of a musth male by backing toward him and urinating. In 191 h of focal sampling on musth males who were with females, males were observed testing females or their urine 431 times. In 1985, in 113 3-h focal samples, musth males contacted a median of three families (or an average of 13.5 adult females) per 3 h (interquartile range = 1.0–5.5; range = 0–23). Lone musth males spent a higher percentage of time walking (median = 43%, interquartile range = 23–47%) than did musth males already associating with females (median = 20%, interquartile range = 17–21%; Wilcoxon $T = 2$, $N = 9$, $P < 0.01$).

Males associating with cow/calf groups typically followed behind the group as it was moving and tested the dung and urine-soaked soil left behind. Lone males also tested the dung and urine that they found. The rate of testing dung or urine found on the ground by a musth male who had located an oestrous female was significantly less than the rate of testing by lone musth males (Mann-Whitney $U = 1$, $N_1 = 6$, $N_2 = 9$, $P < 0.001$; median rate of musth males with an oestrous female = 0.1, interquartile range = 0–0.2; median rate of lone males = 1.2, interquartile range = 0.8–2.3).

Females appeared to announce approaching oestrus. Approximately two and a half weeks prior to behavioural oestrus males began to test the female's urine repeatedly. During 1985–87 nine females were observed both during this brief pre-oestrous period and then again during oestrus proper. The mean interval between the two periods was 17.1 ± 3 days, $N = 9$. In five out of these nine cases the musth male seen showing most interest in the female during pre-oestrus was the same male who was then seen guarding her two and a half weeks later.

Females showed great interest in the urine of musth males. During experiments placing musth male urine in the path of cow/calf groups, females went out of their way to inspect the urine (J.H. Poole unpublished data). Responses included touching the urine with their trunk tips, rumbling, urinating and defecating. In addition, females often followed the natural urine trails of musth males. When a musth male arrived in a group females frequently reached their trunks toward the male's genitals and vocalized with a female chorus.

## Discussion

Male and female elephants live separately in essentially two different social systems. Within these systems musth males and oestrous females are scarce resources in both space and time. Both sexes are confronted with the problem of finding mates.

Females have relatively short and infrequent receptive periods and apparently prefer musth males as mating partners (Moss 1983; Poole in press), but these males are few and often distant. Females in the Amboseli population appear to use several strategies to search for and attract these preferred mates.

We suggest that oestrous females move in large groups so that they can more easily find and be found by large musth males. Since females live in stable family groups and it is assumed that aggregations are formed by the initiation of matriarchs, it is interesting to speculate whether these matriarchs (who may not be available for breeding themselves) position their close female relatives in large groups in order to meet musth males. Females may also search for musth males by following their urine trails and by moving toward the sound of musth rumbles.

We also suggest that oestrous females attract males, by exhibiting conspicuous behaviour (Moss 1983), by calling loudly and frequently, and by producing urine with particular olfactory components (Rasmussen *et al.* 1982).

During early oestrus, females draw attention to themselves by displaying a characteristic posture and running away from the group when pursued by males (Moss 1983). During the peak of oestrus females vocalize loudly (Poole in press). Their female relatives often join in with a pandemonium of calls increasing the sound-pressure level significantly. We suggest that oestrous females and their relatives may call to attract distant preferred mates (Poole *et al.* 1988). Again it is interesting to speculate on the role that these calls may play in increasing the inclusive fitness (Hamilton 1964) of family members.

Oestrous female urine is of particular interest to males who are able to detect the changes in hormone levels associated with oestrus and possibly ovulation (Rasmussen *et al.* 1982). We have some behavioural evidence of a pre-oestrus change in urine which may announce to males a female's approaching oestrous period.

The musth periods of large adult males in the Amboseli population are asynchronous but overlapping and vary in duration from several days to five months (Poole 1987). Individual males attempt to locate, guard and mate with as many oestrous females as possible during their musth periods. Competition between sexually active males for the few receptive females available at any one time is intense. The problem for an individual male is to find and mate with an oestrous female before a higher-ranking male can locate her (Poole 1989). Like females, musth males appear to use a variety of strategies to locate mates.

Males search for females by travelling over long distances contacting many different cow/calf groups, by associating with large aggregations, by listening for the calls of females, and by monitoring female urine and dung.

Lone musth males spent twice as much time walking as did musth males who were in association with females, presumably because the lone males were still searching. In Ruaha, Barnes (1982) also found that sexually active males travelled further when they were alone than when they were with cows. In Kruger, Hall-Martin (1987) found that musth males travelled further and faster and met more cow/calf groups than did non-musth males.

Groups containing musth males were significantly larger than groups without musth males whether or not there was an oestrous female present. We suggest that sexually active males associate with big aggregations so that they can monitor large numbers of females more efficiently.

Musth males listened more often than did non-musth males. In addition, although musth males spent more time with females than alone, lone musth males were seen listening much more frequently than were musth males who were already in association with a group of females. Males may be listening for oestrous calls or for other loud, low-frequency vocalizations made by females (Poole *et al.* 1988) and may locate large aggregations by this higher rate of calling.

In addition to searching, musth males may attract females to them by marking the path they have taken with strong-smelling urine and by calling. The rate of urine dribbling increases significantly while males are walking and they may lose up to 400 litres of urine per 24 h (Poole 1989). Females show considerable interest in these trails and have been observed to follow them. Calling musth males frequently received answers from females and preliminary playback experiments using musth rumbles attracted females to the speaker (J. H. Poole, unpublished data).

Very low-frequency sounds are subject to lower levels of environmental attenuation than are higher-frequency sounds of the same sound-pressure level. Some of the loud, very low-frequency sounds used by elephants may travel several kilometres before becoming inaudible to conspecifics (Payne, Langbauer & Thomas 1986; Poole *et al.* 1988). The reproductive calls made by female African elephants have particularly high sound-pressure levels, which suggests that they may be heard by other elephants at long distances (Poole *et al.* 1988). Elephants may rely on vocal communication in the search for mates, particularly in thick vegetation or where wind direction precludes the use of olfactory communication.

Our study suggests that elephants overcome the potential problem of finding the few and possibly distant mates available by using a variety of strategies. Males and females search for mates by associating with large aggregations, by monitoring urine and by listening for the calls of other elephants. In addition, elephants may attract distant members of the opposite sex to them by leaving urine with particular olfactory components and by loud, very low-frequency calling. Although many species call and mark with urine during oestrus and rut, the behaviour of elephants appears to be particularly well adapted for locating distant mates.

## Acknowledgements

This study was funded by the African Wildlife Foundation, the Harry Frank Guggenheim Foundation, the National Institute of Mental Health (5 F32 MH09277) and the New York Zoological Society, and was carried out in affiliation with the Kenya Rangeland Ecological Monitoring Unit, the Laboratory of Ornithology, Cornell University, Princeton University and Sub-Department of Animal Behaviour, Cambridge University. We thank them for their sponsorship and support. For permission to work in Amboseli National Park we thank the Office of the President of the Republic of Kenya, the National Council for Science and Technology, the Wildlife Conservation and Management Department and the Amboseli Wardens. We are grateful to Chris Clark for allowing us to use his spectrographic equipment and to David Wickstrom for endless technical advice. Thanks are due to Katherine Payne and Bill Langbauer for introducing us to infrasound and for many stimulating discussions. We are grateful to our colleagues, Keith Lindsay, Phyllis Lee and Sandy Andelman, for providing us with their records on births, oestrus and musth. Special thanks are due to Wamaitha Njiraini and Soila Sayialel for their assistance in both data collection and analysis.

## References

Altmann, J. (1974). Observational study of behaviour: sampling methods. *Behaviour* 49:227–67.
Barnes, R.F.W. (1982). Mate searching behaviour of elephant bulls in a semi-arid environment. *Anim. Behav.* 30:1217–23.
Buechner, H.K. (1961). Territorial behavior in Uganda kob. *Science, N.Y.* 133:698–9.

Buss, I.O. (1961). Some observations on food habits and behavior of the African elephant. *J. Wildl. Mgmt* **25**:131–48.

Buss, I.O. & Smith, N.S. (1966). Observations on reproduction and breeding behavior of the African elephant. *J. Wildl. Mgmt* **30**:375–88.

Douglas-Hamilton, I. (1972). *On the ecology and behaviour of the African elephant.* Ph.D. thesis: University of Oxford.

Dunbar, R. & Dunbar, P. (1975). Social dynamics of gelada baboons. *Contr. Primatol.* **6**:1–157.

Grimsdell, J.J.R. (1969). *The ecology of the buffalo,* Syncerus caffer, *in western Uganda.* Ph.D. thesis: University of Cambridge.

Hall-Martin, A.J. (1987). Role of musth in the reproductive strategy of the African elephant (*Loxodonta africana*). *S. Afr. J. Sci.* **83**:616–20.

Hall-Martin, A.J. & van der Walt, L.A. (1984). Plasma testosterone levels in relation to musth in the male African elephant. *Koedoe* **27**:147–9.

Hamilton, W.D. (1964). The genetical evolution of social behaviour. *J. theor. Biol.* **7**:1–52.

Jarman, M.V. (1979). *Impala social behaviour: territory, hierarchy, mating and the use of space.* Verlag Paul Parey, Berlin & Hamburg. (*Adv. Ethol.* No. 21: 1–93.)

Klingel, H. (1965). Notes on the biology of the plains zebra, *Equus quagga boehmi* Matschie. *E. Afr. Wildl. J.* **3**:86–8.

Klingel, H. (1974). A comparison of the social behaviour of the Equidae. In *The behavior of ungulates and its relation to management*: 124–32. (Eds Geist, V. & Walther, F.R.). IUCN, Morges, Switzerland. (*IUCN Publs* (N.S.) No. 24.)

Kruuk, H. (1972). *The spotted hyena: a study of predation and social behavior.* University of Chicago Press, Chicago & London.

Laws, R.M., Parker, I.S.C. & Johnstone, R.C.B. (1975). *Elephants and their habitats: the ecology of elephants in North Bunyoro, Uganda.* Clarendon Press, Oxford.

Lee, P.C. & Moss, C.J. (1986). Early maternal investment in male and female elephant calves. *Behav. Ecol. Sociobiol.* **18**:353–61.

Leuthold, W. (1977). Spatial organization and strategy of habitat utilization of elephants in Tsavo National Park, Kenya. *Z. Säugetierk.* **42**:358–79.

Leuthold, W. & Sale, J.B. (1973). Movements and patterns of habitat utilization of elephants in Tsavo National Park, Kenya. *E. Afr. Wildl. J.* **11**:369–84.

Moss, C.J. (1983). Oestrous behaviour and female choice in the African elephant. *Behaviour* **86**:167–96.

Moss, C.J. (1988). *Elephant memories.* William Morrow, New York.

Moss, C.J. & Poole, J.H. (1983). Relationships and social structure in African elephants. In *Primate social relationships: an integrated approach*: 315–25. (Ed. Hinde, R.A.). Blackwell, Oxford.

Owen-Smith, N. (1975). The social ethology of the white rhinoceros *Ceratotherium simum* (Burchell 1817). *Z. Tierpsychol.* **38**:337–84.

Payne, K.B., Langbauer, W.R., Jr. & Thomas, E.M. (1986). Infrasonic calls of the Asian elephant ( *Elephas maximus*). *Behav. Ecol. Sociobiol.* **18**:297–301.

Perry, J.S. (1953). The reproduction of the African elephant, *Loxodonta africana.* *Phil. Trans. R. Soc.* (B) **237**:93–149.

Poché, R.M. (1974). Ecology of the African elephant (*Loxodonta a. africana*) in Niger, West Africa. *Mammalia* 38:567–80.

Poole, J.H. (1987). Rutting behaviour in African elephants: the phenomenon of musth. *Behaviour* 102:283–316.

Poole, J.H. (1989). Announcing intent: the aggressive state of musth in African elephants. *Anim. Behav* 37:140–52.

Poole, J.H. (In press). Mate guarding, reproductive success and female choice in African elephants. *Anim. Behav.*

Poole, J.H. & Moss, C.J. (1981). Musth in the African elephant, *Loxodonta africana. Nature, Lond.* 292:830–1.

Poole, J.H., Payne, K.B., Langbauer, W.K. Jr. & Moss, C.J. (1988). The social contexts of some very low frequency calls of African elephants. *Behav. Ecol. Sociobiol.* 22:385–92.

Rasmussen, L.E., Schmidt, M.J., Henneous, R. & Groves, D. (1982). Asian bull elephants: flehmen-like responses to extractable components in female elephant estrous urine. *Science, N.Y.* 217:159–62.

Schaller, G.B. (1963). *The mountain gorilla: ecology and behavior.* University of Chicago Press, Chicago & London.

Schaller, G.B. (1972). *The Serengeti lion: a study of predator-prey relations.* University of Chicago, Chicago & London.

Washburn, S.L. & DeVore, I. (1961). The social life of baboons. *Scient. Am.* 204:62–71.

Western, D. (1975). Water availability and its influence on the structure and dynamics of a savannah large mammal community. *E. Afr. Wildl. J.* 13:265–86.

Western, D. & Lindsay, W.K. (1984). Seasonal herd dynamics of a savanna elephant population. *Afr. J. Ecol.* 22:229–44.

Symp. zool. Soc. Lond. (1989) No. 61:127–146

# Ontogeny of female dominance in the spotted hyaena: perspectives from nature and captivity

Laurence G. FRANK
Stephen E. GLICKMAN and
Cynthia J. ZABEL

*Psychology Department*
*University of California*
*Berkeley, CA 94720, USA*

## Synopsis

Among wild spotted hyaenas, females strongly dominate males in most social interactions. In a captive colony, while most social behaviours develop normally, any tendency for females to dominate males is much attenuated. We suggest that, rather than being an independent, sexually dimorphic phenomenon, female dominance among wild hyaenas results from a complex interaction of androgen-mediated female aggressiveness, male dispersal, intolerance of strangers and alliance formation. In captivity, in the absence of maternal intervention on behalf of infants, rank develops as a result of size, age, individual behavioural tendencies, and an individual's ability to form alliances with others. While females are more aggressive, this characteristic is not strong enough to overcome other behavioural factors and produce clearcut female dominance.

## Introduction

Among spotted hyaenas (*Crocuta crocuta*), females strongly dominate males in nearly all social contexts (Kruuk 1972; Frank 1986b). Not only are females more aggressive than males, they display a unique suite of highly masculinized characteristics, apparently related to androgenic influences during embryonic development. The female has no normal female external genitalia (Harrison Matthews 1939; Neaves, Griffin & Wilson 1980). Instead, the labia are fused to form a pseudoscrotum and the clitoris is greatly enlarged, nearly indistinguishable from the male penis in size and shape. This organ is fully erectile and both sexes display an erect phallus as a component of meeting ceremonies. The endocrine events underlying this syndrome have been investigated in wild hyaenas (Racey & Skinner 1979;

Lindeque & Skinner 1982; Frank, Davidson & Smith 1985; Lindeque, Skinner & Millar 1986), and in captive animals (Frank, Smith & Davidson 1985; Glickman, Frank, Davidson, Smith & Siiteri 1987).

### Androgens and sexual differentiation

A large body of literature demonstrates the ontogenetic role of androgens in the development of sexual and aggressive behaviour in mammals. In the contemporary formulation (Goy & McEwen 1980), hormones are seen as having two basic categories of effects on behavioural development. Organizational effects of androgens on behaviour are analogous to their effects on morphological development (J.D. Wilson, George & Griffin 1981): they irreversibly shape the neural substrate early in development, the sensitive period depending upon species and behaviour. Thus, androgens produced by the foetal testis irreversibly masculinize and defeminize the brain, promoting the subsequent display of typically male sexual behaviour or aggressiveness. In the absence of appropriate early exposure to androgen, typically female behaviour will appear in the adult. Activational effects of hormones are reversible, potentiating steroid-dependent behaviours in the post-pubertal animal. Thus, seasonal surges of testosterone produce sexual behaviour and related aggressiveness such as the rut in ungulates. A given sexually dimorphic behaviour may depend on activation or organization or both for its full expression.

The study of sexual differentiation has traditionally depended heavily on experimentally altered animals. Typically, subjects are castrated in early infancy and subsequently exposed to exogenous hormones in varying dosages at varying periods during development. Alternatively, pregnant females are exposed to exogenous hormones, and the behavioural and morphological effects on offspring are studied.

L. Harrison Matthews (1939) first noted that '. . . the striking resemblance of the condition found in the normal female spotted hyaena to the pathological conditions found in adrenal virilism in the human subject should be noted . . . It is unfortunate that the hyaena is not a more convenient laboratory animal, for experimental work on this aspect of its physiology would likely be of considerable interest.' In a broader sense, the spotted hyaena is a natural experiment in sexual differentiation. The suggestion of Racey & Skinner (1979) that masculinization in the hyaena might be due to high levels of testosterone in the foetal female strongly implied that the roots of female aggression and dominance should be sought in the spotted hyaena's endocrine system. Note that aggressiveness and dominance are independent phenomena (Rowell 1974): a dominant individual may not necessarily be particularly aggressive, while a mid-ranking or subordinate animal may exhibit frequent aggression.

### Field studies and research on captive animals

While the focus of this Symposium is on large mammals in their environment, we will present data from both field and laboratory investigations of the spotted hyaena, in a preliminary and somewhat speculative exploration of the behavioural basis of female dominance in this species.

Because of problems in reliably capturing spotted hyaenas repeatedly, it would have been extremely difficult to perform longitudinal physiological studies on free-living animals. Moreover, a long-term field study (Frank 1986b) had demonstrated the importance of maternal behavioural influences on the development of social rank in both male and female hyaenas. In order to have ready access to maturing hyaenas, and to control for the influence of the mother, a colony of captive infant spotted hyaenas was established, to study behavioural and anatomical sexual differentiation.

We first summarize observations of free-living spotted hyaenas to establish the nature of aggression and dominance in this species, and the social influences which seem to be important. We then present data from the study of captives which are consistent with field observations in some dimensions of behaviour, but not in regard to dominance: in captivity, females are not uniformly dominant to males. Finally, we attempt to reconcile the differences between field and captivity in a synthesis which we believe presents a more complete and complex picture of the development of sex-based dominance than could be acquired from either study in isolation.

## Field studies

### Methods

Since 1979, one of us (LGF) has been studying a clan of about 70 spotted hyaenas resident along the Talek River in Kenya's Masai Mara National Reserve.

Basic methodology employed in this study can be found in Frank (1986a,b). Individual animals were identified on the basis of ear notches, natural ear damage and variation in spot patterns. Contrary to widespread belief, the sex of free-ranging hyaenas can usually be determined visually, using techniques summarized in Frank (1986a) and Frank, Glickman & Powch (in prep.).

Behavioural observations were made with binoculars from a vehicle, and recorded on audio tape. Most observations were made in daylight, between the hours of 0600–1000 and 0600–1930; during full moon periods, Night Vision Goggles in combination with 7 × 50 binoculars were used.

**The social system**
## Females
As with many other social mammals, females remain in their natal clan, retaining close relationships with their female kin. Thus, the clan females comprise a group of matrilines. The social rank of the mother is conferred upon young females, with the result that relative rankings of matrilines within the clan are stable over generations; the same matriline that was at the top of the Talek Clan hierarchy in 1979 still held that position in 1987, although with much change in membership through births and deaths. The relative rank of other matrilines has been similarly stable.

## Males
Most males emigrate from the clan at puberty (two years of age), but the sons of top-ranking females stay in the clan much longer, emigrating at three to five years. Males spend part of their lives as nomadic transients that wander among clan ranges, eventually joining a new clan, but usually moving on again after periods of time varying from days to years. Mating is polygynous, high-ranking males obtaining a disproportionate share.

## Intolerance of strangers
As first described by Kruuk (1972), among wild hyaenas in Ngorongoro Crater, clan residents are intolerant of unfamiliar animals. He describes extreme aggression against trespassers, sometimes resulting in their death. In the Mara study, this degree of intolerance has not been witnessed; transient males move through clan ranges, partly in response to wildebeest migrations and local prey availability. These nomadic adult males and dispersing subadults are very subordinate to clan residents and generally avoid contact with them, as they are frequently chased when encountered by resident males. Frank, Davidson *et al.* (1985) found that transient adult males have lower androgen levels than resident males, apparently reflecting social stress associated with nomadism. When a male joins a clan, he is subordinate to all clan residents, and male rank is correlated with length of residence for at least two years (Frank 1986b).

**Female dominance and aggression in nature**
In nearly all social interactions among wild hyaenas, adult males yield to females. This is most evident at kills, where competition among the feeding hyaenas is intense. In the Mara, wildebeest (*Connochaetes taurinus*) and zebra (*Equus burchelli*) are favoured prey. Males frequently make kills or arrive shortly after the prey has been brought down, but they abandon the carcase when adult females arrive, to loiter around the outskirts of the kill site, picking up scraps dropped by the feeding females. Out of 342 5-min

periods during which groups of more than five Talek hyaenas were seen feeding, males fed in only 34 (10%), when no females were feeding (the adult sex ratio in the clan was 0.78 males/female). Except when females are approaching sexual receptivity, males avoid them and tend to be found in all-male groups. Kruuk (1972: 124–125, 234–236) gives other examples of female dominance.

The rank of females is conferred upon their dependent offspring (Frank 1986b) to the degree that even cubs less than five months old frequently threaten and chase adult males which approach the den too closely. While adult females will threaten the offspring of other females, males have never been seen to be aggressive to cubs. Clan subadults can readily displace adult males from food.

Females are also more aggressive among themselves than are males. Brief squabbles over food are common among females, but rare among males. Females frequently threaten each other in non-feeding situations, but relations among males are notably peaceable by comparison: agonistic interactions normally consist of very low-level threats, avoidance and passive displacement, and low-intensity appeasement displays, rather than the more aggressive threats, biting attacks, and more intense appeasement displays seen among females.

## Exceptions: male dominance in nature

In nature, there are two exceptions to the rule of female dominance. The first is the phenomenon of female baiting. As a female approaches sexual receptivity, males, singly and in groups, begin to follow her. When several males are present, they will occasionally attack her vigorously, regardless of her rank. The female defends herself, but is clearly subordinate to the attacking males. The proximate causes and ultimate function of this behaviour are unknown, but its clear correlation with impending oestrus suggests an association with changes in a female's hormonal profile, possibly reducing her normal aggressiveness.

The second exception is the unusual status of male offspring of alpha females, which we refer to as alpha sons (Frank 1986b). Juveniles rank immediately below their mothers in the clan hierarchy, and females attain approximately their mother's rank as adults. Nothing is known of the fates of males after they emigrate from the clan, but alpha sons occupy a unique social niche while still in their natal clan: they are able to dominate all other clan residents, including females that rank below their mother. One apparent consequence of high rank is that they remain in the clan to a much greater age than other immature males before emigrating.

Since 1979, there have been ten sons of alpha females in the Talek Clan. Since an observer has been present in the study area for only 2–4 months each year since 1982, it is not possible to calculate accurately the age at dispersal of males born after 1980. For males born between 1978 and 1981, the mean age of disappearance for alpha sons was 40 months, compared to a mean of 25 months for other males born to the clan in that period. For example, Female N2 was alpha in 1985–86. During October–December 1986, she was constantly accompanied by her three sons born in 1982, 1983 and 1985. By July 1987, N2 and her male offspring had disappeared from the clan range, although her two-year-old daughter was still present. The disappearance of the N2 family was most likely due to the death of N2 followed by dispersal of her sons, which would no longer benefit from the status of their mother. Such dispersal did not, however, always take place: in 1982, an orphaned alpha son remained in the clan range, retaining close associations with kin (but now ranking below offspring of the new alpha female, his older half-sib).

The unusual social status of alpha sons demonstrates that social factors can play a significant role in the development of rank: even in wild spotted hyaenas there are important exceptions to the rule of female dominance, occurring in predictable social contexts. While females are more aggressive than males, it is clear that aggressiveness alone is not sufficient to produce absolute female dominance.

Both these examples are consistent with the hypothesis that males can dominate females if they are able to form appropriate alliances. However, it also seems likely that there are changes in the hormonal state of the oestrous female hyaena which make her less likely to challenge a group of males, and this cannot be disentangled in the field situation.

## Studies of hyaenas in captivity

This project was undertaken with the goal of examining the development of behaviour, morphology and their endocrine substrates in an environment permitting daily observation under controlled conditions. There are potential costs to such an approach, e.g. distortions of behaviour that have no applicability to the natural situation. In general our peer-reared hyaenas developed much as they would have in nature regarding basic patterns of affiliation and aggression. There has been one major departure from the rules of behaviour found in nature. However, we believe that this is a very informative change with substantial implications for understanding the mechanisms of female dominance in nature.

### Assembling, housing and maintaining the hyaena groups

To establish the colony in Berkeley, we collected two cohorts of ten infants each in Narok District, Kenya, in January and December, 1985, under permit from the Kenya Department of Wildlife Conservation and Management. A cohort of ten infants of the same age is typical of hyaena clans in this part of East Africa (Frank 1986a). Infants were between one week and two months of age; the younger ones were bottle-fed until they learned to lap formula from a pan. The first cohort consisted of seven females and three males, the second, five females and five males.

In Berkeley, each cohort was initially housed in indoor-outdoor enclosures measuring 40 × 100 ft. Two females in Cohort One and two females and two males in Cohort Two were bilaterally gonadectomized at 5–6 months of age. Achievement of full sexual maturity (as evidenced by reproductive competence) in this species is generally placed at two years in males and three years in females (Kruuk 1972). However, the period between the beginnings of puberty, as indicated by nipple growth and a small but consistent rise in oestrogen secretions, and attainment of full sexual maturity, covered a long time-span in our female hyaenas. Marked increases in the intensity and consequences of aggression occurred in both cohorts, coincident with the initiation of the female pubertal period at 18–22 months of age (see also Kranendonk, Kuipers & Lensink 1983). Because of this aggression, each cohort was divided into two subgroups before reaching their second birthday, with a concomitant division of the enclosures. Three females from the older group delivered infants at 35–37 months of age. All animals were fed daily with a commercial carnivore diet (ground beef and horse by-products) supplemented with fresh sheep bones.

## Methods and results

### Meeting ceremonies

When two clan members come together after a period of separation, they commonly engage in a 'meeting ceremony' (Kruuk 1972), each presenting its genital region to the other for mutual sniffing and licking. In nature, the ceremony is usually initiated by the subordinate animal, which presents an erect phallus or clitoris for inspection by the dominant. This ceremony also occurs in the context of socially exciting situations, or following an aggressive interaction, and appears to be a crucial means of establishing/ maintaining affiliative relations between members of the clan. We have carried out several experiments on the meeting ceremony within our younger cohort, with the goal of determining whether peer-reared subadults would display the same rules of behaviour as wild adults and subadults. When the ten animals in this group were 15–20 months of age, we examined the effects of social excitement and physical separation of group members on the occurrence of meeting ceremonies (Krusko, Weldele & Glickman in prep.). Social excitement was induced by releasing the group of animals from one part of the enclosure, where they had been confined for 4 h, to another section of the enclosure. This manipulation was sufficient to produce a ten-fold increment in the mean frequency of meeting ceremonies for each animal. If the group was separated into two subsets of five animals, with physical contact prevented for 4 h, we observed a still greater increment in the frequency of meeting ceremonies. Moreover, meetings occurred significantly more often between animals that had been separated, than between those that had been housed together during the brief confinement interval. Finally, there was a strong negative correlation between initiation of meeting ceremonies and dominance ranks (determination of rank is discussed below). We conclude that the meeting ceremonies of peer-reared hyaenas in captivity follow the same basic rules as those described by Kruuk (1972) for hyaenas in nature.

## Aggression

Aggression was assessed in infant and prepubertal hyaenas during the course of nightly time-sampling of spontaneous behaviour in both cohorts (Zabel, Glickman, Weldele, Krusko, & Frank in prep.) Descriptions of behaviour were recorded by the observer on a voice-activated tape recorder, employing a common behavioural inventory. An electronic metronome provided time intervals on each tape.

One major result is presented in Fig. 1 (adapted from data contained in Zabel *et al.* in prep.). These data were derived from more than 100 h of critical incident sampling of both cohorts between 12 and 21 months of age. Aggression was measured as the frequency of threats, bites and attacks, and

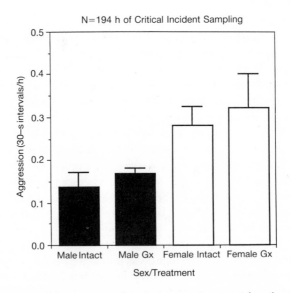

**Fig. 1.** Spontaneous within-sex aggression among captive pre-pubertal spotted hyaenas. Females are significantly more aggressive than males, but gonadectomy at 5–6 months of age does not affect levels of aggression. Gx denotes gonadectomized subjects.

only interactions between pairs of animals were considered. Results were expressed as the number of 30 s intervals in which aggression was seen per potential opponents per hour. There were no significant effects of gonadectomy in either sex on this measure of aggression in prepubertal animals (males: intact vs. gonadectomized, $t = 0.94$, $d.f. = 6$, n.s.; females intact vs. gonadectomized, $t = 0.45$, $d.f. = 10$, n.s.). However, female–female aggression was approximately twice as frequent as male–male aggression ($t = 3.31$, $d.f. = 18$, $P < 0.01$). Male–female aggression occurred at a level intermediate between these values. Comparable data from sexually mature animals are not yet available.

The results presented for prepubertal subjects are in accord with field observations regarding sex differences in aggression among hyaenas. The lack of effect of gonadectomy is not surprising given the time period when most of our data were gathered, i.e. after organizational effects of androgens had probably occurred, but before activating effects are expected, coincident with the onset of puberty. It must also be noted that there may be problems in detecting modulating influences of androgens in the context of a social group (Bernstein, Gordon & Rose 1983). Behavioural or pheromonal effects of hormones circulating in one or two animals may have social consequences that affect all group members. Extended studies of isolated dyads may be required to estimate hormonal effects with greater precision.

### Size

The fact that female hyaenas in nature are heavier than males raises several possibilities. It may be that their increased weight is an indirect result of greater access to food, through greater aggressiveness and dominance. Alternatively, it is possible that the competitive advantage of females in relation to males stems from greater weight. Treatment with androgens during the sensitive period for sexual differentiation has been shown to increase weight (Bell & Zucker 1971) and aggressiveness (Beatty 1979) of female mammals. If this has happened naturally in female hyaenas, we might expect independent actions on weight and aggressiveness, although each might well reinforce the other.

In order to estimate the relationship between gender and body weight in our colony, we weighed all individuals as they matured. The overall results are presented in Fig. 2. Two males (Males 9 and 10) and one female (Female 18) were omitted from this analysis because they were consistently much smaller than their peers or wild hyaenas of comparable age, probably because of nutritional problems as neonates.

There were no detectable sex differences or effects of gonadectomy on weight within the first year of age. However, between 19 and 30 months of age, small but significant sex differences did emerge between males and

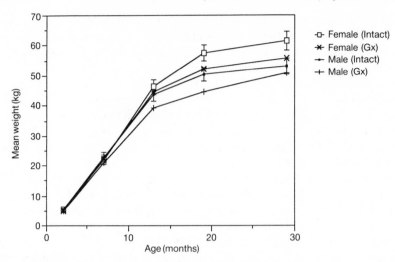

**Fig. 2.** Effects of age, sex and gonadal status on weights of captive spotted hyaenas from infancy through puberty. Gx denotes gonadectomized. By the age of 30 months, females are significantly heavier than males ($P < 0.05$).

females ($t = 2.31$, $d.f. = 15$, $P < 0.05$). The magnitudes of the differences approximate those seen in wild hyaenas (Kruuk 1972; Hamilton, Tilson & Frank 1986). There were no significant effects of gonadectomy, although our sample sizes are small.

When the data are analysed by cohort, including all subjects, in order to determine the relation between weight and dominance, a somewhat different picture emerged. Partly because of the presence of the two very small males in Cohort One, males were significantly lighter than females in that cohort at all ages sampled. However, in Cohort Two, there were no significant differences between males and females. The latter result was influenced by the presence of the exceptionally small female (Female 18).

### Intolerance of strangers

Kruuk (1972) describes the dramatic boundary disputes that arise when members of neighbouring clans meet at their mutual border. Interestingly, we see similar behaviour at the fence which divides subgroups of each cohort of captives. Animals raised together from infancy, when prevented from physically contacting each other, display elevated aggression, postures and vocalizations very similar to those seen during encounters between mutually hostile groups of neighbours in the wild. Thus, intolerance of non-group members is very strong, and only minimal separation (e.g. a wire fence) may within a few days create a distinction between 'us' and 'them', even among animals raised together. This suggests that the distinction between group and non-group may depend in part on close physical contact, specifically the meeting ceremony. Among both wild and captive hyaenas, females appear to be more prominent than males in inter-group aggression.

## Dominance

Despite being heavier on average than males and more prone to engage in within-sex aggression, the females in our captive groups are not uniformly dominant over males. Dominance ranks were derived for each cohort based on a test designed to simulate the highly competitive feeding situation that is typical of wild hyaenas in the dense populations of the Masai Steppe. Each group of animals was allowed to compete for one hour over access to meaty beef neck bones fastened to the fence of the enclosure. Video tapes of the test were subsequently scored for all aggressive interactions (threats, bites, attacks and displacements) as well as appeasement or submissive displays. Dominance matrices (Lehner 1979) based on aggression and submission were highly concordant, but the submission matrices tended to be 'cleaner', showing fewer reversals or circular relationships. Rankings discussed here were based on submission matrices, with the dominant animal in each group or subgroup ranked as number one.

Figure 3 shows the relative ranks of each of the captive animals every six months from infancy through 9–12 months of age, the time when the groups of ten were subdivided. Because animals were shifted between the two subgroups of Cohort One several times, it is not possible to display rankings over time within the subgroups in a meaningful way; membership of the two subgroups of Cohort Two has remained stable, and the rankings of the members as adults are shown in Fig. 4.

In both cohorts, rankings were relatively stable, particularly at the top and middle of the hierarchy. As in wild hyaenas, we observed relatively few interactions among lower-ranking animals, resulting in tied ranks and rank assignments made on the basis of scanty data. Thus, when a lower-ranking animal moved between adjacent ranks in consecutive sampling periods, the apparent rank shift may be due to a single interaction.

Only in Cohort One at the age of 7–8 months was the mean rank of females significantly higher than that of males (Mann-Whitney $U$-test, $U = 19$, $P < 0.05$; Table 1). In each cohort, there was a significant correlation between an individual's weight and its rank in the bone test only in early infancy (Table 2). By the age of seven months, size became a poor predictor of competitive ability.

While there was no overall tendency for females to dominate males, within both male–female sibships (14–15 and 16–17) in Cohort Two the females consistently ranked above the male. In each case, the female was also the larger of the two sibs as adults (47.0 kg vs. 54.6 kg and 56.8 kg vs.

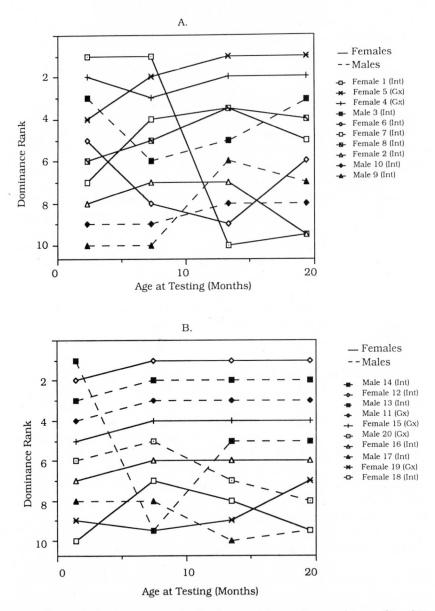

**Fig. 3.** Changes in dominance as measured by the group bone test between the ages of 2 and 20 months in (A) Cohort One and (B) Cohort Two. Females are denoted by open symbols, males by closed symbols. In dominance rank, 1 denotes the top rank, 10 the lowest. Int, intact individuals; Gx, gonadectomized individuals.

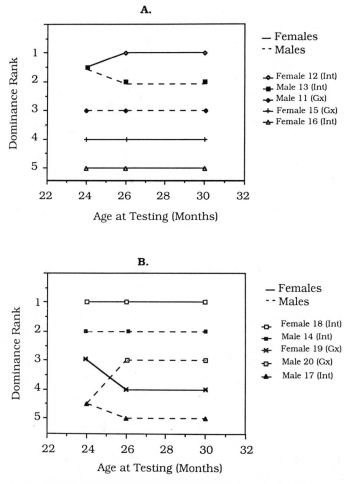

**Fig. 4.** Changes in dominance of post-pubertal spotted hyaenas as measured by the group bone test in the subgroups of Cohort Two after it was subdivided at the age of 20–21 months. Symbols and abbreviations as in Fig. 3. (A) Subgroup 2A; (B) Subgroup 2B.

**Table 1.** Mean social rank, as measured by the group bone test, for males and females in both cohorts. Only in Cohort One at the age of 7–8 months was there a significant difference (Mann-Whitney $U$-test) between the sexes.

| Age (months) | Cohort One Male | Female | $U$ | Cohort Two Male | Female | $U$ |
|---|---|---|---|---|---|---|
| 2–3 | 7.7 | 4.6 | 16 | 4.4 | 6.6 | 18 |
| 7–8 | 8.3 | 4.3 | 19 | 5.5 | 5.5 | 12.5 |
| 13–14 | 6.3 | 5.1 | 13 | 5.4 | 5.6 | 13 |
| 19–20 | 6.0 | 5.3 | 12 | 5.5 | 5.5 | 12.5 |

**Table 2.** Spearman rank correlation co-efficients ($r_s$) between weight and social rank as measured by the group bone test, corresponding to the ages shown in Fig. 3. In both cohorts, the correlation was statistically significant ($P < 0.01$) only in infancy.

| Age (months) | Cohort One | Cohort Two |
| --- | --- | --- |
| 2–3 | 0.85** | 0.81** |
| 7–8 | 0.55 | 0.51 |
| 13–14 | −0.08 | 0.18 |
| 19–20 | 0.16 | 0.49 |

75.0 kg, respectively). Within wild male–female sibships, similar size relations usually pertain (unpublished data), although the differential between sibs 16 and 17 is extreme. Thus, when two animals are evenly matched for age and social history, females may be able to consistently dominate males.

Additionally, at sexual maturity an intact female ranks at the top of each of the four subgroups, and at the bottom of all three subgroups containing more than one intact female. This further suggests that female aggressiveness is in itself a strong influence on dominance relations between the sexes, but not strong enough to produce the absolute female dominance seen in the wild. The tendency for an intact female to rank at the bottom of each group suggests that coalitional aggression within the group may be biased against them.

### Coalition and alliance formation

Kruuk (1972) describes the 'parallel walk', a distinctive display used by two or more hyaenas when jointly threatening another individual, or while performing boundary-related behaviours such as scent-marking or 'patrolling'. In a directed threat, this may escalate into a group attack by several clan members, often close kin, against a lower-ranking animal. Transient males are often chased by groups of clan males.

Group aggression within the clan may take the form of very temporary *ad hoc* coalitions, or may involve stable long-term alliances. Among wild hyaenas, the latter usually involve kin groups, particularly in higher-ranking matrilines. Co-operative aggression and defence among kin presumably contribute to the maintenance and stability of rank relations among individuals and, consequently, among matrilines.

One of the most dramatic aspects of hyaena aggression observed in our colony has been the tendency for dyadic aggressive encounters to escalate into group attacks on a single individual. In the course of nightly critical incident sampling, during the prepubertal period, there were (on average) two or three sequences of group aggression for each hour of observation in each cohort (Zabel *et al.* in prep.). Such attacks commonly lasted for several minutes, with additional animals joining the group until as many as nine hyaenas were arrayed against one animal. Membership in an attacking group fluctuated from one attack to the next, although there were certain pairs of animals which worked very effectively together, and were unusually likely to be found in the same group of aggressors. We prefer to use the term 'coalition' to describe a group of animals with no consistent, reciprocal relationship, while reserving the term 'alliance' for animals with a particular tendency to assist each other.

We believe that such coalitions/alliances have played a major role in determining dominance status within our cohorts. Three case histories illustrate this point.

1. Female 1 had dominated Cohort One from the time of our initial bone test in February 1985 until October of that year. Following an intense fight, she was displaced in the dominance hierarchy by Females 4 and 5, and their supremacy lasted until the cohort was subdivided in October 1986. It is of particular interest that Female 1 did not move down just one or two places in the hierarchy. Rather, she was near the bottom of the group throughout most of the period between her fall from top rank and subdivision of the cohort. Formerly low-ranking animals (e.g. Males 9 and 10) could now challenge her with the assurance that they would receive support from other animals.

2. The second example involves the two ovariectomized females (Females 4 and 5) which supplanted Female 1. These animals had formed a true alliance and continued to dominate their subgroup until they were separated in December 1986. At that time, both fell drastically in rank in their newly constituted subgroups. Deprived of mutual support and compelled to confront an array of hyaenas which coalesced against them, these formerly dominant animals could no longer maintain their status.

3. The low ranks of the twin Males 9 and 10 and Female 2 in their first year of life were due in part to the fact that they were at least a month younger than the rest of the group. The age differential was exaggerated by the unusually small size of the two males, which grew slowly and attained only about 80% of normal adult male size. However, the rise in rank of Male 10 after one year of age is especially noteworthy because, at 39.6 kg, he is the smallest of the captives (mean ± s.e. of all eight captive males is 49.6 ± 2.5 kg) and the smallest of all adult male hyaenas for which measurements are available (V. J. Wilson 1968; Kruuk 1972; Whateley 1980; Smithers 1983; Hamilton *et al.* 1986; L.G. Frank, unpublished data). Moreover, after Cohort One was subdivided, he consistently ranked near the top of his subgroup.

A current subgroup of sexually mature hyaenas consists of two males (Males 9 and 10), one gonadectomized female (Female 5), and two intact females (Females 2 and 6). The group is dominated by Female 2. However, the two males are able to dominate the other two (substantially larger) females. They initially achieved this by aligning themselves with the dominant female, but eventually her direct support was no longer required.

We believe that these observations of coalition/alliance formation are a critical piece in the solution of the dominance puzzle presented by spotted hyaenas. Although our observations of such behaviours in captivity are both more frequent and more dramatic than most instances in nature, they call attention to similar processes that may be at work in the natural setting.

**Summary**
Our captive, peer-reared hyaenas developed and behave like wild hyaenas, in terms of the expected dimorphism in body weight, basic patterns of affiliation and aggression. This result is comparable to that obtained by Goy & Resko (1972) in peer-reared rhesus monkeys. However, on the critical dimension of competitive dominance, female hyaenas in captivity failed to dominate males unless there were weight discrepancies far greater than those occurring in nature.

## Female dominance in nature: a hypothesis

Clearly, the lack of female dominance seen in the captive hyaenas is at variance with the situation described for wild ones. Rather than dimissing the difference as an artifact of captivity, we suggest that the emergence of near-absolute dominance of female hyaenas over males in nature is the result of a complex sequence of events. Instead of being entirely a product of sexually dimorphic behaviour, female dominance may be due to enhanced female aggressiveness, male dispersal, intolerance of strangers and the tendency to form alliances among familiar animals. These interactions are schematically summarized in Fig. 5.

### Enhanced female aggressiveness
Unusual patterns of androgen secretion in female hyaenas (during foetal life and again as adults) predispose them to be somewhat heavier and more prone to engage in aggression than males. This would tend to favour females in dyadic encounters with males, but would not necessarily confer the absolute dominance observed in nature.

### Female philopatry, male dispersal and xenophobia
Additionally, females spend their entire lives within a territory as part of the same social group. Males disperse at puberty and join other groups, as

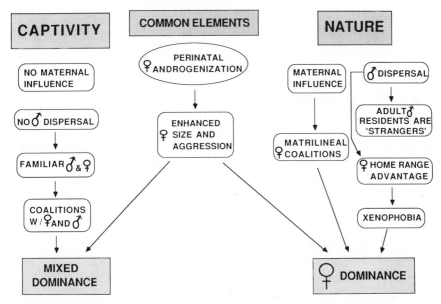

**Fig. 5.** Schematic representation of factors postulated to lead to absolute female dominance in the wild, and the simpler situation in captivity, where the lack of opportunity for males to disperse leads to mixed dominance.

individuals, for varying periods of time. This set of circumstances promotes female dominance in several ways. First, there is a 'home territory' effect. A broad array of information suggests that resident animals have a competitive advantage when dealing with intruders (Figler & Einhorn 1983). This may be attributed to the greater motivation and confidence of the resident, or the insecure, subordinate behaviour of the intruder. Since all female spotted hyaenas are resident in their natal area throughout life, they benefit from that experience. Immigrant males are always highly subordinate and exhibit the reduced androgen levels characteristic of severe social stress (Frank, Davidson *et al.* 1985). With continued residence in an area, a male rises in rank, at least in relation to subsequent immigrants.

### Alliance formation within the clan

Finally, there is a second route through which sexually dimorphic dispersal might be expected to promote female dominance: sustained interactions, particularly between close kin, would facilitate the emergence of alliances within the matrilines that form the social core of a hyaena clan, and more general coalitions among resident females confronting other clans or newly arrived males. Experimental studies with primate groups have shown that dominant individuals in one group become highly subordinate when moved to a different social group, as the result of difficulties dealing with an established cohesive social unit (Bernstein *et al.* 1983). Individual qualities that promote dominance ultimately depend upon the social context for expression.

### The natural situation re-examined

Our studies in captivity call attention to the importance of new field observations. Observations of coalitions/alliances in nature are largely anecdotal. We need to understand the frequency and the conditions under which they occur, the role of kinship and the possible existence of alliances based on reciprocity (Trivers 1971). We also need a better understanding of the causal mechanisms of male dispersal. Are males driven out of the clan, or do they leave for other reasons (e.g. Holekamp 1986)? What happens to alpha females which disappear from a clan? In captivity, we observed a very dramatic decline in rank when a dominant female was displaced. In nature, would she have disappeared from the clan?

Only through the sustained interplay of field and laboratory studies can we gain a proper understanding of the mechanisms underlying the behaviour of hyaenas.

## Acknowledgements

LGF thanks the Office of the President, the Wildlife Conservation and Management Department, and the Narok County Council, Republic of Kenya, for permission to study hyaenas in the Masai Mara National Reserve, and to export infants captured outside the Reserve for the Berkeley study.

Research on the captive colony is supported by grant No. 5R01 MH 39917 from the National Institute of Mental Health. Field work was supported by grant BNS 78–03614 from the National Science Foundation and grants from the National Geographic Society, the Center for Field Research and Earthwatch, and the Harry Frank Guggenheim Foundation.

## References

Beatty, W.W. (1979). Gonadal hormones and non-reproductive behaviors in rodents: organizational and activational influences. *Horm. Behav.* 12:112–63.
Bell, D.D. & Zucker, I. (1971). Sex differences in body weight and eating: organization and activation by gonadal hormones in the rat. *Physiol. Behav.* 7:27–34.
Bernstein, I.S., Gordon, T.P. & Rose, R.M. (1983). The interaction of hormones, behavior and social context in nonhuman primates. In *Hormones and aggressive behavior*: 535–61. (Ed. Svare, B.B.). Plenum Press, New York & London.
Figler, M.H. & Einhorn, D.M. (1983). The territorial prior residence effect in

convict cichlids (*Cichlasoma nigrofasciatum* Gunther): temporal aspects of establishment and retention, and proximate mechanisms. *Behaviour* 85:157–83.

Frank, L.G. (1986a). Social organization of the spotted hyaena (*Crocuta crocuta*). I. Demography. *Anim. Behav.* 34:1500–09.

Frank, L.G. (1986b). Social organization of the spotted hyaena *Crocuta crocuta* II. Dominance and reproduction. *Anim. Behav.* 34:1510–27.

Frank, L.G., Davidson, J.M. & Smith, E.R. (1985). Androgen levels in the Spotted hyaena *Crocuta crocuta*: the influence of social factors. *J. Zool., Lond. (A)* 206:525–31.

Frank, L.G., Glickman, S.E. & Powch, I. (In preparation). *Sexual dimorphism in the spotted hyaena.*

Frank, L.G., Smith, E.R. & Davidson, J.M. (1985). Testicular origin of circulating androgen in the Spotted hyaena *Crocuta crocuta*. *J. Zool., Lond (A)* 207:613–5.

Glickman, S.E., Frank, L.G., Davidson, J.M., Smith, E.R. & Siiteri, P.K. (1987). Androstenedione may organize or activate sex reversed traits in female spotted hyenas. *Proc. natn. Acad. Sci. USA* 84:3444–7.

Goy, R.W. & McEwen, B.S. (1980). *Sexual differentiation of the brain.* MIT Press, Cambridge, Mass.

Goy, R.W. & Resko, J.A. (1972). Gonadal hormones and behavior of normal and pseudohermaphroditic nonhuman female primates. *Rec. Prog. Horm. Res.* 28:707–33.

Hamilton, W.J. III, Tilson, R.L. & Frank, L.G. (1986). Sexual monomorphism in spotted hyenas, *Crocuta crocuta*. *Ethology* 72:63–73.

Harrison Matthews, L. (1939). Reproduction of the spotted hyaena (*Crocuta crocuta* Erxleben). *Phil. Trans. R. Soc.* (B) 230:1–78.

Holekamp, K.E. (1986). Proximal causes of natal dispersal in Belding's ground squirrels (*Spermophilus beldingi*). *Ecol. Monogr.* 56:365–91.

Kranendonk, H.J., Kuipers, J. & Lensink, B.M. (1983). The management of spotted hyaenas, *Crocuta crocuta*, in Artis-Zoo, Amsterdam, The Netherlands. *Zool. Gart. Jena* 53:339–53.

Krusko, N.A., Weldele, M. & Glickman, S.E. (In preparation). *Meeting ceremonies in a colony of juvenile spotted hyenas.* (Paper presented at the 1988 North American Animal Behavior Society Meeting, Missoula, Montana, August 7–13, 1988).

Kruuk, H. (1972). *The spotted hyena: a study of predation and social behavior.* Chicago University Press, Chicago & London.

Lehner, P.N. (1979). *The handbook of ethological methods.* Garland STPM Press, New York.

Lindeque, M. & Skinner, J.D. (1982). Fetal androgens and sexual mimicry in spotted hyaenas (*Crocuta crocuta*). *J. Reprod. Fert.* 65:405–10.

Lindeque, M., Skinner, J.D. & Millar, R.P. (1986). Adrenal and gonadal contribution to circulating androgens in spotted hyaenas (*Crocuta crocuta*) as revealed by LHRH, hCG and ACTH stimulation. *J. Reprod. Fert.* 78:211–7.

Neaves, W.B., Griffin, J.E. & Wilson, J.D. (1980). Sexual dimorphism of the phallus in spotted hyaena (*Crocuta crocuta*). *J. Reprod. Fert.* 59:509–13.

Racey, P.A. & Skinner, J.D. (1979). Endocrine aspects of sexual mimicry in Spotted hyaenas *Crocuta crocuta*. *J. Zool., Lond.* 187:315–26.

Rowell, T.E. (1974). The concept of social dominance. *Behav. Biol.* 11:131–54.

Smithers, R.H.N. (1983). *The mammals of the Southern African subregion.* The University of Pretoria, Pretoria.

Trivers, R.L. (1971). The evolution of reciprocal altruism. *Q. Rev. Biol.* 46:35–57.

Whateley, A. (1980). Comparative body measurements of male and female spotted hyaenas from Natal. *Lammergeyer* 28:40–3.

Wilson, J.D., George, F.W. & Griffin, J.E. (1981). The hormonal control of sexual development. *Science, N.Y.* 211:1278–84.

Wilson, V.J. (1968). Weights of some mammals from eastern Zambia. *Arnoldia (Rhodesia)* 3:1–20.

Zabel, C.J., Glickman, S.E., Weldele, M.L., Krusko, N.A. & Frank, L.G. (In preparation). *Dyadic aggression and dominance in a colony of prepubertal spotted hyaenas: effects of sex, age and gonadectomy.*

Symp. zool. Soc. Lond. (1989) No. 61:147–161

# Assessment of reproductive status of the black rhinoceros (*Diceros bicornis*) in the wild

R.A. BRETT, J.K. HODGES and E. WANJOHI[1]

*The Gallmann Memorial Foundation*
*P.O. Box 45593*
*Nairobi, Kenya*

*Institute of Zoology,*
*The Zoological Society of London*
*Regent's Park*
*London NW1 4RY, UK*

*Research Section*
*Wildlife Conservation and*
*Management Department*
*P.O. Box 40241*
*Nairobi, Kenya*

## Synopsis

There is a need for the development of techniques in monitoring the reproductive performance of the black rhinoceros in the wild, particularly as a tool for improving breeding performance, and in cohesive management of the small protected populations that remain in Kenya. The priorities here for long-term propagation of this endangered species are easy detection of oestrus and pregnancy in females and assessment of breeding status in males. The social organization and reproductive physiology of the indigenous, protected black rhinoceros population of Ol Ari Nyiro ranch, Laikipia, are being monitored to this end. The home range movements and behaviour of resident rhinoceros have been recorded and urine samples have been collected from known individuals and assayed for hormone metabolites in order to provide a non-invasive indicator of breeding status.

Methods of collection of urine from free-living rhinoceros are described. Pregnanediol-3-glucuronide immunoreactivity was measured in urine from females by a simplified enzyme assay procedure, showing the potential of this as a field test for the detection of mid–late pregnancy. Urinary oestrone-3-sulphate did not reflect reproductive status in males, although a positive association of body size and home range size was found. Levels of urinary immunoreactive testosterone in males were extremely low and did not correlate with breeding status. Subsequent chromato-

[1] Present address: Institute of Zoology, Zoological Society of London, London NW1 4RY, U.K.

ZOOLOGICAL SYMPOSIUM No. 61
ISBN 0–19–854009–4

graphic analysis showed testosterone to be completely undetectable but revealed an 'androstane-tiol-one' as the major androgen metabolite. The potential for the use of endocrine monitoring techniques in the black rhinoceros is discussed, particularly in relation to the potential need to be able to transfer genetic material between small rhinoceros populations as part of their long-term management.

## Introduction

The number of black rhinoceros remaining in Kenya is estimated to be in the region of 400. Much money and effort is being spent on measures to protect and manage the few viable black rhinoceros populations remaining in Kenya within special sanctuaries or reserves where their interests are paramount. The need for sanctuaries was spelt out clearly over two decades ago (Ritchie 1963); the Kenya black rhinoceros population has dropped by c. 98% since then.

The remnant black rhinoceros populations in Kenya are small and fragmented, and successful management and breeding in the long term will require the ability easily to assess their reproductive status. Firstly, there is the need for a reliable determination of pregnancy in female rhinoceros. Apart from the difficulty of observing black rhinoceros in their favoured habitat of dense bushland, pregnant females do not become noticeably gravid even during the later stages of pregnancy. Secondly, an indicator of breeding status in mature male rhinoceros would be tremendously useful for long-term management of black rhinoceros under conditions where the movement of breeding animals between populations will be essential for their genetic and demographic health. Breeders need to be easily identified, and the total number of adult rhinoceros breeding in a given population assessed in order to estimate its genetically effective population size (Lande & Barrowclough 1987).

To date, the only available method for identifying and assessing reproductive status has been observation. Even here, reliable information on breeding behaviour, particularly mating, is extremely sparse, largely because of the difficulty of obtaining frequent observations of black rhinoceros in dense bush habitat.

Goddard (1966) and Schenkel & Schenkel-Hulliger (1969) recorded most of the little information available on the breeding behaviour of black rhinoceros. Apart from mating, the breeding and dominance behaviours in males include consorting with oestrous females around the time of mating, mate-guarding, a high frequency of spray urination, scraping hind feet through communal middens in the manner of white rhinoceros (Owen-Smith 1975), redirected activity in the form of destruction of bush (which is particularly associated with urination and defecation), and tracking the scent trails of oestrous females (Schenkel & Schenkel-Hulliger 1969; G.W. Frame pers. comm.).

Female rhinoceros in oestrus urinate frequently, so leaving scent trails (Goddard 1966), dribbling urine down their hind quarters which dries to form a white streaky deposit around and beneath the vulva. Together with a swollen vulva, and the close attentions of males, these are signs of oestrus. It appears that it is the manner of urination rather than a colour change in the urine that results in this white deposit. Previous observers have noted clear seasonal peaks in breeding behaviour, although black rhinoceros appear to be able to mate at any time of year (Ritchie 1963; Schenkel & Schenkel-Hulliger 1969; Mukinya 1973; Hall-Martin 1986).

Since the bush habitat in Laikipia is particularly dense, so that typically only four to six sightings of rhinoceros are made per week, and there is a pressing need to establish reproductive data on this population (only three matings have been observed on the ranch in two years), an alternative method of obtaining data was needed.

A relatively simple method of establishing the identity of breeding animals, particularly males, within any rhinoceros population, based on the measurement of reproductive hormones in urine, would be a potentially useful alternative or addition to chance observations of matings and/or of breeding behaviour, most of which must occur at night (Goddard 1967). Non-invasive assessment of reproductive status through urinary hormone analysis could therefore prove extremely useful as a management tool for determining which animals in a population were breeding, and in future development of methods of transferring genetic material through artificial breeding techniques as an alternative to the difficult, costly and often impractical exchange of breeding animals.

The practical advantages of using urine in reproductive studies on a variety of exotic species have been described by Hodges (1986) and Lasley (1985), although so far most of the available data are from captive animals, largely kept in zoos. The only reports on studies in the wild are from Poole, Kasman, Ramsay & Lasley (1984) on African elephants and from Andelman, Else, Hearn & Hodges (1985) on vervet monkeys. Similar studies in the wild have not previously been attempted on rhinoceroses owing both to the lack of opportunity for sample collection and to the absence of suitable laboratory techniques.

This paper describes initial efforts, firstly, to determine the feasibility of collecting urine from black rhinoceros under field conditions; secondly, to establish the usefulness of urinary hormone analysis for reproductive assessment; and thirdly, to introduce and simplify the methodology for use in field conditions, in particular to sample from the rhinoceros kept in sanctuaries in Kenya, which, given the areas of land enclosed for such purposes, are likely to have a mean holding potential of approximately 50 animals.

## Methods

### Study population

Data and samples were gathered from known individuals among an indigenous population of 47 black rhinoceros resident on Ol Ari Nyiro ranch, Laikipia, Kenya. This is an unfenced cattle and sheep range of *c.* 400 km$^2$ of lower highland bush country at an altitude of 1600–2000 m. The rhinoceros population, which occupies approximately one third of the ranch area, is protected by a private anti-poaching force, and is at medium density (Goddard 1967, 1970: *c.* 1 rhinoceros per 4 km$^2$).

### Identification of rhinoceroses

Because of the difficult visibility in thick bush, rhinoceroses were mostly identified using the measurements and individual features of their footprints, in particular the width of the hind feet between the two side toes, in combination with the patterns of wrinkles in the base of the footpad, which are recorded in footprints in suitably fine, dry soil, and are unique to each animal (Fig. 1). Daily patrols of the extensive road network, water sources and salt licks on the ranch located sets of footprints which were tracked until the animal was sighted or located. Depending on the visibility, the identities of individual rhinoceros were confirmed from sex, horn size and shape, and notches and marks on the margins of the pinnae of the ears (Goddard 1966; Mukinya 1976).

### Home range sizes

All locations of footprints of known animals were plotted on a 1:50 000 map of the ranch to an accuracy of ± 200 m. The data were analysed using the methods of Dixon & Chapman (1980) to yield home range probability contours. The areas within the 90% probability contours were used to measure range sizes: these gave best fit to the home range of each animal, excluding the effects of the occasional distant excursions that black rhinoceroses make from their home ranges.

### Collection of urine

Urine was collected during tracking of a rhinoceros identified from its tracks or from subsequent sightings. Black rhinoceros in bush habitats typically urinate against shrubs that they pass, and may then scrape through the shrubs with their hind feet. Whenever fresh urine was detected during tracking (normally by its strong odour) and found on leaves it was collected. Fresh urine (about 1 h old or less) could be recognized because it was cloudy; a dark-coloured urine with a white precipitate would have been

**Fig. 1.** Photograph of the right hind footprint of a black rhinoceros, showing the wrinkle impressions from the base of the footpad used to identify individuals.

sitting for several hours. The urine was easily drawn up from the leaf axils using a 1 ml syringe (Fig. 2). The average volume collected was *c.* 0.5 ml, although much more could be obtained if a rhinoceros sprayed against a broad-leaved shrub.

The extent to which urine could be collected was influenced by the sex of individual rhinoceroses, creating a bias in sampling. Spray-urination (as described by Schenkel & Schenkel-Hulliger (1969) ) is performed more by dominant males, which project atomized urine for several metres to their rear, and if this is directed onto a bush, and the droplets are deposited all over it, the urine is easy to collect.

Non-ritualized urination, which is the more usual method of females and subdominant males, results in a stream of urine directed down or to the

**Fig. 2.** Photograph of fresh urine being collected for hormone assay from the leaves of a shrub onto which a passing rhinoceros has spray-urinated.

rear, and not necessarily over a bush, and the urine is rapidly absorbed into the ground. This made it particularly difficult to obtain urine from females, which are more reluctant to spray-urinate against bushes than males, and will dribble urine at frequent intervals during oestrus. Of a total of 100 urine samples collected over 18 months in 1986–87, only 17 came from females. All urine samples collected in the field in Kenya were frozen and sent to the Institute of Zoology at London Zoo, where they were kept at −20°C without additives until analysis.

Urine samples were also collected from captive animals for comparative purposes. Samples were collected from a single female during the last six months of pregnancy, and from two females over the oestrous cycle. Samples were also collected from three adult males, all proven breeders. Collections were made during the early morning where possible, urine being drawn up from the floor with a syringe after deposition. Samples were frozen and centrifuged prior to assay.

### Hormone analysis

### Testosterone

Testosterone was measured by radioimmunoassay using a modification of the method previously described by Hodges, Eastman & Jenkins (1983). The antiserum (Steranti Research Ltd) was raised in a rabbit against testosterone-3-CMO-BSA and showed maximum cross-reactivity of 52% against 5α dihydro-testosterone. Conjugated and unconjugated testosterone were measured with and without a hydrolysis step respectively prior to extraction and assay. Hydrolysis was achieved by overnight incubation with glucuronidase arylsulphatase (1000 units) at pH 5 and 37°C. Extraction (0.5–1.0 ml sample) was achieved with methanol: water primed $C_{18}$ Seppak catridges using 100% methanol as eluting solvent. The extract was dried under nitrogen and reconstituted in buffer, ready for assay (mean ± s.e. recovery = 79 ± 3.4%) Sensitivity of the assay was 5 pg/tube or 80 pg/ml and inter- and intra-assay CVs were below 15%. The relationship between serial dilutions of rhinoceros urine and testosterone standards was variable: dilutions of some samples were parallel to the standards whereas dilutions of others were not. In all cases levels of testosterone were very low, producing a maximum displacement of binding of only approximately 30% (1.0 ml sample).

### Conjugated oestrone

Oestrone conjugates were measured using a non-extraction assay according to the method described by Hodges & Eastman (1984) and Eastman, Makawiti, Collins & Hodges (1984). In brief, 0.1 ml of diluted urine (1:5 v/v) was incubated overnight at 4°C with 0.1 ml antiserum raised against oestrone-3-glucuronide-BSA (1:15 000 initial dilution) and 0.1 ml $^3$H oestrone-3-sulphate as tracer (Sp.Act 48 Ci/mmol). Doubling dilutions of oestrone-3-sulphate standards were run over the range 2000–31.3 pg/ tube. Cross-reactivities of the antiserum include 126% oestrone, 84% oestrone-3-glucuronide, 5.4% oestrone-17-sulphate, and 0.1% oestradiol-3-sulphate. Serial dilutions of male rhinoceros urine gave displacement curves parallel to those obtained with oestrone-3-sulphate standards. Assay sensitivity was 18.5 pg/tube or 0.92 ng/ml urine and inter- and intra-assay coefficients of variation were < 10%.

### Pregnanediol glucuronide

Pregnanediol glucuronide immunoreactivity was measured in unextracted samples using a microtitre plate enzyme assay described in detail and validated for use in the rhinoceros by Hodges & Green (in prep.). The assay utilizes an antiserum raised in a rabbit against pregnanediol-3-glucuronide-BSA and alkaline phosphatase conjugated to pregnanediol-3-glucoronide (Sauer, Foulkes, Worsfold & Morris 1986; Hodges, Green, Cottingham, Sauer, Edwards & Lightman 1988) as enzyme conjugate. Serial dilutions of rhinoceros urine gave displacement curves parallel to that obtained with standards. Cross-reactivity of the antiserum has been previously reported (Hodges & Green in prep.). Because of the absence of definitive identification of pregnanediol-3-glucuronide in black rhinoceros urine the true identity of the hormone being measured cannot be verified. Accordingly, the results are expressed as pregnanediol glucuronide immunoreactivity. Assay sensitivity was 10 pg/well or 0.6 ng/ml and inter- and intra-assay CVs were below 15%.

### Creatinine

The creatinine content of each urine sample was estimated to help compensate for differences in urine concentration and volume (Brand 1981; Hodges & Eastman 1984). Sensitivity of the assay was 0.1 mg creatinine and interassay coefficient of variation < 10%. Hormone levels reported in this paper have been divided by urinary creatinine concentration and are thus expressed as mass (ng)/mg creatinine.

## Results

### Home range and body size

The home range sizes of six adult male black rhinoceroses are shown in Table 1, set against the width of their hind footprint, which is used as an index of body size. There is a clear positive association between these two variables, indicating that the home range sizes of males increase as they get larger or more mature. The home ranges of these males were very variable, but all large; the difficulty of attempting to sample from animals covering such areas of dense bushland is apparent, particularly when location of animals and subsequent tracking to obtain urine samples were largely opportunistic.

**Table 1.** Comparison of home range size and hind footprint width in male black rhinoceros.

| Rhinoceros I.D. | Hind foot width (cm) | Home range size (km$^2$) |
|---|---|---|
| L | 21.75 | 54 |
| Z | 21.5 | 31 |
| U | 21.0 | 31 |
| X | 20.5 | 25 |
| M2 | 20.5 | 17 |
| D | 19.75 | 15 |

Males L and X were both known breeders, but did not have the largest home ranges. The levels of oestrone conjugates detected in the urine of these six males showed no relationship to the link between home range size and body size.

## Assessment of reproductive status

## Males

### Testosterone

Because only small amounts of urine could be recovered from free-ranging males, initial attempts to assay urinary testosterone were confined to samples from captive animals. The results shown in Table 2 indicate that levels of total urinary testosterone (i.e. conjugated and unconjugated) in male black rhinoceros were extremely low. Levels, even in samples of 0.5–1.0 ml, often bordered on the limit of sensitivity of the assay, making accurate assessment difficult. Furthermore, the extensive overlap between values obtained in males and females clearly indicated the lack of association of this measurement with testicular function. Because of this lack of potential and the need for samples to be of large volume, the assay of samples from free-ranging males was not attempted.

**Table 2.** Total immunoreactive testosterone (ng/mg Cr) in urine samples from adult male and female captive animals. Mean ± s.e.m. (range).

| Males | Females |
|---|---|
| 0.85 ± 0.22 | 0.19 ± 0.07 |
| (0.11 – 2.6) | (0.06 – 0.45) |
| $n = 12$ | $n = 6$ |

### Conjugated oestrone

Since large amounts of urinary oestrogens are excreted by stallions (Velle 1975), and male Indian rhinoceros (E. Wanjohi & J.K. Hodges, unpubl. data), it was reasoned that it might be possible to assess the reproductive status of male black rhinoceros by measuring conjugated oestrone instead of testosterone. Levels of urinary oestrone conjugates in urine samples from captive males and females and free-ranging male rhinoceros are shown in Table 3.

**Table 3.** Levels of oestrone conjugates (ng/mg creatinine) in urine from male and female captive and male free-ranging animals. Mean ± s.e.m. (range).

| Captive | | Free-ranging |
|---|---|---|
| Males | Females | males |
| 4.74 ± 0.85 | 5.91 ± 1.5 | 2.68 ± 0.16 |
| (0.82 – 16.1) | (0.79 – 24.3) | (0.76 – 6.1) |
| $n = 27$ | $n = 16$ | $n = 41$ |

Conjugated oestrone was detectable in all of 41 urine samples collected from six free-ranging males. Levels were, however, highly variable within individuals (range in one captive male: 0.82–16.1 ng/mg Cr) and there was no difference between values in males and females during the oestrous cycle. Furthermore, there were no differences in levels of oestrone conjugates between free-ranging males, even though at least two males (L and X) were known breeders, and two other males (D and M2) were almost certain non-breeders. We can therefore exclude the measurement of oestrone conjugates in urine as having diagnostic value in identifying reproductive status in black rhinoceros.

In view of the lack of success in using the measurement of either testosterone or conjugated oestrone in assessing reproductive status in male rhinoceros, a more detailed analysis of urinary androgen metabolites was initiated. Preliminary gas chromatographic and mass spectrometric analysis of derivatized black rhinoceros urine collected from two captive males revealed the adrenal androgen androstane-triol-one as the most abundant metabolite present, whilst also detecting small but significant amounts of the testicular androgen androstanediol.

The lack of measurable quantities of testosterone was notable and may offer an explanation for our inability to establish a valid radioimmunoassay for the hormone. Since androstanediol is the major urinary metabolite of testosterone in man, its presence in rhinoceros urine is encouraging, and may also indicate a potential method for assessment of testicular function and breeding status in males of this species. The development of an immunoassay for measurement of androstanediol is currently under way.

## Females

### Pregnanediol-3-glucuronide

Since levels of urinary immunoreactive pregnanediol-3-glucuronide have previously been shown to increase during the later stages of pregnancy in captive black rhinoceros (Ramsay, Kasman & Lasley 1987; Hodges & Green in prep.), its measurement may also provide the basis for a pregnancy test for animals living in the wild.

Figure 3 presents data from pregnant and non-pregnant captive females and from 17 samples collected from free-ranging females in Laikipia. Very large differences between levels in late pregnant and non-pregnant captive animals are seen. Pregnanediol-3-glucuronide was detected in all field samples but in a range consistent with these animals being non-pregnant, or in early pregnancy. This assessment has been confirmed, as none of the females sampled has given birth in the 12 months since sampling.

As described in the methods section, the assay for pregnanediol is a simple micro-titre plate enzyme-based method (Hodges & Green in prep.). The several advantages of ELISA assays, including low cost, ease of performance

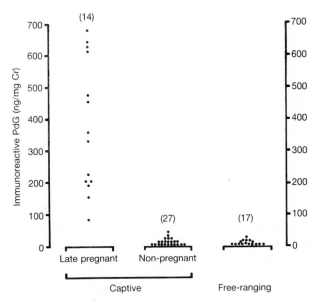

**Fig. 3.** Pregnanediol-3-glucuronide immunoreactivity (ng/mg Cr) in urine from pregnant and non-pregnant captive and free-ranging black rhinoceros. Numbers of samples in brackets.

and stability of reagents, are not only useful in a laboratory setting but also make them amenable to development for use under field conditions. In particular, the end point of the assay (i.e. colour change) is far simpler to quantify than the measurement of radioactivity, and consequently can be read without the need for sophisticated laboratory equipment. In order to assess the potential of the pregnanediol-3-glucuronide ELISA for field use we have evaluated a portable plate reader suitable for use under field conditions, by comparing the results obtained by this and a much larger standard laboratory machine.

A comparison of the results obtained with both plate readers showed a high degree of correlation over the maximum range of pregnanediol-3-glucuronide values normally found. The results gave a regression equation of $y = 1.14\,x - 0.03$ (where $y$ = values were obtained by the portable plate reader, and $x$ = values were obtained by the laboratory reader) and a correlation coefficient $r = 0.98$. The data shown in Fig. 3 were obtained using the portable plate reader.

## Discussion

This paper establishes for the first time the feasibility of collection of urine from black rhinoceros in the field and the potential usefulness of urinary hormone analysis in monitoring reproductive status.

Although the collection of urine from wild rhinoceros is feasible, sampling is limited by the small volumes that can be obtained (< 1.0 ml), the uncertainty of locating the individuals to be sampled, which all occupy large home ranges, and the need to find and collect the urine promptly after it has been deposited by an identified rhinoceros. Furthermore, the differences in urination behaviour make it very much easier to collect urine from males, with the result that most of the urine samples analysed here were from males.

Despite greater access to urine samples from males, we have been unable to establish a method for assessment of breeding status based on the measurement of testosterone or conjugated oestrone. It has been found that matings by male rhinoceros are more likely to be restricted to a few dominant individuals in higher-density populations in fenced sanctuaries (R.A. Brett unpubl.), and it is possible that greater differences in levels of these hormones between breeding and non-breeding males would be seen under these conditions.

The failure to detect conjugated testosterone in male urine was surprising but clearly points to the need to look for alternative metabolites. Androstanediol is one possibility which we are currently investigating further. Thus, although the identification of the major urinary metabolite of testosterone may eventually lead to successful physiological assessment of male breeding status, for the present such assessment must chiefly rely on observation of matings by known males, and of regular consorting by particular males with females which are likely to be in oestrus.

It was found that larger and probably more mature males had larger home ranges. It remains to be determined whether these larger males are necessarily breeders, or occupy a minimum home range size that may be used as another indicator of breeding status. If a correlate of breeding status in the form of a urinary androgen is found, it should be possible to determine which males are breeding, simply from its levels in urine. However, given recent information about the earless male rhinoceros in Addo Elephant National Park that was castrated yet continued to guard females and attempt to mate (Hall-Martin 1986), this is open to doubt. Hall-Martin's observations should be taken into account by managers of rhinoceros sanctuaries: whilst castration might prevent further genetic contributions from males that have sired too many calves in a given sanctuary, it might lead to behavioural problems and obstruction from the castrates.

Results on the assessment of reproductive status in females were more encouraging. Studies on captive female black rhinoceros indicate a variable but significant increase in the levels of immunoreactive pregnanediol-3-glucuronide during mid to late pregnancy (Ramsay et al. 1987; Hodges & Green in prep.), and the present data substantiate this. Given the variability in levels between animals and the limited number of data available, it is difficult to gauge the ultimate value of the pregnanediol-3-glucuronide assay as a pregnancy test in this species.

At the moment, however, we can say that high levels of immunoreactivity are a certain indicator of pregnancy and that the method required to perform the test under field conditions is now available. The value of such a test for pregnancy in the wild would be considerable as a tool in the monitoring and management of rhinoceros sanctuaries, and we are currently hoping to attempt collection from female rhinoceros in more open habitat, where matings are easily recorded and progress of gestation and parturition monitored.

It was noted earlier how difficult it may be to collect urine samples regularly from free-ranging females, and thus allow the possibility of accurate determination of peak oestrus. The fact that urine of oestrous females is deposited frequently in small amounts and is typically not sprayed means that, besides urine being hard to obtain from female rhinoceros outside oestrus, it is most unlikely that it will be possible to collect regular urine samples of sufficient volume from free-living oestrous females.

This should be borne in mind in considering the future possibility of artificial insemination of female black rhinoceros in sanctuaries. Except by observing matings by resident males, it may prove impractical to monitor oestrus accurately in any female black rhinoceros except those in captive conditions. It should also be noted that levels of oestrogens in the urine of captive female black rhinoceros so far show no relationship to the onset or termination of oestrus (Ramsay *et al.* 1987), and at present it is not possible to detect the time of ovulation in this species using any method of urinary hormone analysis.

## Acknowledgements

We thank the Office of the President and the Wildlife and Conservation and Management Department, Kenya, for permission to carry out research in Kenya, Kuki Gallmann for permitting R.A.B. to work on the Ol Ari Nyiro Ranch, which is dedicated to the conservation ideals of the Gallmann Memorial Foundation, and Tim Oloo and Colin Francombe for assistance in the field.

We are grateful to the British Council for support of E.W. at the Institute of Zoology, to Hoechst Pharmaceuticals for financial support of J.K.H. and to Cambridge Life Sciences PLC for generous donation of the portable plate reader.

## References

Andelman, S.J., Else, J.G., Hearn, J.P. & Hodges, J.K. (1985). The non-invasive monitoring of reproductive events in wild Vervet monkeys (*Cercopithecus aethiops*) using urinary pregnanediol-3-α-glucuronide and its correlation with behavioural observations. *J. Zool., Lond. (A)* **205**:467–77.

Brand, H.M. (1981). Urinary oestrogen excretion in the female cotton-topped tamarin (*Saguinus oedipus oedipus*). *J. Reprod. Fert.* **62**:467–73.

Dixon, K.R. & Chapman, J.A. (1980). Harmonic mean measure of animal activity areas. *Ecology* **61**:1040–44.

Eastman, S.-A.K., Makawiti, D.W., Collins, W.P. & Hodges, J.K. (1984). Pattern of excretion of urinary steroid metabolites during the ovarian cycle and pregnancy in the marmoset monkey. *J. Endocr.* **102**:19–26.

Goddard, J. (1966). Mating and courtship of the black rhinoceros (*Diceros bicornis* L.). *E. Afr. Wildl. J.* **4**:69–75.

Goddard, J. (1967). Home range, behaviour, and recruitment rates of two black rhinoceros populations. *E. Afr. Wildl. J.* **5**:133–50.

Goddard, J. (1970). Age criteria and vital statistics of a black rhinoceros population. *E. Afr. Wildl. J.* **8**:105–21.

Hall-Martin, A. (1986). Recruitment in a small black rhino population. *Pachyderm* No. 7:6–8.

Hodges, J.K. (1986). Monitoring changes in reproductive status. *Int. Zoo Yb.* **24/25**:126–30.

Hodges, J.K. & Eastman, S.A.K. (1984). Monitoring ovarian function in marmosets and tamarins by the measurement of urinary oestrogen metabolites. *Am J. Primatol.* **6**:187–97.

Hodges, J.K., Eastman, S.A.K. & Jenkins, N. (1983). Sex steroids and their relationship to binding proteins in the serum of the marmoset monkey (*Callithrix jacchus*). *J. Endocr.* **96**:443–50.

Hodges, J.K. & Green, D.I. (In prep.). *The development of an enzymeimmunoassay for urinary pregnanediol-3-glucuronide and its application to non-invasive reproductive assessment in exotic mammals.*

Hodges, J.K., Green, D.I., Cottingham, P.G., Sauer, M., Edwards, C. & Lightman, S.L. (1988). Induction of luteal regression in the marmoset monkey (*Callithrix jacchus*) by a gonadotrophin-releasing hormone antagonist and the effects on subsequent follicular development. *J. Reprod. Fert.* **82**:743–52.

Lande, R. & Barrowclough, G.R. (1987). Effective population size, genetic variation, and their use in population management. In *Viable populations for conservation*: (Ed. Soulé, M.). Cambridge University Press, Cambridge.

Lasley, B.L. (1985). Methods of evaluating reproductive function in exotic species. *Adv. vet. Sci. comp. Med.* **30**:209–28.

Mukinya, J.T. (1973). Density, distribution, population structure and social organization of the black rhinoceros in Masai Mara Game Reserve. *E. Afr. Wildl. J.* **11**:385–400.

Mukinya, J.G. (1976). An identification method for black rhinoceros (*Diceros bicornis* Linn. 1758). *E. Afr. Wildl. J.* **14**:335–8.

Owen-Smith, R.N. (1975). The social ethology of the white rhinoceros *Ceratotherium simum* (Burchell 1817). *Z. Tierpsychol.* **38**:337–84.

Poole, J.H., Kasman, L.H., Ramsay, E.C. & Lasley, B.L. (1984). Musth and urinary testosterone concentrations in the African elephant (*Loxodonta africana*). *J. Reprod. Fert.* **70**:255–60.

Ramsay, E.C., Kasman, L.H. & Lasley, B.L. (1987). Urinary steroid evaluations to monitor ovarian function in exotic ungulates: V. Estrogen and pregnanediol-3-glucuronide excretion in the black rhinoceros (*Diceros bicornis*). *Zoo Biol.* 6:275–82.

Ritchie, A.T.A. (1963). The black rhinoceros (*Diceros bicornis* L.). *E. Afr. Wildl. J.* 1:52–62.

Sauer, M.J., Foulkes, J.A., Worsfold, A. & Morris, B.A. (1986). Use of progesterone 11-glucuronide-alkaline phsophatase conjugate in a sensitive microtitre-plate enzymeimmunoassay of progesterone in milk and its application to pregnancy testing in dairy cattle. *J. Reprod. Fert.* 76:375–91.

Schenkel, R. & Schenkel-Hulliger, L. (1969). *Ecology and behaviour of the black rhinoceros* (Diceros bicornis L.). *A field study* (*Mammalia Depicta* 5). Verlag Paul Parey, Berlin & Hamburg.

Velle, W. (1975). Endogenous anabolic agents in farm animals. In *Anabolic agents in animal production*:159–70. (Eds Lu, F.C. & Rendel, J.). Georg Thieme, Stuttgart.

Symp. zool. Soc. Lond. (1989) No. 61:163–180

# Locomotion of African mammals

R.McN. ALEXANDER and
G.M.O. MALOIY

Department of Pure & Applied Biology
University of Leeds
Leeds LS2 9JT, UK

Department of Animal Physiology
University of Nairobi
P.O. Box 30197
Nairobi, Kenya

## Synopsis

Much research on mammalian running has used domestic species, but wild African species have also been studied extensively.

Maximum speeds of horses, dogs and humans are well known but those of wild mammals are not. Horses and dogs can maintain oxygen balance at higher running speeds than wild mammals. Lions accelerate faster than the prey they are pursuing.

The dynamic similarity hypothesis predicts that animals of different sizes will tend to move in dynamically similar fashion, when travelling at speeds that give them equal Froude numbers. Mammals travelling with equal Froude numbers tend to use the same gait, as the hypothesis predicts. However, the relationships between relative stride length and Froude number are different for primates (which take remarkably long strides), non-cursorial mammals and cursorial mammals.

Primates tend to have longer legs than fissipedes of the same mass, and have large flexor muscles with long fibres in the distal parts of their legs. Antelopes and other advanced artiodactyls also have long legs but this is because they have very long, fused metapodials. They have small distal flexor muscles with short fibres. They are highly adapted to save energy in running by exploiting tendon elasticity. Peak stresses in muscles and bones, in fast running or jumping, have been calculated for a few species of mammals.

## Introduction

Much of our knowledge of mammal locomotion comes from studies of domestic animals, especially dogs, cats, horses and humans. Dogs and horses have been bred for running ability but, in the breeding of most other domestic animals, locomotion has been largely ignored. Thus the selection for locomotor performance that acts on wild species has been enhanced in the breeding of some domestic mammals and relaxed in the breeding of

ZOOLOGICAL SYMPOSIUM No. 61
ISBN 0–19–854009–4

others. Information about the locomotion of domestic mammals may be misleading if applied to wild ones, even to those wild species that most closely resemble the domestic ones.

African species have figured very largely in research on locomotion of wild mammals, for two reasons. First, Africa is unrivalled for the diversity of its mammals and is especially rich in artiodactyls, fissipedes and primates. It also has a remarkable group of graviportal mammals (elephant, rhinoceros and hippopotamus). Second, the short grasslands of the East African plains offer exceptional opportunities for observing the locomotion of the many large species that live there. They can be seen from far off, and closely approached or followed by vehicles across the easy terrain.

## Performance

Speed, endurance, acceleration and manoeuvrability are all important aspects of running performance.

We will start by examining the maximum speeds that mammals can reach in short sprints. These are well known from race results for racehorses, greyhounds and humans, but much less well known for other species. Most horse races over distances of 1 mile (1600 m) or less are won at $15-17 \text{ m s}^{-1}$, and most greyhound races over 460 m at $15-16 \text{ m s}^{-1}$ (Alexander, Langman & Jayes 1977; Jayes & Alexander 1982). Good male human athletes reach $11 \text{ m s}^{-1}$ in the fastest part of a 100 m sprint (Ballreich & Kuhlow 1986).

Maximum sprinting speeds have been published for many other species (Garland 1983) but in many cases no information is given about how they were measured. Since it is impossible to assess the reliability of these measurements we feel obliged to reject them. Some other data are subjective estimates, presumably based on the observer's experience of road traffic. These also we reject. The remaining data, which are more acceptable, come from speedometer readings, from film analysis and from measurements of the time required to run measured distances.

Speedometer readings are not wholly reliable, even when the vehicle drove close behind or alongside the animal. Most speedometers overestimate speed because it is illegal (at least in Britain) for them to underestimate speed. Also, if an animal and a vehicle travel in a curve with the vehicle on the outside, the vehicle must travel faster than the animal to keep abreast of it. Measurements from films avoid these errors, but it is generally difficult to establish the scale of length for films taken in the field. Timing over measured distances is generally possible only for captive animals, and it may be difficult to persuade the animal to run at maximum speed.

The speeds for wild animals, shown in Fig. 1, are therefore less reliable than we would wish. Most of them lie between the maximum speeds of men

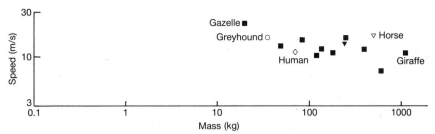

**Fig. 1.** Maximum speeds and body masses of ◇, humans; ●○, fissipedes; ■□, artiodactyls; and ▼▽, other mammals. Filled symbols represent wild African species and hollow symbols represent domestic mammals (from any continent). The data are from Garland (1983, omitting subjective estimates and speeds published without information about how they were measured); from Alexander, Langman *et al.* (1977); and from Ballreich & Kuhlow (1986).

(11 m s$^{-1}$) and horses (17 m s$^{-1}$), but there is a record of 23 m s$^{-1}$ for Thomson's gazelle (*Gazella thomsoni*). We have omitted the even higher speed of 31 m s$^{-1}$, given by Garland (1983) for the cheetah (*Acinonyx jubatus*) because it seems to be based on a record that was discredited by Hildebrand (1959).

All the mammals in Fig. 1 are fairly large. Garland's (1983) data for non-African species show that small mammals cannot run as fast. Elephants and other graviportal mammals also seem to be slower than antelopes etc., but we have no reliable data. Garland's (1983) text makes it clear that the speed he gives for *Loxodonta* is little better than a guess.

The speeds in Fig. 1 are sprint speeds, that could be sustained only for a short time. They presumably require anaerobic metabolism. Figure 2 shows the (lower) maximum speeds at which mammals can maintain oxygen balance. They were measured in experiments in which captive animals ran on motorized treadmills. Their rates of oxygen consumption were measured and the lactate concentration in their blood was monitored. At the speeds shown in Fig. 2, the rate of oxygen consumption reached its maximum value and lactate began to accumulate in the blood. The speeds for horses, dogs and eland (*Taurotragus oryx*) were not measured directly, but have been estimated from maximum rates of oxygen consumption measured on sloping treadmills by extrapolation of the relationships between speed and oxygen consumption given by Taylor, Heglund & Maloiy (1982).

Figure 2 shows a trend from low maximum aerobic speeds for small mammals to higher speeds for large ones, but there are no data for animals of masses above 250 kg. The domestic mammals selected for running ability (dog and horse) have higher maximum aerobic speeds than any of the other mammals.

The maximum aerobic speed of an animal depends both on how economically it can run (i.e. on the relationship between oxygen consumption and speed) and also on how fast it can take up oxygen. Dogs and horses are

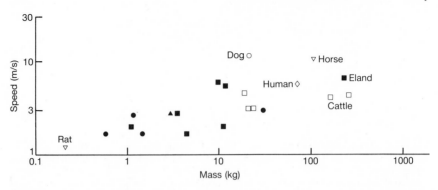

**Fig. 2.** Maximum speeds that can be supported by aerobic metabolism, and body masses, for wild African mammals and domestic mammals. The data are from Seeherman, Taylor, Maloiy & Armstrong (1981), Taylor, Maloiy *et al.* (1981) and (for humans) Margaria, Cerretelli, Anghemo & Sassi (1963). Other details in in Fig. 1.

not particularly economical runners, in comparison with other mammals of similar size (Taylor, Heglund *et al.* 1982) but they are capable of remarkably high rates of oxygen consumption (Taylor, Maloiy *et al.* 1981).

Speeds have been measured for many mammals but accelerations for only a few. Elliott, Cowan & Holling (1977) filmed lion attacks on various prey and obtained graphs of speed ($v$) against time ($t$), both for the predator and for the prey. Speed increased rapidly at first but approached a maximum value ($v_{max}$) asymptotically. They fitted an equation to their data

$$v = v_{max} \left(1 - e^{-Kt}\right) \qquad (1)$$

and evaluated $v_{max}$ and $K$ for each species (Table 1). The values of $v_{max}$ for lion, zebra and wildebeest lie in the range that Fig. 1 would lead us to expect, but the value for Thomson's gazelle is remarkably high. We suspect that it may have been obtained from the speed-time curve by extrapolation from lower speeds. Elliott *et al.* (1977) show no points on their speed-time curves so we cannot tell whether the highest speeds were actually observed. For this reason we prefer to use their paper as a source for accelerations only, and not for maximum speeds.

Differentiation of Equation 1 gives the acceleration

$$dv/dt = Kv_{max}e^{-Kt}$$

which has its maximum value, $Kv_{max}$, when $t = 0$. Table 1 shows maximum accelerations calculated in this way. They are about 5 m s$^{-2}$ for all three prey species but almost 10 m s$^{-2}$ for the lion. We cannot tell whether this is

**Table 1.** Accelerations of athletes and wild mammals, starting from rest. The accelerations of the wild mammals were calculated from values of $K$ and $v_{max}$ (Equation 1) given by Elliott *et al.* (1977). The accelerations for humans are those of male athletes, given by Ballreich & Kuhlow (1986).

|  | $K$ $(s^{-1})$ | $v_{max}$ $(m\ s^{-1})$ | Initial acceleration $(m\ s^{-2})$ |
|---|---|---|---|
| Lion, *Panthera leo* | 0.68 | 13.9 | 9.5 |
| Zebra, *Equus burchelli* | 0.31 | 16.0 | 5.0 |
| Wildebeest, *Connochaetes taurinus* | 0.39 | 14.3 | 5.6 |
| Gazelle, *Gazella thomsoni* | 0.17 | 26.5 | 4.5 |
| Man, *Homo sapiens* |  |  | 10.0 |

due to some physiological difference between the species or merely to their different roles in the chase: the lion was able to prepare itself for the attack and to choose the starting time, but the prey may have been surprised by the attack.

Table 1 also shows the acceleration of human sprinters leaving the starting blocks. They were fully prepared for the start, and the blocks probably enabled them to accelerate faster than they could have done on smooth ground, so it is perhaps not surprising that they accelerated as fast as the lion.

Elliott *et al.* (1977) stressed the importance of forward acceleration for the ability of lions to capture prey, and for prey to escape, but pursued gazelles swerve in their efforts to escape. Howland (1974) argued that swerving ability is critically important. In his model, the animal that succeeds in the chase is the one that is capable of the higher value of $v(r)^2/r$, where $v(r)$ is the speed of running in a curve of radius $r$. A standard proof in dynamics shows that a body moving with speed $v$ around a circle of radius $r$ has acceleration $v^2/r$ towards the centre of the circle, so $v(r)^2/r$ is the animal's *transverse* acceleration.

There seem to be no published data for transverse accelerations of animals other than man. McMahon (1984) asked human subjects to run as fast as possible along circular paths of various radii. His graph shows that they could run at $4.7\ ms^{-1}$ with a radius of $3.7\ m$, which implies a transverse acceleration of $(4.7)^2/3.7 = 6.0\ m\ s^{-2}$. No other point on the graph requires so large an acceleration.

Transverse acceleration is an important aspect of animal running performance, about which we are regrettably ignorant. Manoeuvrability depends largely upon it. In some circumstances, forward and transverse acceleration may be limited by the animal's strength, but in others they may be limited by the friction between the feet and the ground: an animal that attempts to accelerate on slippery ground may skid.

## Gait

Mammals use different gaits at different speeds. For example, horses walk at low speeds, trot at moderate speeds, canter at higher speeds and gallop at their highest speeds. Other mammals use some or all of these gaits.

Each gait is recognizable, whatever mammal performs it. There are clear similarities between, for example, the gallops of a mouse and of a horse, despite the great difference in size, speed and stride frequency. These similarities become especially clear when examined in the light of the dynamic similarity hypothesis of Alexander & Jayes (1983. See also Alexander 1976).

The concept of dynamic similarity is closely related to the more familiar concept of geometric similarity. Two bodies A and B are geometrically similar if A could be made identical in form to B by multiplying all its linear dimensions by the same factor. Similarly, two motions are dynamically similar if one could be made identical to the other by multiplying all lengths by one factor, all times by a second factor and all forces by a third factor.

Dynamic similarity is possible only between geometrically similar systems, and there are other restrictions that depend on the nature of the forces involved. If gravity is important (as it is in walking and running) dynamic similarity is possible only between systems that have equal values of the Froude number $v^2/gl$ (Duncan 1953) where $g$ is the gravitational acceleration and $v$ and $l$ are a speed and a linear dimension. Alexander & Jayes (1983) defined $v$ as the mean forward speed and $l$ as the height of the hip joint from the ground in normal standing, and we use these definitions again here, but we would like to point out that they are arbitrary. It would be permissible, for example, to use the peak vertical speed of the centre of mass and the length of the humerus, provided that this were done consistently for all the animals being compared.

The dynamic similarity hypothesis (Alexander & Jayes 1983) postulates that animals tend to move in dynamically similar fashion when travelling at speeds that make their Froude numbers equal. Different species cannot show precise dynamic similarity, because they are not geometrically similar, so we expect to see only approximations to dynamic similarity. The section of this paper on structure will show what the principal deviations from geometric similarity are, between and within major groups of mammals.

There is good reason for expecting animals to behave as predicted by the dynamic similarity hypothesis. It seems likely that they adjust their gaits to minimize the energy cost of locomotion (except in gaits with a display function, such as stotting: Caro 1986). Imagine two animals of different sizes moving in dynamically similar fashion. The forces that their muscles exert will be proportional to their body weights and the amounts that their muscles shorten will be proportional to the linear dimensions of their bodies. Thus the work that is done by the muscles in the course of a stride is proportional to (weight × linear dimensions). The distance travelled in the stride is also proportional to the linear dimensions so the mechanical cost of transport (the work required to move unit weight of animal through unit distance) is the same for both animals. If one has adjusted its gait to minimize the mechanical cost of transport, the other, moving in dynamically similar fashion, has also minimized this cost.

We will examine the gaits of wild African mammals and of domestic mammals, to test the dynamic similarity hypothesis.

One of the predictions of the hypothesis is that mammals will tend to use the same gait, when travelling with equal Froude numbers. Figure 3 shows the Froude numbers at which various mammals used named gaits. In some mammals (for example, horses and dogs) the transition from walking to trotting involves abrupt changes in duty factors and in the phase relationships of the feet. In others (for example, sheep: Jayes & Alexander 1978) the transition is more gradual and it is not always clear whether a particular gait should be regarded as a walk or a trot. The transition from trotting to cantering generally involves an abrupt change in phase relationships (Alexander & Jayes 1983: Fig. 1) but cantering generally merges into galloping, so we make no distinction, in Fig. 3, between cantering and galloping.

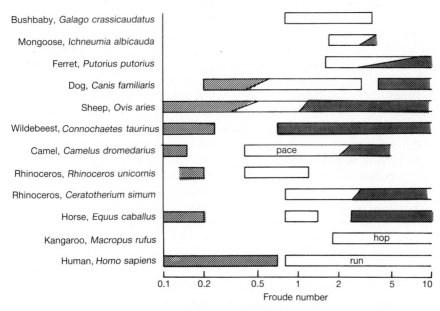

**Fig. 3.** The Froude numbers at which various mammals use each of their gaits. Hatched bars indicate walking, white bars indicate trotting (unless otherwise labelled) and stippled bars indicate cantering or galloping. This diagram is based on the data of Hayes & Alexander (1983). Alexander & Jayes (1983) and Alexander & Maloiy (1984).

Most mammals change from walking to trotting at Froude numbers close to 0.5 and from trotting to galloping at Froude numbers between 2 and 3. Figure 3 shows that these general rules apply to animals as different as ferrets, horses and rhinoceroses, which are obviously not geometrically similar to each other and so cannot move in perfect dynamic similarity. They nevertheless use gaits similar enough to be given the same name, when travelling with equal Froude numbers.

Though this rule holds widely, there are exceptions. Wildebeest, giraffe and hyaena change directly from walk to canter without the usual intervening trot. Pennycuick (1975) points out that all of them have sloping backs. Camels and some long-legged breeds of dog pace instead of trotting, possibly to keep the fore and hind legs out of each other's way (Hildebrand 1980). Elephants amble instead of trotting, and do not gallop even at their highest speeds. Large African elephants have hip heights of about 2.3 m, so they would reach a Froude number of 3 (by which other mammals would be galloping) at a speed of 8 m s$^{-1}$, which is well within their reported range of speeds (Garland 1983). Lemurs and monkeys move their feet in a different order, when they walk, from most other mammals (Hildebrand 1967). Other peculiarities of gait among mammals have been surveyed by Hildebrand (1976, 1977).

When different-sized animals move in dynamically similar fashion their Froude numbers must be equal, so their speeds must be proportional to the square roots of their linear dimensions. Also their stride lengths must be proportional to their linear dimensions, and stride frequency (speed divided by stride length) should be inversely proportional to the square root of the linear dimensions.

Pennycuick (1975) observed the stride frequencies of large mammals moving undisturbed in Serengeti National Park. Figure 4 shows mean stride frequencies for each gait plotted against shoulder height on logarithmic co-ordinates. The argument of the previous paragraph suggests that the points for each gait should lie along a line of slope −0.5, and Fig. 4 shows that they do approximately that, despite the diversity of the species.

If mammals were geometrically similar to each other their linear dimensions would be proportional to (body mass)$^{0.33}$. The next section of this paper will show that they tend to be quite close to geometric similarity in many respects. This implies that if stride frequencies are proportional to (hip height)$^{-0.5}$ they are also about proportional to (body mass)$^{-0.17}$. Heglund, Taylor & McMahon (1974) measured the galloping stride frequencies of laboratory and domestic mammals (mouse, rat, dog and horse) and found them to be proportional to (body mass)$^{-0.14}$.

Figure 5 shows galloping stride frequency plotted against body mass for a different selection of mammals, including many wild African species. The non-primates, ranging from small rodents to rhinoceros, show a relationship

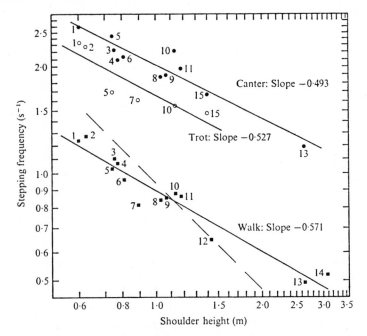

**Fig. 4.** Graphs on logarithmic co-ordinates of stride frequency against shoulder height for African mammals ■, walking, ○, trotting and ●, cantering or galloping.

1. Thomson's gazelle (*Gazella thomsoni*)
2. Warthog (*Phacochoerus aethiopicus*)
3. Juvenile gnu (*Connochaetes taurinus*)
4. Spotted hyaena (*Crocuta crocuta*)
5. Grant's gazelle (*Gazella granti*)
6. Impala (*Aepyceros melampus*)
7. Lion (*Panthera leo*)
8. Hartebeest (*Alcelaphus buselaphus*)
9. Topi (*Damaliscus korrigum*)
10. Zebra (*Equus burchelli*)
11. Adult gnu (*Connochaetes taurinus*)
12. Black rhinoceros (*Diceros bicornis*)
13. Giraffe (*Giraffa camelopardalis*)
14. Elephant (*Loxodonta africana*)
15. Buffalo (*Syncerus caffer*)

From Pennycuick (1975).

indistinguishable from that given by Heglund *et al.* (1974) for domestic mammals. The primates, however, show a quite different relationship. They use much lower stride frequencies than other mammals of the same body mass (Vilensky 1980; Alexander & Maloiy 1984). Notice that this is true of all the primates in Fig. 5, despite their diversity: in order of body mass they are *Galago*, two species of *Macaca*, *Cercopithecus*, *Papio*, juvenile *Gorilla* and *Pan*.

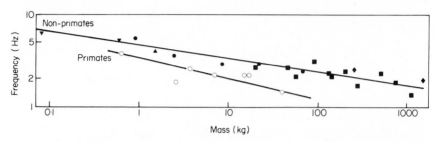

**Fig. 5.** A graph on logarithmic co-ordinates of galloping stride frequency against body mass for ○, Primates; ●, Fissipedia; ■, Artiodactyla; ◆, Perissodactyla; ▼, Rodentia and ▲, Lagomorpha. The line for non-primates shows stride frequency proportional to (body mass)$^{-0.14}$ and the line for primates shows it proportional to (body mass)$^{-0.22}$ but neither exponent is significantly different from the theoretical value of $-0.17$. From Alexander & Maloiy (1984).

This analysis distinguishes two main groups of quadrupedal mammals (primates and non-primates) but another analysis distinguishes three. Mammals generally increase the length of their strides as they increase speed. The dynamic similarity hypothesis predicts that, at any particular Froude number, they will have stride lengths proportional to their hip heights. Alexander & Jayes (1983) and Alexander & Maloiy (1984) found that this was approximately true within each of three groups of mammals, but not between the groups. These groups are:

(a)  Cursorial non-primates. Cursorial mammals are those that walk and run on relatively straight legs (Jenkins 1971). They include the large fissipedes (cats, dogs etc.) and all ungulates. Most mammals with body masses of 5 kg or more are cursorial in this sense.

(b)  Non-cursorial non-primates. Non-cursorial mammals run on strongly bent legs (Jenkins 1971). They include nearly all the rodents, and most other mammals of less than 5 kg.

(c)  Primates.

At any particular Froude number, group (c) generally use longer relative stride lengths than group (b), which use longer relative stride lengths than group (a). The long stride lengths of primates correlate with their low stride frequencies. Imagine a primate and a non-primate with equal hip heights, running at the same speed. The primate takes fewer, longer strides than the non-primate.

Most mammals are quadrupeds but a few, including some African ones, are bipeds. Kangaroos and some rodents hop bipedally and tend to move in dynamically similar fashion to each other, at any particular Froude number. Film of a springhare (*Pedetes*) hopping on a treadmill showed that it used about the same relative stride length and duty factor as a kangaroo at the same Froude number (Hayes & Alexander 1983). Chimpanzees and gorillas occasionally walk bipedally, keeping their knees bent and the trunk sloping forwards (Jenkins 1972). The movements of their hind legs are not grossly different from those of quadrupedal walking and (at least in the case of the chimpanzee) stride lengths are about the same as for quadrupedal walking at the same Froude number (Alexander & Maloiy 1984) or a little shorter (Reynolds 1987). Humans also walk bipedally, but with very different movements. We keep our trunks erect, and our knees are almost straight while the feet are on the ground. Alexander (in press) has reviewed human walking and running, comparing them with the gaits of other mammals.

## Structure

The varied feet of mammals are conveniently described by using the adjectives plantigrade, digitigrade and unguligrade. Plantigrade mammals put the whole of each foot (including the metapodial region) on the ground. Non-cursorial mammals (Jenkins 1971) are generally plantigrade but the only plantigrade mammals heavier than about 10 kg are primates and bears. The great apes have plantigrade hind feet but walk on the bent knuckles of their hands (Larson & Stern 1987). Digitigrade mammals put the digits on the ground but keep their metapodials sloping at a steep angle, so that the wrist and ankle are well off the ground. The larger African fissipedes (Canidae, Felidae and Hyaenidae) are all digitigrade. Unguligrade mammals stand and run on the tips of their toes, which have hooves instead of claws. Artiodactyls, perissodactyls and elephants are all unguligrade. The more advanced artiodactyls (including antelopes, giraffes and camels) have only two functional digits on each foot, with their metapodials fused to form a single cannon bone. Zebras and other horses have only one functional toe on each foot.

Alexander, Jayes, Maloiy & Wathuta (1979) measured the lengths and diameters of the principal leg bones of a wide variety of mammals, most of them African. They ranged in size from 3 g shrews (*Sorex*) to a 2.5 t elephant (*Loxodonta*). Figure 6 shows the dimensions of the femur plotted against body mass. The continuous lines represent allometric equations calculated from all the data. Both have gradients of 0.36, showing that both the length and the diameter of the femur tend to be proportional to (body mass)$^{0.36}$. Similar exponents were found for all the major limb bones: the mean values were 0.35 for lengths and 0.36 for diameters. These are only a little different from the exponents of 0.33 that would have been obtained if the mammals were geometrically similar to each other.

Figure 6 shows that the femurs of primates are generally about 50% longer, and also slightly thicker, than those of other mammals of equal mass. Small Bovidae (*Rhynchotragus*, *Gazella*) have femurs of about the same size as those of other mammals of equal mass but large Bovidae

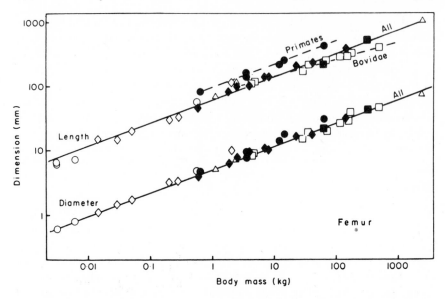

**Fig. 6.** A graph on logarithmic co-ordinates showing the lengths and diameters of the femurs of mammals plotted against body mass. ○, Insectivora; ●, Primates; ◇, Rodentia; ◆, Fissipedia; □, Bovidae; ■, other Artiodactyla; △, other orders (Lagomorpha and Proboscidea). The lines represent allometric equations fitted to the data. From Alexander, Jayes *et al.* (1979).

(*Connochaetes, Oryx, Syncerus*) have relatively short femurs. The springhare (*Pedetes*, the 2.1 kg rodent in Fig. 6) hops bipedally on its hind legs and has longer, thicker femurs than most other mammals of the same mass.

Graphs like Fig. 6 show similar differences between taxa for the tibia, humerus and ulna (Alexander, Jayes *et al.* 1979). These bones are longer in primates, and shorter in large bovids, than in other mammals of equal mass. The springhare has a long tibia as well as a long femur, but its humerus and ulna are short.

Figure 7 shows the dimensions of the metacarpals. Primates have metacarpals (and metatarsals) of about the same length as in other mammals of equal mass, despite having relatively long bones more proximally. Bovidae have metacarpal (and metatarsal) lengths about proportional to (body mass)$^{0.2}$. Their cannon bones are much longer and thicker than the metapodials of other mammals of equal mass. Two of the artiodactyls in Fig. 7 are not bovids. *Camelus* has its metapodials fused to form cannon bones which are long and thick like those of bovids, but *Phacochoerus* (the warthog) has unfused metapodials similar in length and diameter to those of non-artiodactyls of equal mass. The springhare has very short metacarpals in its reduced fore legs but long metatarsals in its strong hind legs.

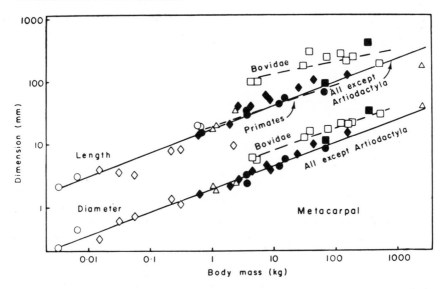

**Fig. 7.** The lengths and diameters of the longest metacarpals of mammals ploted against body mass. Other details as for Fig. 6. From Alexander, Jayes *et al.* (1979).

The data of Alexander, Jayes *et al.* (1979) show that both bovids and primates have long legs for their body masses, but that bovids owe this to their long metapodials and primates to more proximal bones. (See also Alexander 1985a.) They also show that the allometric exponents relating bone lengths to body mass for mammals in general are close to the value of 0.33 required for geometric similarity but that the exponents for Bovidae are close to the value of 0.25 predicted by McMahon's (1973) theory of elastic similarity. Economos (1983) pointed out that most of the large mammals in the data set are bovids and suggested that large and small mammals scale according to different rules, large ones approximating to elastic and small ones to geometric similarity. Alexander (1982) argued that the premises of the theory of elastic similarity are unrealistic for mammalian running. If he is right, the exponents for bovid bone lengths require a different explanation.

Alexander, Jayes, Maloiy & Wathuta (1981) measured the masses of the principal leg muscles of (mainly African) mammals and the lengths of their muscle fibres. They found as expected that the hind leg muscles of the springhare and other bipedal hoppers were more massive than those of quadrupedal mammals of equal body mass. Apart from that, the most marked differences between groups of mammals were in the muscles of the lower leg and forearm. Figure 8 summarizes the data for the flexor muscles of the wrist and of the digits of the fore foot. It shows that these muscles are

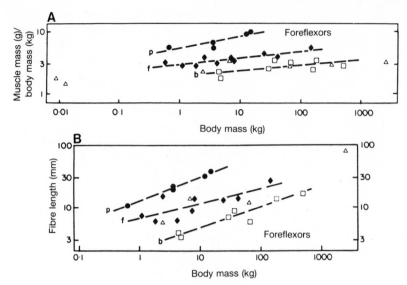

**Fig. 8.** Graphs showing properties of the flexor muscles of the wrist and of the digits of the fore foot, of various mammals. (A) shows (muscle mass/body mass) and (B) shows muscle fibre length, plotted in each case against body mass. Muscle masses are totals and fibre lengths are weighted harmonic means for the whole group of muscles. ●, Primates; ◆, Fissipedia; □, Bovidae and △, other mammals. The lines represent allometric equations for p, Primates; f, Fissipedia and b, Bovidae. From Alexander, Jayes *et al.* (1981).

remarkably large in primates, with remarkably long muscle fibres. They are small in bovids, with remarkably short muscle fibres. The fibres are about six times as long in a typical primate, as in a typical bovid of the same mass. The corresponding muscles of the hind leg show similar differences between primates, fissipedes and bovids, but they are less dramatic, and humans have rather small deep digital flexor muscles to operate their toes. Primates presumably need long muscle fibres, capable of large length changes, to operate their prehensile hands and (except in humans) feet. Their style of climbing trees requires the ability to exert large forces with the hands and feet in a wide variety of positions. In contrast, running involves much more predictable movements, requiring large forces in only a restricted range of foot positions. The long fibres of the muscles of primate hands and feet may be regarded as part of their adaptation to climbing.

Large mammals save energy in running by using leg tendons as springs (Alexander 1984). Each leg exerts a braking action in the first half of its stance phase, removing kinetic and potential energy from the body. Some of this is stored as elastic strain energy in tendons and returned by elastic recoil in the second half of the stance phase. Thus mechanical energy is conserved, reducing the work required of the muscles. The movements of distal joints that occur while the foot is on the ground may be accommodated largely by stretching and recoil of tendons, so that the muscle fibres need make only small length changes, and short fibres suffice. The Bovidae and other advanced ungulates are particularly well adapted to save energy in this way. The more extreme adaptations are found in camels and horses, in which the muscle fibres of several distal leg muscles have almost disappeared, leaving tendon running virtually uninterrupted from the origin to the insertion (Alexander, Maloiy, Ker, Jayes & Warui 1982; Dimery, Alexander & Ker 1986). Such adaptations require very long tendons, capable of stretching enough to allow the requisite leg movements. Dimery *et al.* (1986) studied the geometry of horse legs and the movements they make in galloping and calculated that various tendons must stretch by 4–9%. They could not have stretched much more without breaking: indeed, mammal tendons seldom stretch more than 8% before breaking in tensile tests (data of Bennett, Ker, Dimery & Alexander 1986). The main significance of the long metapodials of ungulates may be that they allow the tendons of the digital flexor muscles to be very long.

McMahon's (1973) theory of elastic similarity makes predictions about muscle as well as bone dimensions, but it seems unprofitable to apply it to mammalian muscles for two reasons. First there is the doubt (already mentioned) about the relevance of its premises to running. Secondly, the exponents of the allometric equations for muscle dimensions (Alexander, Jayes *et al.* 1981) are not known precisely: the confidence limits are too wide to exclude any plausible similarity model. It seems more useful to try to understand muscle and bone dimensions in relation to each other and to the animal's running performance.

Muscles and bones have limited strength. Running speed and jumping ability are probably generally limited by the strengths of leg muscles. Bones must be strong enough not to be broken in the animal's most strenuous activities, and indeed should be built with appropriate factors of safety (Alexander 1981). The forces that various mammals exert on the ground, in running and jumping, have been measured by means of force plates or calculated from films (Alexander 1985b). The mammals investigated by filming include various Bovidae, running in their natural habitats in Kenya, and an African elephant running in a zoo (Alexander 1977; Alexander, Maloiy, Hunter, Jayes & Nturibi 1979). The forces have been combined with anatomical measurements to obtain muscle and bone stresses. Stresses in bones have also been calculated from the output of strain gauges cemented to leg bones of living animals (Biewener, Thomason & Lanyon 1983). Peak muscle stresses in running and jumping seem generally to lie between 0.15 and 0.3 MPa, and may be compared to the maximum stress of about 0.3 MPa obtainable in isometric contractions of excised mammalian muscle (Alexander, Maloiy, Hunter *et al.* 1979; Alexander 1985b). Most of the peak bone stresses indicate factors of safety between 2 and 5 (Alexander 1981; Biewener *et al.* 1983, 1988). Alexander (1977) considered the scaling of African Bovidae, which span a wide range of sizes. He calculated that the peak stresses in their leg bones, muscles and tendons, when running at maximum speed, and the work required per unit mass of muscle for each stride, were all more or less independent of body mass. The dimensions of the bones, muscles and tendons seemed to be nicely matched to each other, throughout the range of body sizes, in accordance with the principle of symmorphosis (Taylor & Weibel 1981).

## References

Alexander, R.McN. (1976). Estimates of speeds of dinosaurs. *Nature, Lond.* **261**:129–30.

Alexander, R.McN. (1977). Allometry of the limbs of antelopes (Bovidae). *J. Zool., Lond.* **183**:125–46.

Alexander, R.McN. (1981). Factors of safety in the structure of animals. *Sci. Progr., Oxf.* **67**:109–30.

Alexander, R.McN. (1982). Size, shape and structure for running and flight. In *A companion to animal physiology*: 309–24. (Eds Taylor, C.R., Johansen, K. & Bolis, L.). Cambridge University Press, Cambridge.

Alexander, R.McN. (1984). Elastic energy stores in running vertebrates. *Am. Zool.* **24**:85–94.

Alexander, R.McN. (1985a). Body size and limb design in primates and other mammals. In *Size and scaling in primate biology*: 337–43. (Ed. Jungers, W.L.). Plenum Press, New York & London.

Alexander, R.McN. (1985b). The maximum forces exerted by animals. *J. exp. Biol.* **115**:231–8.

Alexander, R.McN. (In press). Characteristics and advantages of human bipedalism. In *Biomechanics in evolution*. (Ed. Rayner, J.M.V.). Cambridge University Press, Cambridge.

Alexander, R.McN. & Jayes, A.S. (1983). A dynamic similarity hypothesis for the gaits of quadrupedal mammals. *J. Zool., Lond.* **201**:135–52.

Alexander, R.McN., Jayes, A.S., Maloiy, G.M.O. & Wathuta, E.M. (1979). Allometry of the limb bones of mammals from shrews (*Sorex*) to elephant (*Loxodonta*). *J. Zool., Lond.* **189**:305–14.

Alexander, R.McN., Jayes, A.S., Maloiy, G.M.O. & Wathuta, E.M. (1981). Allometry of the leg muscles of mammals. *J. Zool., Lond.* **194**:539–52.

Alexander, R.McN., Langman, V.A. & Jayes, A.S. (1977). Fast locomotion of some African ungulates. *J. Zool., Lond.* **183**:291–300.

Alexander, R.McN. & Maloiy, G.M.O. (1984). Stride lengths and stride frequencies of Primates. *J. Zool., Lond.* **202**:577–82.

Alexander, R.McN., Maloiy, G.M.O., Hunter, B., Jayes, A.S. & Nturibi, J. (1979). Mechanical stresses in fast locomotion of buffalo (*Syncerus caffer*) and elephant (*Loxodonta africana*). *J. Zool., Lond.* **189**:135–44.

Alexander, R.McN., Maloiy, G.M.O., Ker, R.F., Jayes, A.S. & Warui, C.N. (1982). The role of tendon elasticity in the locomotion of the camel (*Camelus dromedarius*). *J. Zool., Lond.* **198**:293–313.

Ballreich, R. & Kuhlow, A. (1986). *Biomechanik der Leichtathletik*. Enke, Stuttgart.

Bennett, M.B., Ker, R.F., Dimery, N.J. & Alexander, R.McN. (1986). Mechanical properties of various mammalian tendons. *J. Zool., Lond. (A)* **209**:537–48.

Biewener, A.A., Thomason, J. & Lanyon, L.E. (1983). Mechanics of locomotion and jumping in the forelimb of the horse (*Equus*): *in vivo* stress developed in the radius and metacarpus. *J. Zool., Lond.* **201**:67–82.

Biewener, A.A., Thomason, J.J. & Lanyon, L.E. (1988). Mechanics of locomotion and jumping in the horse (*Equus*): *in vivo* stress in the tibia and metatarsus. *J. Zool., Lond.* **214**:547–65.

Caro, T.M. (1986). The functions of stotting in Thomson's gazelles: some tests of the predictions. *Anim. Behav.* **34**:663–84.

Dimery, N.J., Alexander, R.McN. & Ker, R.F. (1986). Elastic extension of leg tendons in the locomotion of horses (*Equus caballus*). *J. Zool., Lond. (A)* **210**:415–25.

Duncan, W.J. (1953). *Physical similarity and dimensional analysis*. Arnold, London.

Economos, A.C. (1983). Elastic and/or geometric similarity in mammalian design? *J. theor. Biol.* **103**:167–72.

Elliott, J.P., Cowan, I. McT. & Holling, C.S. (1977). Prey capture by the African lion. *Can. J. Zool.* **55**:1811–28.

Garland, T. (1983). The relation between maximal running speed and body mass in terrestrial mammals. *J. Zool., Lond.* **199**:157–70.

Hayes, G. & Alexander, R.McN. (1983). The hopping gaits of crows (Corvidae) and other bipeds. *J. Zool., Lond.* **200**:205–13.

Heglund, N.C., Taylor, C.R. & McMahon, T.A. (1974). Scaling stride frequency and gait to animal size: mice to horses. *Science, N.Y.* **186**:1112–3.

Hildebrand, M. (1959). Motions of the running cheetah and horse. *J. Mammal.* **40**:481–95.

Hildebrand, M. (1967) Symmetrical gaits of primates. *Am. J. phys. Anthrop.* **26**:119–30.

Hildebrand, M. (1976). Analysis of tetrapod gaits: general considerations and symmetrical gaits. In *Neural control of locomotion*: 203–36. (Eds Herman, R.M., Grillner, S., Stein, P.S.G. & Stuart, D.G.). Plenum Press, New York.

Hildebrand, M. (1977). Analysis of asymmetrical gaits. *J. Mammal.* **58**:131–56.

Hildebrand, M. (1980). The adaptive significance of tetrapod gait selection. *Am. Zool.* **20**:255–67.

Howland, H.C. (1974). Optimal strategies for predator avoidance: the relative importance of speed and manoeuvreability. *J. theor. Biol.* **47**:333–50.

Jayes, A.S. & Alexander, R.McN. (1978). Mechanics of locomotion of dogs (*Canis familiaris*) and sheep (*Ovis aries*). *J. Zool., Lond.* **185**:289–308.

Jayes, A.S. & Alexander, R.McN. (1982). Estimates of mechanical stresses in leg muscles of galloping Greyhounds (*Canis familiaris*). *J. Zool., Lond.* **198**:315–28.

Jenkins, F.A. (1971). Limb posture and locomotion in the Virginia opossum (*Didelphis marsupialis*) and in other non-cursorial mammals. *J. Zool., Lond.* **165**:303–15.

Jenkins, F.A., Jr. (1972). Chimpanzee bipedalism: cineradiographic analysis and implications for the evolution of gait. *Science, N.Y.* **178**:877–9.

Larson, S.G. & Stern, J.T. (1987). EMG of chimpanzee shoulder muscles during knuckle-walking: problems of terrestrial locomotion in a suspensory adapted primate. *J. Zool., Lond.* **212**:629–55.

McMahon, T.A. (1973). Size and shape in biology. *Science, N.Y.* **179**:1201–4.

McMahon, T.A. (1984). *Muscles, reflexes and locomotion.* Princeton University Press, Princeton.

Margaria, R., Cerretelli, P., Anghemo, P. & Sassi, G. (1963). Energy cost of running. *J. appl. Physiol.* **18**:367–70.

Pennycuick, C.J. (1975). On the running of the gnu (*Connochaetes taurinus*) and other animals. *J. exp. Biol.* **63**:775–99.

Reynolds, T.R. (1987). Stride length and its determinants in humans, early hominids, primates, and mammals. *Am. J. phys. Anthrop.* **72**:101–15.

Seeherman, H.J., Taylor, C.R., Maloiy, G.M.O. & Armstrong, R.B. (1981). Design of the mammalian respiratory system. II. Measuring maximum aerobic capacity. *Respir. Physiol.* **44**:11–23.

Taylor, C.R., Heglund, N.C. & Maloiy, G.M.O. (1982). Energetics and mechanics of terrestrial locomotion. I. Metabolic energy consumption as a function of speed and body size in birds and mammals. *J. exp. Biol.* **97**:1–21.

Taylor, C.R., Maloiy, G.M.O., Weibel, E.R., Langman, V.A., Kamau, J.M.Z., Seeherman, H.J. & Heglund, N.C. (1981). Design of the mammalian respiratory system. III. Scaling maximum aerobic capacity to body mass: wild and domestic mammals. *Respir. Physiol.* **44**:25–37.

Taylor, C.R. & Weibel, E.R. (1981). Design of the mammalian respiratory system. I. Problem and strategy. *Respir. Physiol.* **44**:1–10.

Vilensky, J.A. (1980). Trot-gallop transition in a macaque. *Am. J. phys. Anthrop.* **53**:347–8.

*Symp. zool. Soc. Lond.* (1989) No. 61:181–196

# The reproductive biology of the male hippopotamus

Frederick I.B. KAYANJA

*Makerere University*
*P.O. Box 7062*
*Kampala, Uganda*

## Synopsis

Queen Elizabeth National Park in Uganda carried so many hippopotamuses during the early 1960s that it was essential to crop some. But as a result of turbulence and civil war in Uganda, the hippopotamus population has been drastically reduced during the past ten years.

This study, which confirmed that the reproductive strategy of hippopotamus bulls is mating territoriality, investigated their reproductive system. The hippopotamus bull has no pendulous scrotum. All adult bulls in this study had testes with active spermatogenesis. The intertubular connective tissue of the testis has varying amounts of fibrous tissue, depending on the location. Leydig cells are mainly lodged in the areas with the smallest amount of collagenous fibres, in which lymphatic and blood vessels are also found.

The head, body and tail of the epididymus form distinct morphological units. The findings indicate that the head has absorptive capacity, while the tail is a reservoir of semen with a large capacity. The body appears to have intermediate function.

A fibro-elastic penis with a well-developed sigmoid flexure resembles in many features that of the domestic ruminants.

## Introduction

The hippopotamus (*Hippopotamus amphibius* Linn.) is a unique ungulate, feeding mostly during the night on dry land, and spending the rest of the time, largely the hours of daylight, partly or completely submerged in the water. Its habits, including feeding, have been reported upon by several workers (Asdell 1946; Hediger 1951; Verheyen 1954; Curry-Lindahl 1961; Laws & Clough 1966; Olivier & Laurie 1974; Klingel in press).

Hippopotamuses devour large amounts of grass, and when they are present in large numbers (beyond the carrying capacity of the area), there is rapid deterioration of the habitat. They digest most of this food during the daylight hours when they are resting in the water. Large amounts of faeces are voided in the water, which becomes heavily manured and constitutes a vital component in a chain that supports a large biomass of fish-life. This is a highly significant point, since fish is a vital source of protein in Africa where protein for the human population is generally in short supply.

Figure 1 shows the very large number of hippopotamuses in Queen Elizabeth National Park (QENP) during the late 1950s. During the early 1960s it was found necessary to crop 7000 hippopotamuses in QENP since the carrying capacity had been exceeded and the deterioration of the habitat had to be arrested. The 1970s and 1980s (up to 1986) have been times of considerable turbulence in Uganda. The civil unrest and civil war have drastically affected the hippopotamus population. It is no longer relevant to talk about culling: it is now more realistic to strictly enforce the laws and by-laws that protect the remaining population.

When a population is drastically reduced, its reproductive potential should be fully assessed. This is one of the aims of this project, which has been slow in taking off because the restoration of law and order and security have been slow, as was to be expected. This is the first report and it focuses on the adult male.

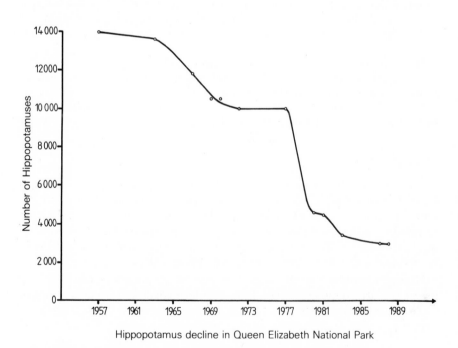

Hippopotamus decline in Queen Elizabeth National Park

**Fig. 1.** Hippopotamus population decline in Queen Elizabeth National Park.

## Materials and methods

Queen Elizabeth National Park is situated within the southern part of the Western Rift Valley (Albertine Rift) of East Africa. The valley runs along the length of Uganda on the west; it comprises the basins of Lakes Edward, George and Albert and the Rwenzori Mountain block. The park is bounded to the west by Lake Edward and extends northwards to enclose Lake George. The two lakes within the rift are linked by Kazinga Channel, which is the heart of the park, and the site of this research project.

Ten adult male hippopotamuses have so far been used in this study. Two identified territorial bulls were shot during the day. Each bull was driven out of the water onto the bank where it was shot. Three other males were also shot during the day but in wallows. The latter are small pools of water, about 6 m in diameter, which form during the rainy season and are almost completely surrounded by thickets. Five males were shot on land during the night.

Immediately after death the reproductive organs were dissected and recovered. Tissues for electron microscopy were fixed in buffered glutaraldehyde, post-fixed in osmium tetroxide, and embedded in epon araldite. Tissues for histology were fixed in Bouin's fluid or formol saline, and the paraffin sections were stained with haematoxylin eosin or Masson's trichrome stain. Semi-thin sections 0.5–1.0 μm in thickness, cut from the epon araldite tissue blocks, were stained with toluidine blue in borax and were also used for light microsocopy. Ultra-thin sections from the epon araldite tissue blocks were stained with uranyl acetate and lead citrate (Venable & Coggeshall 1965) and were examined with a Zeiss EM9S electron microscope.

## Results

Mating territorality, as described by Klingel (in press), working on the same population, was confirmed. Adult bulls occupy territories along the shoreline in which they are dominant over all conspecifics. In the territory, the dominant male has exclusive mating rights but these are open to challenge. The territories are deserted during the night when the entire hippopotamus population invades the land for feeding. The same dominant bull, however, returns to the same territory when the population returns to water before dawn.

During the night, when the hippopotamuses are grazing on land, there are no apparent social groups, but during the day two kinds of social group are seen. One is the 'school', dominated by an adult male bull that holds a territory, and containing not only females and calves, but also adult and subadult males, which behave subordinately to the dominant, territorial bull. The other is a bachelor group, consisting mainly of males.

During the wet season, some males can be found alone in wallows. Occasionally more than one male can be found in these rain-filled pools during the daylight hours. At the onset of the dry season, when the pools dry up, the solitary males may either join a 'school' and behave subordinately to the territorial bull, or they may join a bachelor group.

Neither of the groups is a stable social unit: they appear to be loose associations in a suitable resting-place (Klingel in press).

## The testis

In the hippopotamus, there is no pendulous scrotal sac. The skin covering the testes does not show the usual thin and delicate form associated with the scrotum. With the animal lying on its back, the testes can be easily palpated in the inguinal region.

The two territorial bulls (1 and 2 on Table 1) were the largest animals according to body weight and also had the heaviest testes. The average weight of the testis was 275 g and all ten bulls showed active spermatogenesis typical of the adult testis.

The hippopotamus testis is covered by a tough and fibrous tunica albuginea up to 1.6 mm thick. As in other mammals, septula testis originate from this tunic to subdivide the organ into lobuli testis in which the seminiferous tubules are lodged. Near the periphery of the testis, the connective tissue which surrounds the seminiferous tubules contains the largest amounts of collagenous fibres. Moving away from the tunica albuginea, the content of collagenous fibres is reduced so that loose connective tissue mainly supported by reticular fibre networks surrounds

**Table 1.** Measurements of hippopotamuses used in the study.

| Male No. | Body weight (kg) | Testis weight (g) | Epididymis weight (g) | Bulbo-urethral gland weight (g) | Penis weight (g) | Penis length (cm) |
|---|---|---|---|---|---|---|
| 1 | 1 550 | 300 | 150 | 115 | 2 240 | 120 |
| 2 | 1 420 | 290 | 146 | 115 | 2 080 | 115 |
| 3 | 1 250 | 280 | 140 | 110 | 1 400 | 108 |
| 4 | 1 276 | 285 | 142 | 112 | 1 500 | 110 |
| 5 | 1 290 | 280 | 145 | 110 | 1 500 | 110 |
| 6 | 1 207 | 220 | 80 | 100 | 1 300 | 104 |
| 7 | 1 222 | 282 | 120 | 100 | 1 308 | 110 |
| 8 | 1 280 | 280 | 138 | 110 | 1 450 | 110 |
| 9 | 1 346 | 287 | 145 | 118 | 2 000 | 115 |
| 10 | 1 150 | 250 | 140 | 110 | 1 385 | 110 |
| Average | 1 299 | 275 | 135 | 110 | 1 616 | 111 |

Males 1 and 2 were identified territorial bulls; males 3, 4 and 5 from wallows; males 6 to 10 obtained at night on land.

the seminiferous tubules. Towards the centre of the testis where the septula come together, the intervening connective tissue once again becomes more fibrous with increasing amounts of collagenous fibres. The central connective tissue is, however, less fibrous than that in the periphery.

Arterioles, capillaries and venules as well as lymphatic vessels are abundant, especially in the connective tissue carrying small amounts of collagenous fibres between the seminiferous tubules. Leydig cells do not occur in areas having large concentrations of collagenous fibres. Even in the loose intertubular connective tissue, Leydig cells do not occur in large numbers. They occur in small groups of two to five cells lying close to thin-walled vascular structures and especially lymphatic vessels (Fig. 2). Larger associations of Leydig cells are seldom found.

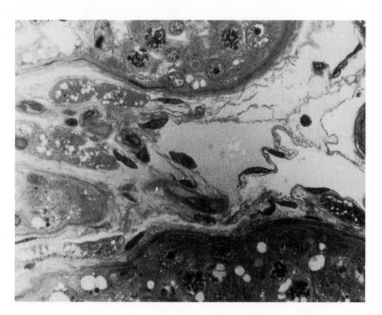

**Fig. 2.** Plastic semi-thin section from the testis. Note the Leydig cells, capillaries and lymphatic vessels between the seminiferous tubules. Toluidine blue, × 5872.

Leydig cells measure between 25 and 30 μm in diameter but are difficult to measure accurately since they are rarely spherical. Small lipid droplets can be seen in Leydig cells even in semi-thin toluidine blue-stained sections. These droplets are much smaller than those present in the seminiferous epithelium. As expected, Leydig cells contain abundant smooth endoplasmic reticulum. This organelle occurs in the form of random tubules, cisternae and whorls (Fig. 3).

Leydig cells contain moderate numbers of rather pleomorphic mito-

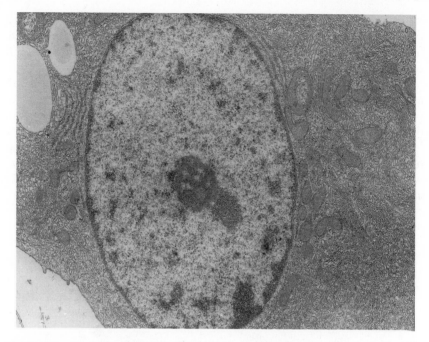

**Fig. 3.** Section through a Leydig cell. Note the smooth endoplasmic reticulum, mitochondria, granulated endoplasmic reticulum and lipid droplets, × 9094.8.

chondria carrying tubular cristae. The granulated or rough endoplasmic reticulum is not a prominent organelle but occurs in isolated areas. The nuclei of Leydig cells usually contain little heterochromatin and this is largely peripherally deployed. Usually one prominent nucleolus was identified. Some dense granules were present in the Leydig cells and these tended to occur to one side of the nucleus.

The observations made above were applicable to all ten hippopotamuses used in this investigation.

### Seminiferous epithelium

Only preliminary remarks can be made here since further studies are continuing on spermatogenesis in the hippopotamus. Cell divisions were obviously taking place in the seminiferous epithelium of all ten animals examined. The presence of lipid droplets in the seminiferous epithelium has already been mentioned. There were areas of the epithelium apparently free of lipid droplets. Where they occurred, the droplets were evenly and deeply stained in the semi-thin sections. The seminiferous epithelium of the two hippopotamuses obtained from wallows (Nos 4 and 5 in Table 1) contained slightly more lipid droplets than that of all the other bulls.

## Epididymis

The head, body and tail of the epididymis are relatively easy to identify morphologically. Histological sections show ducts surrounded by vascularized connective tissue in all parts of the epididymis.

The ductuli efferentes have a simple columnar epithelium 25 μm in height (Fig. 4). They are easily differentiated from the ducts of the initial segment which have an epithelium 41.67 μm in height (Table 2). This epithelium is largely of pseudostratified columnar appearance (Fig. 5). Two cell types form this epithelium. One has a clear appearance, while the other is darker-staining and its nucleus is richer in small clumps of heterochromatin. Vacuolation of the cytoplasm is an apparent and characteristic feature of the epithelial cells of the initial segment, as is the accumulation of mitochondria at the base of the cells (Fig. 6). The vacuoles are large and extend to all parts of the cell including the base of some of the cells, as seen in electron micrographs. A few cells were noted to contain a few granules that stained intensely with toluidine blue in semi-thin sections. A few smooth muscle cells are already assembled around the ducts.

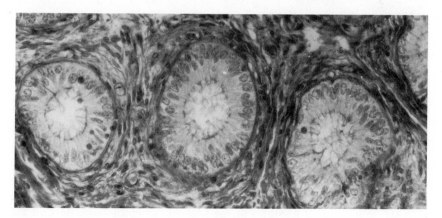

**Fig. 4.** Plastic section through the ductuli efferentes. Masson trichrome, × 2584.

**Table 2.** Epithelial height and duct diameter of the epididymis.

| Epididymal region | Mean epithelial height (μm) | Mean duct diameter (μm) |
|---|---|---|
| Initial segment | 41.67 ± 1.97 | 171.66 ± 20.41 |
| Head | 147.50 ± 26.41 | 485.00 ± 146.26 |
| Body | 114.66 ± 11.78 | 473.33 ± 37.24 |
| Tail | 66.00 ± 17.11 | 1210.00 ± 349.28 |

Data are mean ± S.D. N = 10 animals.

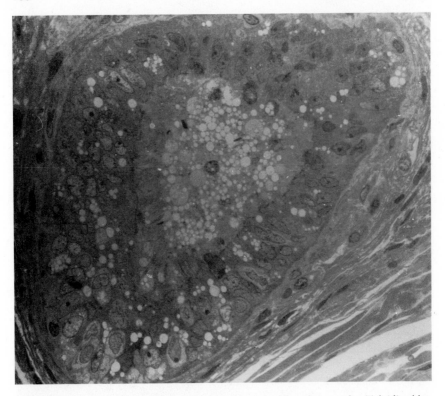

**Fig. 5.** Plastic section through a duct of the initial segment. Note the vacuoles. Toluidine blue, × 5872.

The epithelial height rises steeply to 147.50 μm in the rest of the head of the epididymis (Table 2). Basal and principal cells constitute this epithelium (Fig. 7). The apical third is largely free of nuclei and in electron micrographs vacuolation is observed (Fig. 8). Granules staining intensely in toluidine blue are seen in small numbers at the base of the epithelium. Beyond the smooth muscle cells, which surround the ducts, lies a well-vascularized connective tissue.

There is a slight fall in the average height of the epithelium in the body of the epididymis to 114.66 μm (Fig. 9). The largest ducts have the lowest epithelium. Vacuolation of the apical cytoplasm is greatly reduced and is absent in the terminal part of the body of the epididymis. Intraepithelial cells with the appearance of a lymphocyte are occasionally seen. The epithelial surface is often undulating and small intraepithelial canals opening into the lumen are most obvious in the body of the epididymis. Only a few cells contain basal granules that stained deeply with toluidine blue. In electron micrographs this part of the epididymal cells contains

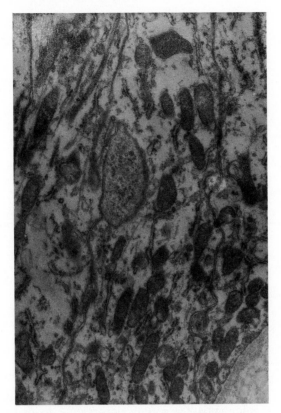

**Fig. 6.** Section through the base of cells of the initial segment. Note the mitochondria, endoplasmic reticulum and microtubules. × 11154.

cisternae of granular or rough endoplasmic reticulum. The smooth muscle layer around the ducts consists of fewer cell layers than it does in the head region. However, the connective tissue between the ducts of the body of the epididymis is less richly vascularized than in the head.

The epithelium of the tail of the epididymis measures only 66.0 μm in height although the greatest diameter is attained in this region (Table 2). The large dimension of the duct is associated with occasional infolding of the duct wall that involves the connective tissue. There is generally no apical vacuolation of the epithelial cells. The largest deposits of granules staining intensely in toluidine blue are found in tissues taken from this region (Fig. 10). The granules accumulate in the basal cytoplasm of the epithelial cells. This area is also rich in small rod-shaped mitochondria. In electron micrographs, tubules of smooth and granular endoplasmic reticulum are present. The toluidine blue-staining granules are membrane-bound bodies with a slight variation in the electron-density of the content in electron

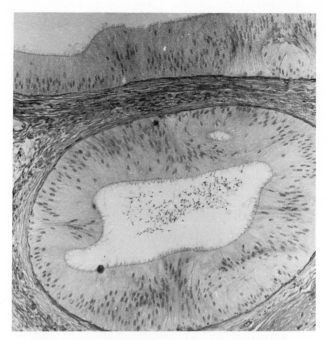

**Fig. 7.** Paraffin section through the ducts of the head of the epididymis. Masson trichrome, × 1175.

micrographs (Figs 11, 12). The smooth muscle coat is best developed in the tail region of the epididymis. The region appears to be divided into two parts. The first part of the tail consists of the widest ducts, while the second, smaller part, the terminal part, has an even thicker muscular wall but its ducts have a reduced luminal size.

### The bulbo-urethral gland

This is the only accessory gland reported upon so far in this study because the others remain under investigation. The gland is remarkable because of its large size (8 cm diameter, 110 g average weight). It is almost completely covered by striated muscle and the interior contains mucus-secreting elements (Fig. 13).

### The penis

The penis of the hippopotamus resembles that of the pig and domestic ruminants in having a sigmoid flexure and also in being fibro-elastic. It averages 111 cm in length and 1616 g in weight, with a cross-sectional diameter of about 3 cm. The s-bend of the sigmoid flexure averages 40 cm

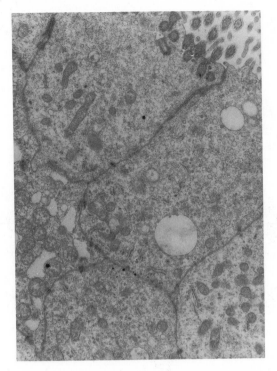

**Fig. 8.** Section through the apical region of the epithelial cells of the head of the epididymis. Note the microvilli, the cilia and the vacuoles, × 8085.

in length. In life the penis is sometimes extruded and when this happens on dry land the protruding terminal part has a curved, almost spiral shape.

The urethra, surrounded by a prominent cavernous tissue (corpus spongiosum urethrae), occupies a position slightly below the centre in a cross-section of the penis. Above the urethra are prominent vascular spaces of the cavernous tissue of the penis (corpus cavernosum penis) and the entire structure is enclosed in a tunica albuginea.

## Discussion

The difficult years, from 1971 to 1986, have taken a heavy toll of wildlife in Uganda. During the wars, soldiers and others turned their guns on mainly the large mammals, most species of which have declined drastically (Kayanja & Douglas-Hamilton 1982). The turbulence also left many guns in the hands of the population, and these continue to threaten the populations of large mammals in the country. The hippopotamus is not one of the species threatened by the international trade in valuable trophies. It is

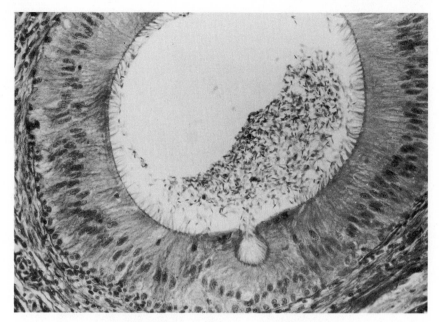

**Fig. 9.** Paraffin section through a duct from the body of the epididymis. Masson trichrome, × 2554.

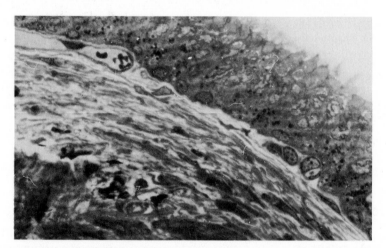

**Fig. 10.** Plastic section through part of a duct from the tail of the epididymis. Note the intensely stained granules. Toluidine blue, × 5872.

threatened because the difficult times have brought about economic hardship, responsible for high prices for meat from domestic stock. The turbulence has also caused a notable decline in the numbers of cattle, sheep

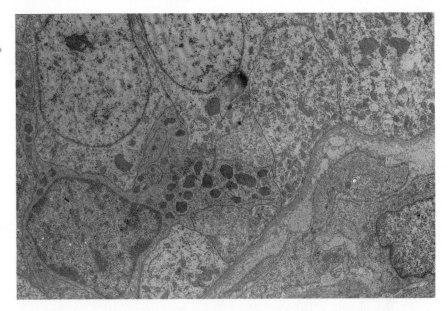

**Fig. 11.** Section through the base of the epithelial cells from the tail of the epididymis. Note the dense membrane-bound granules and the mitochondria, × 4180.

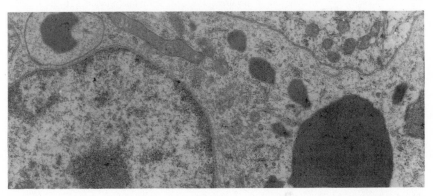

**Fig. 12.** Section through epithelial cells from the tail of the epididymis. Note the electron-dense granules which are membrane-bound, × 9438.

and goats in the country. Hippopotamus meat has therefore become an attractive alternative and this is the reason why the remaining population is threatened. Stabilizing the population and allowing it to recover would eventually allow a sustainable use of this source of meat for the people.

Klingel (in press) studied the same population but before the drastic decline shown in Fig. 1 had taken place. He recorded mating territoriality in this species. This investigation confirms his findings and also reveals that the

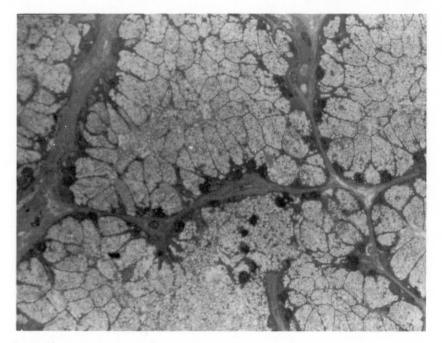

**Fig. 13.** Plastic section through the bulbo-urethral gland. Toluidine blue, × 5167.

strategy of mating territoriality has been retained after the population has come under tremendous pressure in which about 70% has been lost.

The findings of this investigation show that adult hippopotamus bulls are, at any time, able to take over the mating role of the territorial bull. Laws & Clough (1965) showed that there is no seasonal reproductive cycle in the adult male hippopotamus. There is, however, evidence that some of the bulls found in the wallows are the older animals and perhaps some displaced territorial bulls. If this is ultimately confirmed, it will be proved that successors are usually within the 'schools' or bachelor groups and are strategically placed to challenge the dominant or territorial bull.

It can be argued that a pendulous scrotum would interfere with the streamline effect of the body in the water. Whether this is true or not, the absence of the scrotum is apparently adequately compensated for. The testes still lie just under the skin and the hours of daylight, the hottest part of the day, are spent in the water.

According to comparative observations on intertubular lymphatics and the organization of the interstitial tissue of the mammalian testis, Fawcett, Neaves & Flores (1973) recognized three patterns. The hippopotamus belongs to the second category defined by these authors. To this category belong species in which clusters of Leydig cells are widely scattered in a loose connective tissue stroma drained by conspicuous lymphatic vessels.

Different morphological characteristics have been ascribed to the various segments of the epididymis. Dym (1977) suggested that each region of the epididymis had its own particular function. The vacuolation noted in the supra-nuclear cytoplasm of epithelial cells in the head of the epididymis indicates activity related to fluid absorption. This activity begins early in the initial segment, continues over the head and falls off where the body of the epididymis begins. The tail of the epididymis is a reservoir of spermatozoa and it has a large capacity in the hippopotamus. The first part of the tail seems uniquely adapted to this function by having ducts with the widest luminal dimensions. In fact the ducts are so wide that they are prone to artefactual folding of the wall during processing, as observed in some histological materials in this study. The second part or terminal portion of the tail has ducts with a narrower lumen and a thicker muscular wall. These structural attributes indicate that this portion plays a more active role in getting the spermatozoa into the ductus deferens during ejaculation. The exact role the body of the epididymis plays in the hippopotamus, although under investigation, remains elusive.

Membrane-bound dense bodies have been recorded as present in the basal part of the epithelium in all segments of the hippopotamus epididymis. The bodies are present in greatest numbers in the tail of the epididymis. This interesting observation warrants further investigation.

## Acknowledgements

Funds supporting much of the work so far done in this continuing project were obtained from the Research Grants Committee of Makerere University. I am indebted to the Board of Trustees of Uganda National Parks for granting permission for this work to be done in Queen Elizabeth National Park.

## References

Asdell, S.A. (1946). *Patterns of mammalian reproduction.* Comstock, New York.

Clough, G. (1966). *Reproduction in the hippopotamus Hippopotamus amphibius, Linn.* PhD thesis: University of Cambridge.

Curry-Lindahl, K. (1961). *Contribution à l'étude des vertébrés terrestres en Afrique tropicale.* Institut des Parcs Nationaux du Congo et du Ruanda, Bruxelles.

Dym, M. (1977). The male reproductive system. In *Histology* (4th edn): (Eds Weiss, L. & Greep, R.O.). McGraw-Hill Book Co., New York.

Fawcett, D.W., Neaves, W.B. & Flores, M.N. (1973). Comparative observations on intertubular lymphatics and the organization of the interstitial tissue of the mammalian testis. *Biol. Reprod.* 9:500–32.

Hediger, H. (1951). *Observations sur la psychologie animale dans les Parcs Nationaux du Congo Belge.* Institut des Parcs Nationaux du Congo Belge, Bruxelles.

Kayanja, F.I.B. & Douglas-Hamilton, I. (1982). The impact of the unexpected: the case history of Uganda National Parks. In *National parks: conservation and development. The role of protected areas in sustaining society*: 87–92. (Eds McNeely, J.A. & Miller, K.A.). IUCN & Smithsonian Institution Press, Washington.

Klingel, H. (In press). The social organisation and behaviour of *Hippopotamus amphibius.* In *African wildlife: research and management.* (Eds Kayanja, F.I.B. & Edroma, E.). (Symp. Uganda Institute of Ecology, Kampala, Uganda, December 1986). ICSU Publication.

Laws, R.M. & Clough, G. (1965). Observations on reproduction in the hippopotamus, *Hippopotamus amphibius. J. Reprod. Fert.* 9:369–70.

Laws, R.M. & Clough, G. (1966). Observations on reproduction in the hippopotamus *Hippopotamus amphibius* Linn. *Symp. zool. Soc. Lond.* No. 15:117–40.

Olivier, R.C.D. & Laurie, W.A. (1974). Habit utilization by hippopotamus in the Mara River. *E. Afr. Wildl. J.* 12:249–71.

Venable, J.H. & Coggeshall, R.E. (1965). A simplified lead citrate stain for use in electron microscopy. *J. Cell Biol.* 25:407–8.

Verheyen, R. (1954). *Monographie éthologique de l'hippopotame* (Hippopotamus amphibius *Linné*). Institut des Parcs Nationaux du Congo Belge, Bruxelles.

Symp. zool. Soc. Lond. (1989) No. 61:197–215

# Chemosensory investigation, flehmen behaviour and vomeronasal organ function in antelope

Benjamin L. HART,
Lynette A. HART and
John N. MAINA

*School of Veterinary Medicine
Department of Physiological
Sciences
University of California
Davis, CA 95616, USA*

*Department of Veterinary Anatomy
University of Nairobi
PO Box 30197
Nairobi, Kenya*

## Synopsis

During sexual encounters, ungulate males commonly investigate recently voided urine or the genitalia of females and often perform flehmen following such investigations. Flehmen behaviour is presumably involved in transferring fluid-borne chemical stimuli from the oral cavity to the vomeronasal organs (VNOs) for detection of chemical cues (sex pheromones) indicating a female's reproductive status. Field studies of Thomson's and Grant's gazelles (*Gazella thomsoni*; *G. granti*) and eland (*Taurotragus oryx*) quantitatively confirmed observations by previous investigators that females often urinate when being sexually pursued by males and that following their investigation and flehmen, the males terminate their sexual pursuit, presumably because the females have communicated their state of anoestrus. A difference between impala (*Aepyceros melampus*) females and the other antelope is that they do not urinate when being sexually pursued, perhaps because impala females can easily escape sexual pursuit of males by dashing into the tightly clustered herd of females or losing the male in light woodland brush. Eland devote about 90% of flehmen responses to genital contact or the urine stream of females they are pursuing, as contrasted with gazelle males devoting less than 10% of flehmen responses to these stimuli. The particularly close attention that eland males pay to females may reflect a response of herd bulls to possible competition from other males in the herd.

The transfer of fluid-borne stimuli to the VNOs from the oral cavity during flehmen takes place through the incisive ducts. There is evidence that the incisive papilla, located on the hard palate just behind the dental pad, is an important part of the mechanical process in propelling fluid into the incisive ducts. Although the papilla and ducts are generally found in ruminants, all species of alcelaphine

ZOOLOGICAL SYMPOSIUM No. 61
ISBN 0–19–854009–4

antelope examined, including topi (*Damaliscus lunatus*), Coke's hartebeest (*Alcelaphus buselaphus*), blesbok (*Damaliscus dorcas*), Hunter's hartebeest (*Beatragus hunteri*), blue wildebeest (*Connochaetes taurinus*) and black wildebeest (*Connochaetes gnou*), were found to lack the incisive papillae. All species except the two wildebeest species lacked patent incisive ducts. The ducts in wildebeest were very small. Observations on topi and hartebeest revealed they did not exhibit flehmen behaviour and, in fact, showed evidence of little, if any, chemosensory interest in females during sexual encounters, which suggests that some alcelaphines may be unique among ungulates in not using sex pheromones for communication of sexual status. The blue wildebeest, unlike the topi and hartebeest, did perform flehmen to urine from females. However, intermittent nostril-licking during flehmen apparently delivers the stimulus material to the VNOs by way of the nasal route, compensating for reduced oral access to the VNOs. Further research should be done on alcelaphine antelopes to understand the evolutionary implications of the anatomical and behavioural alterations in their chemosensory investigatory behaviour.

## Introduction

The descriptions of sexual encounters in various ungulates often refer to olfactory investigation by the male of the female genital area or recently voided urine. The implication is that the male is learning something about the reproductive status of the female in performing the behaviour. There is a common reference to the occurrence of lip curl, or what is now called flehmen, following such investigations. It is implied that this behaviour also is involved in evaluating the reproductive status of females. The impression often conveyed from such accounts is that while male courtship displays, female pursuit and even copulatory behaviour patterns may vary according to species, the behaviours of urine investigation and flehmen are virtually the same across species and have undoubtedly the same functional role in all species.

The work reported in this paper was undertaken with the goal in mind of documenting possible species differences in urine investigation and flehmen behaviour patterns in some common East African antelopes, namely Thomson's gazelle (*Gazella thomsoni*), Grant's gazelle (*G. granti*), impala (*Aepyceros melampus*), and eland (*Taurotragus oryx*), and to relate these findings to what differences had been known with regard to social structure. A second goal was to explore behavioural and anatomical questions arising from the reports that some alcelaphine antelope species did not display flehmen behaviour.

Our observations were conducted on the plains of Kenya and Rwanda where there is little interference by vegetation with visual observation of the subjects and where one can readily follow a herd of antelope for several hours in a four-wheel-drive vehicle. One can, in fact, drive onto a plain and choose among two or three species just which one to observe that day.

## Urine investigation and flehmen: the general picture

In a typical ungulate species, the male, after making contact with recently voided urine or the genital surface of females, elevates his head and curls or lifts his upper lip, retaining this posture for as little as 5 s to as long as 30 s or more. From recent work on goats it is known that this behaviour is involved in the transport of urine or vaginal secretions from the oral cavity through incisive ducts and into the vomeronasal organs (VNOs), presumably for the detection of chemical cues indicating a female's state of oestrus (Ladewig & Hart 1980; Melese d'Hospital & Hart 1985). We shall assume this concept holds for other ruminants as well as various species of antelopes.

At the present time, it is not possible to ascribe a particular role to the lip curl and head elevation during flehmen, but apparently the incisive papilla, located just behind the dental pad, plays a role in the movement of fluid from the oral cavity into the VNOs. Surrounding the incisive papilla there are moat-like grooves which lead to the incisive ducts (nasopalatine ducts) which communicate with the openings to the VNOs. Bulls stroke their tongues against the incisive papilla just prior to flehmen, and these tongue compressions are considered effective in forcing fluid along the grooves and into the incisive ducts where there is access to the VNOs (Jacobs, Sis, Chenoweth, Klemm, Sherry & Coppock 1980). There is a highly developed venous complex surrounding the lumen of the VNOs and, extrapolating from research on hamsters (Meredith & O'Connell 1979) and cats (Eccles 1982), it would appear that there is a vascular pumping apparatus surrounding the mucus-filled vomeronasal lumen that causes emptying and refilling of the lumen, much as squeezing and releasing a rubber bulb empties and fills a tube. The vascular morphology surrounding the VNO lumen of ruminants (Adams 1986; Kratzing 1971; Ladewig & Hart 1980) is as extensive as that of rodents and cats, strongly suggesting that a similar pumping mechanism operates during flehmen.

It has been proposed that flehmen behaviour and VNO chemosensory analysis might enable a male to detect impending oestrus (pro-estrus) in a female so that he could guard or tend her until she is in oestrus (Estes 1972). The VNOs could have the role of detection of small or rising concentrations of chemical cues (sex pheromones) of impending oestrus, thus allowing a male to discriminate between females approaching oestrus and those in anoestrus, a discrimination that is perhaps too difficult for main olfaction (B.L. Hart 1983). Also, through flehmen behaviour and vomeronasal chemosensory analysis, the detection of falling concentrations of sex pheromones could give a male information about how much longer a female will be in oestrus (B.L. Hart 1983).

## Urine investigation and flehmen in gazelles, impala and eland

The antelopes observed in this study all overlap in distribution and are found in plentiful numbers in East Africa. Most conceptions occur during one or two annual peaks, with some births throughout the year. Our work was carried out during one of the peaks of mating activity in the months of June and July. In the case of gazelles and impala, subjects selected for observation were territorial males which were in the company of one or more adult females. With eland, the dominant and subordinate males accompanying herds of females were observed. Male subjects were observed for 5–120 min, depending upon whether the females moved away or observations were arbitrarily terminated. Observation periods were counted only if the males displayed some sign of sexual activity towards females such as courtship, pursuit, attempted mounting, mounting, or copulation. The results of the study are reported in detail elsewhere (L.A. Hart & Hart 1987).

Female urinations occurring in the immediate vicinity of the focal male were categorized as provoked or unprovoked. Provoked urinations were those which occurred while a male pursued and displayed courtship postures towards females. What we refer to as provoked urinations are termed response urinations by Estes (1967) or demand urinations by Walther (1984). Unprovoked urinations were those made by females not being courted and occurring within the immediate vicinity of a male (estimated as being within 7 m of a male and within the frontal 180° of the visual field).

Provoked urinations in the male's vicinity are frequently followed by the performance of flehmen as depicted, for example, in eland in Fig. 1. Flehmen is also commonly displayed by a male after naso-oral contact with the vulva of a female he is pursuing. As illustrated in impala in Fig. 1, flehmen may be seen following olfactory investigation of unprovoked urinations, or investigation of ground material unidentified by the observer. Presumably, most unidentified ground material that is investigated comprises unobserved urinations. We have also occasionally noticed a male investigate an area where a female has been lying down for an hour or more, so that the stimulus was probably from the female's hair coat or skin rather than recently voided urine.

### Social significance of provoked urinations

One of the interesting species contrasts among antelopes involves the phenomenon of provoked urination. These urinations were common in gazelles and eland, but virtually absent in impala (Fig. 2). The significance of provoked urinations becomes apparent when one notices that gazelle and eland males usually do not continue to pursue females after investigating a

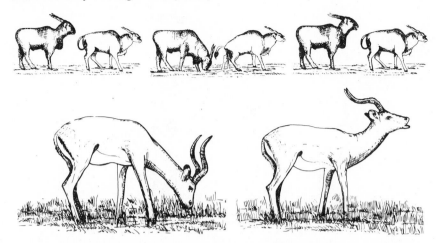

**Fig. 1.** Top, typical behavioural sequence of a male investigating and performing flehmen to a provoked urination from a female he is pursuing, as illustrated by eland. Bottom, flehmen to recently voided, but unprovoked, urine, as illustrated by impala.

provoked urination. In our study we found that only 0–20% of provoked urinations that were followed by flehmen resulted in the male continuing to pursue the female (Fig. 3). In other words, provoked urinations usually have the effect of terminating a male's pursuit. How can one interpret these data? On any particular day most of the females in a male's territory are not going to be in oestrus or near oestrus. But it is useful for a male to know if a female is nearing oestrus even though she acts in a non-receptive manner, and the

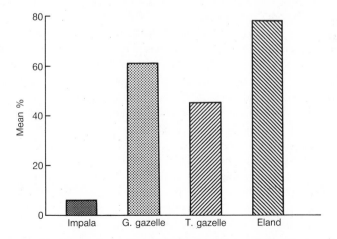

**Fig. 2.** Percentage of all observed urinations in the male's vicinity which were provoked. Impala differed markedly from Grant's (G.) gazelle, Thomson's (T.) gazelle and eland.

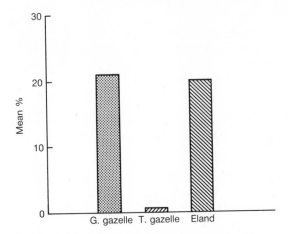

**Fig. 3.** Percentage of provoked urinations, followed by flehmen, in which the female was subsequently pursued, for Grant's (G.) gazelle, Thomson's (T.) gazelle and eland.

performance of flehmen and use of the VNOs presumably enable a male to discriminate impending oestrus from anoestrus. When a female is not near oestrus the provoked urinations allow her a method of repelling the male short of constantly butting him away or leaving his territory. Female urination terminates a male's sexual harassment, leaving the female freer to graze without interruption. For the male the information from provoked urination allows him more time to graze or watch for predators without missing the onset of oestrus. Although it has been previously reported that a male ungulate frequently stops pursuing a female after he has provoked her to urinate (Geist 1971; O'Brien 1982; Walther 1963), the sexual turn-off function of such urinations and the specific advantages to both the male and female of the chemosensory communication have not been highlighted before.

According to the above line of reasoning, one should expect to see occasionally some female urinations that initially turn on a male's sexual pursuit, as has been noted by others (Walther 1964). In most of our observations of mating behaviour, a mounting series was already in progress when the observations were initiated so we seldom observed the initial turn-on effect of a female urination. However, we did note that of the 20–30% of provoked urinations which were not investigated by gazelles, 60–70% occurred while gazelle males were already pursuing the females. To the observer, it appeared as though the male, by virtue of prior investigation, was responding to the female as if he had already determined she was in pro-oestrus and had decided to continue to pursue her, even ignoring further urinations.

Why do impala females not display provoked urinations when they are being sexually pursued by a male if this behaviour helps both the female and male by suppressing a male's unnecessary sexual pursuit? One of the things that has been noted in impala social behaviour is that female herds are relatively large and, compared to the gazelle and eland herds, quite tightly clustered (L.A. Hart & Hart 1987; Jarman & Jarman 1974; Walther 1972). Impala females can easily terminate a sexual pursuit by dashing into the tightly clustered group of females, a pattern we frequently observed. This contrasts with gazelles, where the herd size is typically smaller and much more spread out, and females would have a much harder time shaking off a male in hot pursuit. In impala, as with gazelles and eland, the male would still be able to obtain information by finding and flehmen-testing recently voided urine, which would alert him to the presence of a female worthy of watching.

### Social significance of stimuli that evoke investigation and flehmen

A fundamental difference between eland and the other antelopes in the style of investigating females is presented in Fig. 4 which depicts the percentage of flehmen episodes that are evoked by the different stimuli. The eland male distinguishes himself by very close attention to the females he is pursuing. A

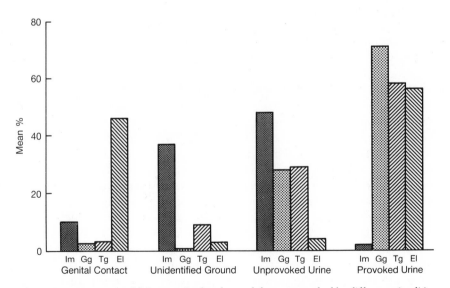

**Fig. 4.** Mean percentage of flehmen episodes observed that were evoked by different stimuli in impala (Im), Grant's gazelle (Gg), Thomson's gazelle (Tg) and eland (El). Impala differed from other antelope in percentage of flehmen responses to unidentified ground stimuli. Eland differed in percentage of flehmen responses to genital contact.

much higher percentage of total flehmen responses are evoked by genital contact than in any of the other species. Also, a high percentage of flehmen responses to provoked urinations were to the urine stream as it was being voided. Over 90% of flehmen responses in eland were directed to the urine stream or to genital contact. Less than 10% of flehmen responses in the gazelles were devoted to these stimuli. In the impala flehmen responses were not directed towards provoked urination, but a disproportionately higher percentage were evoked by unidentified ground stimuli than in other species. The impala were observed to devote occasional flehmen responses to genital contact.

Eland differ from the other antelope species in their social structure. Although there is customarily just one dominant herd bull in a large herd, additional smaller bulls of breeding age are common. The other bulls test urine and perform flehmen and do engage in mounting females when not directly inhibited by the herd bull. Since a female in impending oestrus, or in oestrus, could get lost in a large herd, and since there are potential competing males around, there is an advantage to the dominant male in detecting and closely following such females. One might speculate that stimulation of the VNOs in eland acts to reinforce sexual interest in a way perhaps not unlike that in guinea pigs, where reinforcing powers of VNO stimulation seem especially prominent (Beauchamp, Wysocki & Wellington 1985). The maintenance of sexual interest by pheromones would help in keeping the eland herd bull in pursuit of the female. Where competing males are not close, as in the other antelope, the reinforcement function of VNO stimulation (in addition to the chemosensory detector function) may not be adaptive.

### Chemosensory investigation and social structure: conclusions

The differences in chemosensory investigation between the species of antelopes described here can be related to constraints of social structure. Provoked urinations appear to be not only a means by which females communicate oestrus or impending oestrus, but also a mechanism that females use to turn off a male's sexual harassment when they are not in or near oestrus. This behaviour is common in gazelles and eland. However, in impala provoked urinations do not occur, and we have suggested that this reflects the ease with which impala females can shake off a male's sexual pursuit by dashing into the tightly clustered herd of other females. Females of gazelles and eland do not cluster tightly. Whereas the breeding males of gazelle and impala are territorial, breeding eland males reside in a large herd of non-territorial females where there are competing, albeit subordinate, males. We have pointed out that the eland male distinguishes himself from males of the other antelopes studied by his high percentage of flehmen responses to genital contact and the urine stream. This extremely close chemosensory attention to females may be related to the presence of potentially competing males in the same herd.

## Male antelope that do not perform flehmen: anatomical alteration of the vomeronasal system in alcelaphine antelopes

In pursuing the role of chemical cues in co-ordinating sexual interactions, we were fascinated with the prospect of studying a group of antelope that were reported not to display flehmen behaviour. It had been noted that flehmen was not performed by topi (*Damaliscus lunatus*) (Jewell 1972; Joubert 1975; Monfort-Braham 1975), bontebok (*Damaliscus dorcas dorcas*) (David 1973, 1975), blesbok (*Damaliscus dorcas phillipsi*) (Lynch 1971), and Coke's hartebeest (*Alcelaphus buselaphus*) (Backhaus 1959; Gosling 1975), all members of the alcelaphine tribe of antelopes. Along with the blue wildebeest (*Connochaetes taurinus*), the topi and hartebeest are the most common species of alcelaphine antelopes that one sees on a typical safari to East Africa. However, the male wildebeest, according to Estes (1969), displays flehmen, not only to females that he sexually pursues, but also to the urine of conspecific males during agonistic encounters.

There are a number of interesting questions that arise regarding alcelaphine antelopes and use of the VNOs. If topi and hartebeest do not perform flehmen, and presumably do not use the VNOs for detection of sex pheromones, do they display increased olfactory interest in females? Alternatively, could the absence of flehmen indicate that chemosensory information is less important in the sexual behaviour of these species? Another question is whether there are any anatomical characteristics of the alcelaphine VNO system that reflect the absence of flehmen behaviour in several of the species? If there is an anatomical correlate to the absence of flehmen, what about the VNO system in the wildebeest, which does perform flehmen? The results of this study are reported in detail elsewhere (B.L. Hart, Hart & Maina 1988).

### Vomeronasal organ system in alcelaphine antelopes

In the typical ruminant, access to the VNOs from the oral cavity is provided through the incisive papilla and incisive ducts. In most ungulates there is also a nasal connection to the VNOs, leading from the ventral-rostral extent of the nasal cavity. The incisive papillae of non-alcelaphine antelopes, represented by impala, Grant's gazelle, Thomson's gazelle and eland are shown in Fig. 5. Alcelaphine specimens from the East African plains and the San Diego Zoo were examined. Photographs were taken of the same areas of the hard palate of alcelaphines, where one would expect to see incisive papillae, and these specimens (previously unpublished) are presented in Figs 5 and 6. The blue wildebeest and black wildebeest (*Connochaetes gnou*) are

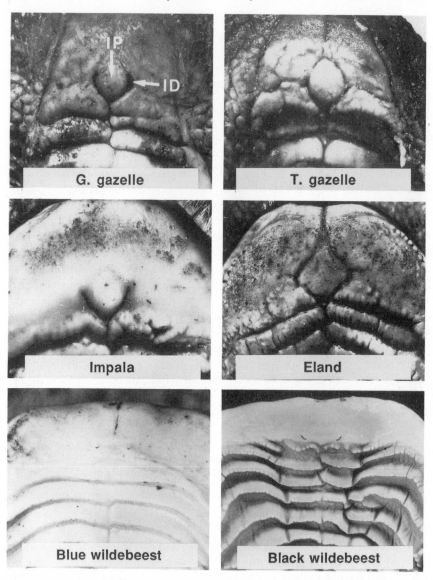

**Fig. 5.** Photographs of the dental pad and hard palate showing normal incisive papilla (P) and incisive ducts (ID) as represented by Grant's (G.) gazelle, Thomson's (T.) gazelle, impala and eland. The absence of an incisive papilla but presence of small incisive ducts are shown in blue wildebeest and black wildebeest.

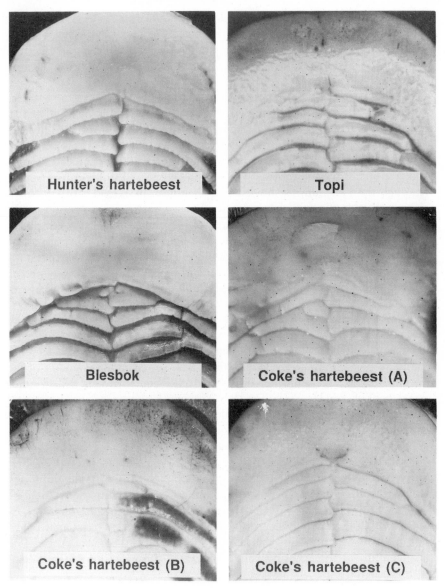

**Fig. 6.** Photographs of the dental pad and hard palates showing absence of incisive papilla and incisive ducts in Hunter's hartebeest, topi and blesbok. In Coke's hartebeest photographs illustrate the variability in the incisive papilla area showing no signs of incisive papilla or ducts in Specimen A, small indentations where incisive ducts would be in Specimen B and small grooves suggestive of the moat around the incisive papilla in Specimen C.

shown in Fig. 5. In Fig. 6 are shown Hunter's hartebeest (*Beatragus hunteri*), blesbok and topi. Coke's hartebeest was the only species found to have some variability in the incisive papilla area and three specimens from this species are shown in Fig. 6. Our survey of hard palates included all species of alcelaphines, according to most classifications, except Lichtenstein's hartebeest (*Alcelaphus lichtensteini*).

None of the alcelaphine antelopes was found to have an incisive papilla. Gross examination of the hard palate of Coke's hartebeest revealed in some individuals no signs of incisive ducts or a papilla, in others small indentations where the incisive ducts would be and, in still others, grooves suggestive of the moat around the papilla (Fig. 6). Histological sectioning through the indentations showed that they did not extend through to the VNOs. In the two wildebeest species, small canals in the position of the incisive ducts were seen and histological sectioning revealed small patent ducts in most specimens, communicating with the VNOs. The size of the canals was smaller than in non-alcelaphine antelopes. In the wildebeest the transportation of fluid-borne stimuli to the VNOs during flehmen would be very minimal in comparison to non-alcelaphine species, not only because of the small size of the ducts, but also because of the absence of a papilla.

From decalcified histological sections through the middle of the VNOs of topi, Coke's hartebeest and blue wildebeest, the VNOs looked quite comparable to the other non-alcelaphine antelopes examined, including eland, impala, Grant's gazelle and Thomson's gazelle. Our estimates of the lengths of the VNOs did not differ in alcelaphine antelopes and in other antelopes of similar size. The VNOs would appear to be functional chemosensory organs in the alcelaphine antelopes, since there was a clear connection between the ventral-rostral end of the nasal cavities and the VNOs, as was found in non-alcelaphine antelopes. The anatomical alteration in the alcelaphine VNO system corresponds to the reports that topi, hartebeest, blesbok and bontebok do not display flehmen. If the animals have no incisive papillae and no incisive ducts, then there would be no reason to perform flehmen. There remained the difficulty, however, of understanding why the wildebeest, which share the same basic anatomical arrangement as topi and hartebeest, do perform flehmen. A possible explanation of this disparity became evident during our field studies of flehmen behaviour.

### Flehmen behaviour and olfactory investigation in alcelaphine antelopes

Behavioural observations focused upon territorial topi, hartebeest and wildebeest males that displayed signs of sexual interest in females. Observations of the male's responses in investigating and performing flehmen to female urinations occurring near the male were noted as in previous observations of impala, gazelle and eland. With topi and hartebeest, provoked urinations did not occur and topi and hartebeest males never displayed flehmen to female urinations nearby. However, wildebeest males did investigate and perform flehmen at a rate typical of the gazelle, impala and eland studied previously (Fig. 7). These observations confirm, in a quantitative manner, the comments by others that topi and hartebeest appeared never to perform flehmen behaviour.

What is perhaps surprising is that not only did topi and hartebeest males not perform flehmen to urinations, but they showed very little inclination to perform olfactory investigation of recently voided female urine, even if the urination occurred immediately in front of them. This, too, was very different from the wildebeest and other antelope, as seen in Fig. 7. The lack of interest in female urine by male topi and tsessebe (*D. lunatus lunatus*) and by bontebok was previously alluded to by Jewell (1972), Joubert (1975) and David (1973, 1975) respectively.

Typically a non-alcelaphine male antelope will engage in genital investigation and perform flehmen at least once either before or after mounting or copulating with a female. In our observations of copulating hartebeest and topi, on no occasion were males found to nuzzle, lick, or investigate the genital area prior to mounting. Some genital sniffing and licking were seen after copulation, but flehmen was never performed. Joubert (1975) also mentions that he did not see tsessebe males muzzle the base of the tail or lick the vulva but Jewell (1972) did note some nosing and

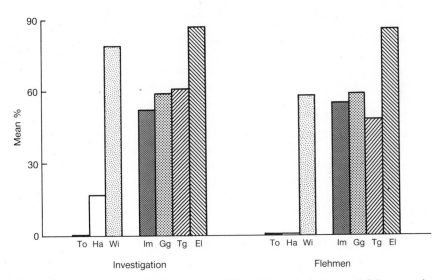

**Fig. 7.** Mean percentage of female urinations followed by investigation and flehmen in the alcelaphines, topi (To), hartebeest (Ha) and wildebeest (Wi) and the non-alcelaphines, impala (Im), Grant's gazelle (Gg), Thomson's gazelle (Tg) and eland (El).

nibbling of the tail base and vulva of females in oestrus by male topi. There appears to be no compensatory enhancement of olfactory investigation in topi and hartebeest to make up for the absence of VNO chemosensory function associated with flehmen. Certainly there is no indication that topi and hartebeest males are using the nasal route to expose the VNOs to female genital or urinary stimuli. The qualitative and quantitative observations point to the absence of flehmen in topi and hartebeest during sexual encounters as a reflection of an overall reduction in chemosensory communication. Alcelaphine antelope do use chemical communication in other contexts such as territorial scent marking (Gosling 1987). At times, we have observed topi and hartebeest to show olfactory interest in females. Males will frequently sniff areas where females have been lying down. Similar patterns have been seen by others in tsessebe (Joubert 1975), topi (Jewell 1972) and bontebok (David 1973).

The behaviour of the wildebeest represents perhaps what one would expect to see in an antelope that has the alcelaphine VNO system anatomy but retains a chemosensory interest in females. Wildebeest fall in the mid-range of non-alcelaphine antelopes in frequency of investigation of female urine and performance of flehmen (Fig. 7). Additionally, wildebeest males do nuzzle and lick the genitalia of females they are sexually pursuing. However, during flehmen we have recorded that males put their tongues into the external nares about 80% of the time. This tonguing occurs intermittently with the flehmen lip curl. Presumably these tongue movements deliver urine samples or vaginal secretions to the nasal cavity for access to the VNOs. Although such tongue movements occasionally occur in non-alcelaphine antelopes, the nostril-licking follows the flehmen episodes rather than occurring intermittently. These intermittent tongue movements are an aspect of flehmen behaviour in the wildebeest that differs from flehmen behaviour in other antelope that have no incisive papilla and gives the wildebeest a way to bypass the normal oral route with an efficiency that compensates for the lack of incisive papillae.

Another perspective on the interpretation of chemosensory investigatory behaviour in topi and hartebeest comes from observations by others on equids. Like topi and hartebeest, equids lack the incisive duct openings and thus have no oral access to the VNOs. However, stallions closely sniff female urine and routinely perform flehmen; but, unlike flehmen in ruminants, it is apparently not preceded by direct lip or tongue contact with urine (Crowell-Davis & Houpt 1985). It would appear as though stallions take volatiles into the nasal cavity during close investigation and the stimuli subsequently move into the VNOs from the nasal cavity during flehmen. Thus, in both equids and wildebeest, there is close and conspicuous nasal investigation of urine as well as flehmen, providing further evidence that chemosensory investigation is of little or no importance in male topi and hartebeest. One would expect that female topi and hartebeest do not produce in urine or vaginal secretions the classical sex pheromones that are associated with mammalian chemosensory communication.

### Visual and tactile communication during sexual encounters in topi and hartebeest

How do male topi and hartebeest determine the state of oestrus in females that enter their territories? We would propose that visual and tactile signals are particularly important. Males typically approach females with a pronounced neck-stretch or low-stretch posture. This posture is seen in topi (B.L. Hart *et al.* 1988; Jewell 1972; Joubert 1975), Coke's hartebeest (Backhaus 1959; Gosling 1975; B.L. Hart *et al.* 1988), bontebok (David 1973, 1975), and blesbok (Lynch 1971) as well as in a number of non-alcelaphine antelopes (Walther 1084). We have observed that upon seeing the male neck-stretch, females usually move a metre or so and the male then terminates his approach. The neck-stretch posture may be presented to several females successively. Unless receptive, each female moves a short distance in response to the posture. If a female does not move upon the initial display of the neck-stretch, the male may nudge his muzzle against the rear quarters of the female, in which case the female moves away a bit more suddenly unless she is receptive. These nudges are not necessarily to the perineal or ano-genital area. When females are sexually receptive, the male can be observed to either mount directly with no genital investigation or to briefly nudge the female's hindquarters with his muzzle and immediately mount.

The neck-stretch posture could be interpreted as a sign that the male is making an attempt to sniff the female, and this has been the interpretation given to this behaviour by some investigators (Jewell 1972; Joubert 1975). Although males may acquire some olfactory information from this behaviour it is not followed by conspicuous genital sniffing or nuzzling as seen in other ruminants. The interpretation offered here is that the neck-stretch is not related to olfaction but rather to visual testing of a female's state of oestrus, a low-pressure method which carries minimal risk of driving her off the male's territory.

The main point to be made in concluding this section is that behavioural observations by a variety of investigators on alcelaphine antelopes of the *Damaliscus* and *Alcelaphus* genera are consistent with the picture that these antelopes do not employ a means to deliver stimuli to the VNOs during sexual encounters such that the VNOs could be used to detect oestrus or pro-oestrus. Thus, the VNOs in these animals are perhaps primarily used in a non-sexual context. Some functional and evolutionary implications of this are discussed below.

## Evolutionary significance of the alcelaphine chemosensory system

The peculiar evolution of the alcelaphine VNO system, the elimination of flehmen behaviour, and the de-emphasis on chemosensory investigation in antelopes of the *Damaliscus* and *Alcelaphus* genera deserve some speculation. It seems likely that there was a remote ancestor that had the typical ruminant incisive papilla and ducts leading to the VNOs, since alcelaphine speciation is rather recent (Gentry 1978; Vrba 1979, 1984). The impala, which is reported to be a descendant of the early alcelaphine stem line (Kingdon 1982; Vrba 1984), has the incisive papilla and ducts and displays investigation of urine and flehmen. In fact, the existence of the incisive papilla in impala is reason to classify the impala as a non-alcelaphine, such as a member of the antilopine tribe, rather than as a member of the alcelaphine tribe as Gentry (1978) and Kingdon (1982) classify it. However, in one respect, that of the absence of provoked urinations, impala are more like topi and hartebeest than gazelles and other antelope.

Hunter's hartebeest is considered to be the most closely related of all extant alcelaphines to the stem ancestor of alcelaphine antelope (Kingdon 1982) and, like the topi, hartebeest and blesbok, has no incisive papilla or ducts. Perhaps somewhere in alcelaphine evolution there was an advantage in anatomically restricting VNO function to analysis only of nasal cavity fluid. The fluid film that lines the nasal cavity epithelium undoubtedly contains dissolved volatiles from environmental chemicals and these chemicals would have access to the VNOs through the nasal route. One hypothesis is that there was some disadvantage to early alcelaphines in allowing oral fluids to contaminate nasal-borne stimuli because of enhanced reliance on the regulation of some neuroendocrine processes by changes in environmentally-produced chemicals. During the rapid radiation and speciation of alcelaphines 2–3 million years ago (Vrba 1979, 1984), there could have been selection against an oral access to the VNO. If this was a distinct advantage, those individuals with the least oral contamination would be the most appropriately regulated. East African day-length changes are relatively minor and probably of no particular value in regulating reproductive function. Seasonal changes in vegetation growth are brought on by changes in rainfall. An individual's reproductive cycle could be modulated by changes in certain chemicals produced by different stages of plant growth. To cue these changes very accurately by using the VNOs might give an animal a particular advantage over one that does not use the VNO chemosensory system for such an exclusive function. The precedent for such regulation exists in rodents where reproductive cycles have been shown to be altered by VNO-mediated chemical signals (Vandenbergh 1983; Wysocki 1979).

Alcelaphine antelopes were in the past among the most successful ungulate species in the evolution of the East African fauna, as documented by palaeontological findings. Representation of these species at the present time is only a fraction of their distribution and numbers in preceding ages (Kingdon 1982). The wildebeest is said to represent one of the more recently derived species in alcelaphine evolution (Kingdon 1982; Vrba 1979, 1984). This species could have evolved the need to detect the chemical cues related to a female's sexual status and acquired the behaviour of using the tongue to transfer fluid-borne urine or genital stimuli to the nasal access of the VNOs. Alternatively, the wildebeest may have been a transitional stage between non-alcelaphine and alcelaphine antelopes, as suggested more recently by E.S. Vrba (personal communication).

Finally, to truly understand the function of the VNO system in alcelaphine antelopes, and to extrapolate these findings to some aspects of VNO function that have not yet been explained in other ungulates throughout the world, requires that we study these animals in the habitat in which they evolved. A concerted conservation effort is needed to maintain both adequate numbers of the alcelaphine antelopes and protection of their diminishing natural habitat.

## Acknowledgements

Financial support was provided by NSF grant BNS81–03574. Participants of the University Research Expeditions Program, University of California, Berkeley, provided field and financial assistance. We appreciate the co-operation of the Kenyan Wildlife Conservation and Management Department in the Ministry of Tourism and Wildlife, and the sponsorship of Professor G.M.O. Maloiy of the University of Nairobi. Further assistance was provided by Warden John Muema of Hell's Gate National Park, John Wright of Oserian, and Cesare Bellingeri of Kongoni Farms. Rwanda arrangements were hosted by Nicole and Alain Monfort and were made possible by the Director of the Rwandan Department of Tourism and National Parks. Anatomical material from the Hunter's hartebeest, blesbok and black wildebeest was supplied by Dr Marilyn Anderson of the San Diego Zoo from animals that had died of natural causes.

## References

Adams, D.R. (1986). The bovine vomeronasal organ. *Arch. Histol. Jpn* **49**:211–55.
Backhaus, D. (1959). Beobachtungen über das Freileben von Lelwel-Kuhantilopen (*Alcelaphus buselaphus lelwel*, Heuglin 1877) und Gelengenheitsbeobachtungen an Sennar-Pferdeantilopen (*Hippotragus equinus bakeri*, Heuglin 1863). *Z. Säugetierk.* **24**:1–34.

Beauchamp, G.K., Wysocki, C.J. & Wellington, J.L. (1985). Extinction of response to urine odor as a consequence of vomeronasal organ removal in male guinea pigs. *Behav. Neurosci.* **99**:950–5.

Crowell-Davis, S. & Houpt, K.A. (1985). The ontogeny of flehmen in horses. *Anim. Behav.* **33**:739–45.

David, J.H.M. (1973). The behaviour of the bontebok, *Damaliscus dorcas dorcas*, (Pallas 1766), with special reference to territorial behaviour. *Z. Tierpsychol.* **33**:38–107.

David, J.H.M. (1975). Observations on mating behaviour, parturition, suckling and the mother-young bond in the bontebok (*Damaliscus dorcas dorcas*). *J. Zool., Lond.* **177**:203–23.

Eccles, R. (1982). Autonomic innervation of the vomeronasal organ of the cat. *Physiol. Behav.* **28**:1011–5.

Estes, R.D. (1967). The comparative behavior of Grant's and Thomson's gazelles. *J. Mammal.* **48**:189–209.

Estes, R.D. (1969). Territorial behavior of the wildebeest (*Connochaetes taurinus* Burchell, 1823). *Z. Tierpsychol.* **26**:284–370.

Estes, R.D. (1972). The role of the vomeronasal organ in mammalian reproduction. *Mammalia* **36**:315–41.

Geist, V. (1971). *Mountain sheep: a study in behavior and evolution.* University of Chicago Press, Chicago and London.

Gentry, A.W. (1978). Bovidae. In *Evolution of African mammals*: 540–72. (Eds Maglio, V.J. & Cooke, H.B.S.). Harvard University Press, Cambridge, Mass. & London.

Gosling, L.M. (1975). *The ecological significance of male behaviour in Coke's hartebeest*, Alcelaphus buselaphus cokei, *Günther.* Ph.D. Thesis: University of Nairobi.

Gosling, L.M. (1987). Scent marking in an antelope lek territory. *Anim. Behav.* **35**:620–2.

Hart, B.L. (1983). Flehmen behavior and vomeronasal organ function. In *Chemical signals in vertebrates* **3**:87–103. (Eds Silverstein, R.M. & Muller-Schwarze, D.). Plenum Press, New York.

Hart, B.L., Hart, L.A. & Maina, J. (1988). Alteration in vomeronasal system anatomy in alcelaphine antelopes: correlation with alternation in chemosensory investigation. *Physiol. Behav.* **42**:155–69.

Hart, L.A. & Hart, B.L. (1987). Species-specific patterns of urine investigation and flehmen in Grant's gazelle (*Gazella granti*), Thomson's gazelle (*G. thomsoni*), impala (*Aepyceros melampus*) and eland (*Taurotragus oryx*). *J. comp. Psychol.* **101**:299–304.

Jacobs, V.L., Sis, R.F., Chenoweth, P.J., Klemm, W.R., Sherry, C.J. & Coppock, C.E. (1980). Tongue manipulation of the palate assists estrus detection in the bovine. *Theriogenology* **13**:353–6.

Jarman, P.J. & Jarman, M.V. (1974). A review of impala behavior and its relevance to management. In *The behaviour of ungulates and its relation to management*: 871–81. (Eds Geist, V. & Walther, F.). IUCN, Morges, Switzerland. (*IUCN Publs* (N.S.) No. 24.)

Jewell, P.A. (1972). Social organisation and movements of topi (*Damaliscus korrigum*) during the rut, at Ishasha, Queen Elizabeth Park, Uganda. *Zool. Afr.* 7:233–55.

Joubert, S.C.J. (1975). The mating behaviour of the tsessebe (*Damaliscus lunatus lunatus*) in the Kruger National Park. *Z. Tierpsychol.* 37:182–91.

Kingdon, J. (1982). *East African mammals. An atlas of evolution in Africa* 3 parts C and D (Bovids). Academic Press, London, New York & San Francisco.

Kratzing, J. (1971). The structure of the vomeronasal organ in the sheep. *J. Anat.* 108:247–60.

Ladewig, J. & Hart, B.L. (1980). Flehmen and vomeronasal organ function in male goats. *Physiol. Behav.* 24:1067–71.

Lynch, C.D. (1971). *A behavioral study of blesbok*, Damaliscus dorcas phillipsi, *with special reference to territoriality*. M.Sc. Thesis: University of Pretoria.

Melese d'Hospital, P.Y. & Hart, B.L. (1985). Vomeronasal organ cannulation in male goats: evidence for transport of fluid from oral cavity to vomeronasal organ during flehmen. *Physiol. Behav.* 35:941–4.

Meredith, M. & O'Connell, R.J. (1979). Efferent control of stimulus access to the hamster vomeronasal organ pump. *J. Physiol.* 286:301–16.

Monfort-Braham, N. (1975). Variations dans la structure sociale du topi, *Damaliscus korrigum* Ogilby, au Parc National de l'Akagera, Rwanda. *Z. Tierpsychol.* 39:332–64.

O'Brien, P.H. (1982). Flehmen: its occurrence and possible functions in feral goats. *Anim. Behav.* 30:1015–9.

Vandenbergh, J.G. (Ed.) (1983). *Pheromones and reproduction in mammals.* Academic Press, New York.

Vrba, E.S. (1979). Phylogenetic analysis and classification of fossil and recent Alcelaphini. Mammalia: Bovidae. In *Living fossils*:62–79. (Eds Eldredge, N. & Stanley, S.M.). Springer Verlag, New York, Berlin etc.

Vrba, E.S. (1984). Evolutionary pattern and process in the sister-group Alcelaphini-Aepycerotini (Mammalia: Bovidae). *Biol. J. Linn. Soc.* 11:207–28.

Walther, F.R. (1963). Einige Verhaltensbeobachtungen am Dibatag (*Ammodorcas clarkei* Thomas, 1891.) *Zool. Gart. Lpz.* 27:233–61.

Walther, F.R. (1964). Verhaltensstudien an der Gattung *Tragelaphus* de Blainville, 1816 in Gefangenschaft, unter besonderer Berücksichtigung des Sozialverhaltens. *Z. Tierpsychol.* 21:393–467.

Walther, F. (1972). Social grouping in Grant's gazelle (*Gazella granti* Brooke 1827) in the Serengeti National Park. *Z. Tierpsychol.* 31:348–403.

Walther, F.R. (1984). *Communication and expression in hoofed mammals.* Indiana University Press, Bloomington.

Wysocki, C.J. (1979). Neurobehavioral evidence for the involvement of the vomeronasal system in mammalian reproduction. *Neurosci. Biobehav. Rev.* 3:301–41.

*Symp. zool. Soc. Lond.* (1989) No. 61:217–240

# African trypanosomiasis in wild and domestic ungulates: the problem and its control

Max MURRAY
and A.R. NJOGU

*Department of Veterinary Medicine*
*Veterinary School*
*University of Glasgow*
*Glasgow G61 1HQ, UK*

*Kenya Trypanosomiasis Research*
*Institute*
*P.O. Box 362*
*Kikuyu, Kenya*

## Synopsis

One of the major constraints on the expansion of the livestock and agricultural industries in Africa is tsetse-transmitted trypanosomiasis. The disease affects all domestic ungulates, as well as man, and direct and indirect losses are many billions of dollars annually. On the other hand, while wild ungulates can become infected with trypanosomes, most appear to be resistant to the effects of infection and act as asymptomatic carriers of both the domestic animal and human infective forms of the parasite, as well as providing an important source of blood meals for tsetse. Because of the phenomenon of antigenic variation, no vaccine is available and, at present, control is limited to tsetse control or to the use of trypanocidal drugs. Although these approaches can be successful, the costs can be high and the logistics complex, with the result that their use is limited and the threat of infection to both domestic animals and humans is not diminishing and, in some areas, is increasing. Nevertheless, recent advances in our understanding of tsetse and trypanosome biology and of genetic resistance to the disease, and in the strategic use of drugs, have meant that new efforts are being and will be undertaken to control tsetse-transmitted trypanosomiasis and permit better utilization of the vast areas of Africa held captive by the tsetse. Integrated multidisciplinary planning and research will be required to ensure that domestic and wild ungulates are not competing for resources but play a complementary role in the future socio-economic development of Africa.

## Introduction

The current world human population is estimated at 4.2 billion and is expanding by about 2% each year, i.e., by about 70 to 80 million people. To

ZOOLOGICAL SYMPOSIUM No. 61
ISBN 0–19–854009–4

deal with this increase, an additional 30 million tonnes of staple foods will be needed annually. One of the major ways in which livestock production could contribute to this need is by reduction of losses due to disease. It is estimated that a 2% reduction in disease loss would provide food for an additional 80 million people (Gavora 1982).

The situation is most serious in developing countries where 75% of the world's population live. It is estimated tht 65 to 70% of the world's livestock resources exist in these regions, yet they account for only 30% of meat output (Murray & Gray 1984). Nowhere is the problem greater than in Africa where the plight of the people, particularly in the Sahelian regions of West Africa, in Sudan, in Ethiopia and Mozambique, is so pitiful. In Africa over the last five years, the human population has increased at twice the rate of food production and by the year 2000 will have risen from the current estimated 470 million to 877 million (Trail, Sones, Jibbo, Durkin, Light & Murray 1985). In fact, food production on the continent is falling, whereas in Asia and Latin America there is an overall increase (ILCA 1984). Production of animal protein is lower in Africa than in any other continent. For Africa, the estimated production of animal protein from livestock farming per 1000 ha is 542 kg, for Latin America it is 4113 kg and for Europe, 38 083 kg (reviewed by Murray & Gray 1984). This is despite the fact that Africa has vast agricultural resources and could readily feed herself.

The causes of this critical situation in Africa are complex but one of the most significant factors is tsetse-transmitted animal trypanosomiasis. Currently, vast humid and semi-humid areas of Africa are held captive by tsetse flies and the trypanosomes, *Trypanosoma congolense*, *T. vivax*, *T. brucei* and *T. simiae*, which they transmit. Tsetse infest 10 million $km^2$ of Africa, representing 37% of the continent, about half the habitable land, and affecting 37 countries (FAO-WHO-OIE 1982). It is considered that 7 million $km^2$ of this area would otherwise be suitable for livestock and mixed agriculture without stress to the environment, if trypanosomes could be controlled (MacLennan 1980). While in certain regions there are indications that as human population densities increase, tsetse density falls (Jordan 1986), it would appear that at the continental level the threat of tsetse is not diminishing: tsetse have shown a remarkable capacity to adapt to peridomestic habitats and at the same time have reinfested many areas as a result of political instability disrupting community structure.

## The problem in domestic ungulates

It could be said that there is no other continent dominated by one disease to the same extent as Africa is by tsetse-transmitted trypanosomiasis. About 30% of the 147 million cattle in countries affected by tsetse are exposed to infection. The situation with regard to sheep, goats, pigs, horses, donkeys and camels is probably as serious but is less well documented (FAO-WHO-OIE 1982; Table 1).

**Table 1.** Domestic livestock population in the 37 countries in Africa infested with tsetse fly (millions).

| Cattle | Sheep | Goats | Pigs | Horses | Mules | Camels |
|--------|-------|-------|------|--------|-------|--------|
| 147 | 104 | 125 | 8 | 3 | 9 | 12 |

Countries in West and Central Africa are particularly severely affected by trypanosomiasis. In the 18 countries from Senegal to Zaire live 26% of Africa's human population, but this vast area maintains only 9% of Africa's cattle, sheep and goats (ILCA 1979). As a result, the livestock biomass per inhabitant in West and Central Africa is only 26 kg, in contrast to 146 kg for the remainder of Africa south of the Sahara, and 79 kg for the continent as a whole. The average potential carrying capacity in this region has been estimated at 20 cattle/km$^2$, as compared with the current 3.4/km$^2$. Equivalent increases would also be possible for sheep and goats. The situation in southern and eastern Africa is no better, with up to 70% of the land in some countries, e.g. Tanzania, being infested by tsetse. It should be emphasized that approximately 50% of the domestic livestock in the 37 countries affected by tsetse are confined to six countries in East Africa, namely Ethiopia, Sudan, Tanzania, Kenya, Uganda and Somalia, and that this is only possible because a large proportion of these animals inhabit highland areas, above the limits of tsetse infestation.

As well as domestic ruminants, the impact of trypanosomiasis on domestic pigs needs emphasis. Domestic pigs usually develop a relatively mild disease when infected with *T. congolense* or *T. brucei*, but *T. simiae* infections produce fulminating outbreaks of fatal trypanosomiasis with the result that there are only eight million pigs in the 37 countries in Africa infested with tsetse.

Non-tsetse-transmitted trypanosomiasis caused by *T. evansi* is a major problem, particularly in Africa's 12 million camels, 86% of the world's total (Mukasa-Mugerwa 1981). *T. evansi* infections are found outside the limits of tsetse infestation in Africa, as well as in South America and in Asia, where it is recognized that this parasite can be pathogenic not only in camels but also in cattle, water buffalo and equines. Non-tsetse-transmitted *T. vivax* infections have long been known to occur in cattle and water buffalo in South America and, more recently, have been found in Asia. The prevalence and economic importance of *T. vivax* in these continents are unknown.

It is also important to appreciate that human African trypanosomiasis (*T.*

*gambiense, T. rhodesiense*) is an important constraint on rural development in Africa, acting as a major factor in the depopulation of large areas. It causes disruption of communities with the resultant depletion of the human resources upon which viable agricultural communities depend. Today it is estimated that some 50 million people in some 36 countries are at risk (Molyneux 1986). Prior to 1979, there were approximately 10 000 new cases every year. At present, there are several serious active foci of the disease in Africa, e.g. in Uganda, Sudan, Zambia, Ivory Coast, Zaire, Angola and Mozambique, and the number of new cases reported every year has increased to 20 000 (Molyneux 1986). There is no doubt that the true prevalence is underestimated owing to inadequate surveillance and reporting, inaccessibility of the population at risk and the difficulties of diagnosing the disease.

The presence of tsetse not only results in severe losses in production in domestic livestock due to poor growth, weight loss, low milk yield, reduced capacity for work, infertility and abortion but excludes domestic animals from a massive area of Africa. The annual loss in meat production alone is estimated at US $ 5 billion (Murray & Gray 1984). This figure excludes milk and mixed agriculture where draught power and manure play a vital role. In Africa, 80% of traction power is non-mechanized. It has been calculated that the availability of a draught ox to a family unit can increase agricultural output six-fold (McDowell 1977). Furthermore, the manure provided by livestock is essential for the production of food and cash crops and is a potential source of energy in the form of biogas. If all these factors are taken into consideration, it has been estimated that livestock and agricultural development of tsetse-infested Africa could generate a further US $ 50 billion annually.

## The problem in wild ungulates

Africa has a vast array of wild ungulates, all of which have the reputation for being highly resistant to trypanosomiasis, a conclusion based largely on the fact that they are able to survive in areas heavily infested with tsetse. However, because they are wild and undomesticated, there are only limited data on the prevalence and pathogenicity of trypanosome infections. The scientific names of animals under discussion are given in Table 2.

### Prevalence of trypanosomes

Natural infections of oribi (Bruce *et al.* 1914) and waterbuck (Bruce *et al.* 1915; Kinghorn & Yorke 1912) with *T. congolense* were reported early this century. Ashcroft (1959) reviewed the results of several parasitological surveys, which showed that of 1242 wild animals examined in various locations at various times, 19.5% were infected with trypanosomes. *T.*

**Table 2.** Scientific names of wild animals discussed.

| | | | |
|---|---|---|---|
| Aardvark | *Orycteropus afer* | Hyrax | *Dendrohyrax* sp. |
| Antelope, roan | *Hippotragus equinus* | Impala | *Aepyceros melampus* |
| Antelope, sable | *Hippotragus niger* | Jackal | *Canis* sp. |
| Baboon | *Papio* sp. | Kob | *Kobus kob* |
| Buffalo | *Syncerus caffer* | Kudu, greater | *Tregelaphus strepsiceros* |
| Bushbaby | *Galago crassicaudatus* | Lechwe | *Kobus leche* |
| Bushbuck | *Tragelaphus scriptus* | Leopard | *Panthera pardus* |
| Bushpig | *Potamochoerus porcus* | Lion | *Panthera leo* |
| Dikdik | *Madoqua (Rhynchotragus)* sp. | Monkey | *Cercopithecus* sp. |
| Duiker, blue forest | *Cephalophus monticola* | Oribi | *Ourebia ourebi* |
| | | Ostrich | *Struthio camelus* |
| Duiker, bush | *Sylvicapra grimmia* | Porcupine | *Hystrix cristata* |
| Eland | *Taurotragus oryx* | Puku | *Kobus vardoni* |
| Elephant | *Loxodonta africana* | Reedbuck | *Redunca* sp. |
| Fox, bat-eared | *Otocyon megalotis* | Reedbuck, Bohor | *Redunca redunca* |
| Gazelle, Thomson's | *Gazella thomsoni* | | |
| | | Rhinoceros | *Diceros bicornis* |
| Giraffe | *Giraffa camelopardalis* | Serval | *Felis serval* |
| Hartebeest | *Alcelaphus buselaphus* | Topi | *Damaliscus korrigum* |
| Hippopotamus | *Hippopotamus amphibius* | Tsessebe | *Damaliscus lunatus* |
| | | Warthog | *Phacochoerus aethiopicus* |
| Hyaena | *Crocuta crocuta* | | |
| | | Waterbuck | *Kobus ellipsiprymnus* |
| | | Wildebeest | *Connochaetes taurinus* |
| | | Zebra | *Equus burchelli* |

*brucei* accounted for 4.2%, *T. vivax* for 6.3% and *T. congolense* for 10.4%. These early surveys showed that the animals with the greatest incidence of trypanosomes were waterbuck (52%), kudu (45%), reedbuck (44%), giraffe (37%), bushbuck (31%), and eland (29%). Buffalo, bushpig, duiker, hartebeest, impala, oribi, puku, roan antelope, topi and warthog had a similar incidence (between 10% and 16%). The incidence in zebra was low (6%) and trypanosomes were not found in sable antelope, rhinoceros, Thomson's gazelle or wildebeest. In his own study attempting to isolate human-infective *T. brucei* from wild animals, Ashcroft (1958) found trypanosomes in only 11 of 74 animals in Tanzania.

Improvements in techniques of parasite detection and identification over the last two decades have extended the list of infected animals. Baker, Sachs & Laufer (1967) demonstrated a relatively high incidence of trypanosomes in wildebeest (27%) and also found infected Thomson's gazelle and lion. No trypanosomes were detected in zebra. *T. brucei* was identified in the blood of two lions in the Serengeti (Sachs, Schaller & Baker 1967) and the overall incidence of *Trypanosoma* species in lions in this region was very high (68.8%). Reedbuck were the most frequently infected of eight species examined in southern Tanzania (Geigy, Kauffmann & Beglinger 1967), two infected zebra were reported by McCulloch (1967) and in Baker's (1968) summary of surveys in the Serengeti, only buffalo, leopard and jackal were free from infection. D.P. Kariuki, L. Boyce & R. Injairo, in an unpublished survey in 1981 of wild animals in Lambwe Valley National Park, Kenya, isolated *T. brucei* sp. by subinoculation into mice from the blood of oribi and reedbuck. Dräger & Mehlitz (1978) investigated the prevalence of trypanosome infections and antibodies in 605 buffalo, 60 lechwe, 23 kudu, 23 impala, 15 tsessebe, 22 sable and two reedbuck in Botswana. The overall prevalence of trypanosomes in buffalo and lechwe was 12% and 42%, respectively, with *T. vivax* and *T. congolense* predominating in buffalo and *T. brucei* in lechwe. Both reedbuck were infected with *T. vivax* and *T. congolense*. The other antelopes were not examined parasitologically but 19 kudu (69%), 14 impala (61%), eight tsessebe (53%) and eight sable (36%) were serologically positive.

Geigy, Mwambu & Kauffmann (1971) surveyed 13 species of wild mammal for possible reservoirs of human trypanosomiasis in Tanzania. Of the *T. brucei* strains isolated, human-infective parasites were found in hyaena, lion and hartebeest (Geigy, Jenni, Kauffmann, Onyango & Weiss 1975). Bushbuck, the first wild animal species shown to harbour human-infective *T. rhodesiense* trypanosomes (Heisch, McMahon & Manson-Bahr 1958), waterbuck and warthog (Dillmann & Townsend 1979), giraffe (Awan 1979) and reedbuck (Rickman 1984) are also potential reservoirs for human infections. At the same time, it is now recognized that at least some wild Bovidae such as hartebeest and kob can be infected with trypanosomes, likely to be *T. gambiense* (reviewed by Molyneux 1986).

## The effects of infection

In the main, it would appear that clinical illness due to trypanosome infection is rare in wild animals, but may occur under conditions of stress. Thus, in all the reports summarized by Baker (1968), only three animals were observed to be clinically ill. Following necropsy, McCulloch (1967) attributed the lethargy and poor condition of two zebra to trypanosomiasis, but the cause of illness in one lion was not determined (Sachs *et al.* 1967). Burridge, Reid, Pullan, Sutherst & Wain (1970) stressed that there was no sign of disease in bushbuck, waterbuck or elephant, in which they detected trypanosomes.

As part of the study by Geigy, Mwambu *et al.* (1971), post-mortem examinations of warthog, impala, Thomson's gazelle, hartebeest, topi, zebra and lion were carried out (Losos & Gwamaka 1973). Significant histological lesions were found in two impalas, one gazelle, three hartebeest and two lions, but they were mild and suggested that the infections were probably asymptomatic. High parasitaemias in wounded buffaloes, buffaloes in poor condition, and bushbuck rams injured in the course of territorial conflicts also emphasized the importance of stress (Awan &

Dillmann 1973). In this context, Lumsden (cited by Baker 1968) concluded that stress contributed to the death of a rhinoceros which showed a high parasitaemia when captured. An account of an outbreak among cattle in Kenya (Wijers 1969) indicated that the wild animal population also appeared to be affected; dead or very thin and weak bushpigs were found in the forests after advances of tsetse.

Only a very limited number of investigations involving experimental infection of wild ungulates have been carried out. The results of many of these studies could be open to question because the history of previous exposure to trypanosomes was not known and only small numbers of animals were tested. Carmichael (1934) found that several wild species were either not infectible with *T. congolense* or developed only transient infections and then recovered. However, there was a spectrum of susceptibility to *T. brucei* and some fatalities occurred. Needle challenge with *T. brucei* and *T. rhodesiense* was examined by Ashcroft, Burtt & Fairbairn (1959), who recognized two broad categories of response to infection. Gazelle, dikdik, blue forest duiker, jackal, bat-eared fox, aardvark, hyrax, serval and monkey usually died, whereas other species showed a range of tolerance to infection. Bush duiker, eland, Bohor reedbuck, hyaena, oribi, bushbuck and impala were all infectible and remained parasitaemic for a considerable time; very slight parasitaemias were exhibited by warthog, bushpig and porcupine, while baboons were apparently refractory to infection.

A study of duiker and gazelle (Roberts & Gray 1972) showed that duiker were very resistant to *T. vivax* and developed only sporadic parasitaemia with no anaemia or signs of disease. The parasitaemia in gazelle was similar to that of cattle; the animals were prostrate and febrile at peak parasitaemia, but this effect was transient: parasitaemia became slight and recovery followed. The disease syndromes induced by *T. congolense* in duiker and gazelle were similar to those caused by *T. vivax*. More recently, a series of experiments has been carried out on several species of wild Bovidae, including buffalo, eland, wildebeest and waterbuck, which had never previously been exposed to trypanosome infection. These animals were subjected to tsetse-transmitted challenge or to needle inoculation with bloodstream forms of different serodemes[1] of *T. congolense, T. vivax* or *T. brucei* (Murray, Grootenhuis, Akol, Emery, Shapiro, Moloo, Dar, Bovell & Paris 1981; Dwinger, Grootenhuis, Murray, Moloo & Gettinby 1986; Rurangirwa, Musoke, Nantulya, Nkonge, Njuguna, Mushi, Karstad & Grootenhuis 1986; J.G. Grootenhuis personal communication). The trypanosome isolates used had been shown to be virulent in susceptible cattle and goats. All wild Bovidae infected in this way exhibited marked resistance

[1] Serodemes are populations of trypanosomes each of which can express the same repertoire of variable antigen types.

to the effects of infection. There were no clinical manifestations of illness, no weight loss, no anaemia and the parasitaemias that developed were less intense and more transient than in corresponding susceptible cattle and goats.

The limited data there are would suggest that wild animals may remain infected for a long time. Thus, Hornby (1952) reported the case of an antelope which died in London Zoo in 1945, 11 years after its capture in Nigeria. Although it had been in apparent good health during its captivity, *T. congolense* was found in its blood at necropsy. Similarly, *T. brucei* was found in a hyaena after it died in London Zoo in 1942. This animal had been captured six years previously in Uganda (Hoare, cited by Baker 1968). Long-term transmission experiments conducted at the Tinde Laboratory in Tanzania also demonstrated that trypanosomes could be found in the blood of wild animals for long periods after fly infection (Willett 1970).

### Hosts of tsetse

Wild ungulates play an important role in the maintenance of tsetse populations as they provide a major source of blood meals. Weitz (1963) showed that the genus *Glossina* has a wide range of hosts, of which pigs of different species are generally favoured. There are also hosts which are attractive to only one or two species of tsetse, e.g. rhinoceros, elephant (*G. longipennis*), reptiles (*G. palpalis* and *G. tachinoides*), hippopotamus (*G. fuscipleuris* and *G. brevipalpis*), porcupine (*G. tabaniformis*), kudu and elephant (*G. morsitans* in Central and East Africa). Certain animals accounted for 8 or 9% of the identified feeds of only one species of tsetse: duiker (*G. austeni*), giraffe (*G. swynnertoni*), aardvark (*G. fusca*), and ostrich (*G. longipennis*). In addition, there are some animals on which flies rarely, if ever, feed. Although the zebra occupies large areas of *G. morsitans* country in East Africa, its blood has not been identified in any meal. Impala, wildebeest, waterbuck, gazelle, dikdik, monkey, baboon, oribi, eland, dogs, cats and other carnivores are also of negligible importance as hosts for the tsetse.

These results indicate a discrepancy between the species of wildlife which are preferred hosts of the tsetse and those shown to be naturally infected with trypanosomes in the field; for example, Ashcroft (1959) found a high incidence of infection in waterbuck, while Weitz (1963) concluded that these species were largely ignored by all tsetse flies as a source of food. It would appear that if enough pressure is exerted by the absence of preferred hosts, tsetse are sufficiently adaptable in feeding habits to find an alternative host. At the same time, the high incidence of infection in lions, which are rarely used by tsetse as a source of food, could be due to killing and eating infected herbivores. Cats (Duke, Mettam & Wallace 1934) and bushbabies (Heisch 1952) have been experimentally infected by eating infected animals. It is likely that trypanosomes penetrate small lesions of the buccal mucosa which must often be damaged by the bones of the prey.

In conclusion, while there are differences in infection rates among wild animals, most species can become infected with both domestic animal and human infective trypanosomes. The limited number of studies carried out on experimentally infected animals has confirmed the widely held belief that wild ungulates possess a remarkable degree of resistance to the disease. By acting as asymptomatic carriers of the infection and as a source of blood meals for tsetse, wild ungulates play a major role in the epidemiology of the disease.

## Complexity of the disease

Many factors contribute to the magnitude of the problem of African trypanosomiasis. One of the major ones is the complexity of the disease itself. In cattle, for example, three species of trypanosome, *T. congolense*, *T. vivax* and *T. brucei*, cause the disease, either individually or jointly. These trypanosomes are transmitted cyclically by several different species of tsetse, each of which is adapted to different climatic and ecological conditions (Ford 1971). While tsetse are not the only vectors of African trypanosomes, cyclical transmission of infection represents the most important problem because, once the tsetse fly becomes infected, it remains infective for a long period, in contrast to the ephemeral nature of non-cyclical transmission. At the same time, trypanosomes infect a wide range of hosts including wild and domestic animals. As discussed, the former do not suffer severe clinical disease but become carriers and constitute an important reservoir of infection. The success of the trypanosome as a parasite is to a large extent due to the ability to undergo antigenic variation, i.e., change a single glyco-protein (Cross 1975) which covers the pellicular surface, thereby enabling evasion of host immune responses and the establishment of persistent infections. Added to the complexity of multiple variable antigen types expressed during a single infection, each trypanosome species comprises an unknown number of different serodemes, all capable of elaborating a different repertoire of variable antigen types (Van Meirvenne, Magnus & Vervoort 1977). As a result, no vaccine is available for use in the field.

## Control

Lack of a vaccine means that current control is reliant on tsetse control or on the use of trypanocidal drugs. While both have been shown to be effective if properly applied, the net effect to date at the continental level has been limited. However, recent developments in the understanding of tsetse biology, antigenic variation and host susceptibility, and in the strategic use of drugs, have identified promising new approaches.

**Tsetse control**

Attempts to control tsetse have been made for over 60 years. Initially, they included eradication of wildlife, clearing of fly barriers to prevent the advance of the vector, and widespread bush clearing to destroy breeding habitats. Following the introduction of modern insecticides, the principal method employed to control tsetse populations has been the use of insecticides, alone or in conjunction with traps and screens.

### Insecticides

This approach is the main method employed at present for tsetse control. The insecticides used fall into two categories, residual and non-residual. Residual insecticides (DDT and more recently dieldrin) are usually applied by hand-operated sprays that deliver the insecticide to sites where resting tsetse are known to alight. Ideally, the persistence of the insecticide should be sufficient to make only one application necessary. Non-residual insecticides require several applications; at present, endosulphan is the insecticide of choice. Non-residual treatments are applied mainly by fixed-wing aircraft or helicopters. Recently, advances in the technology of aerial spraying have resulted in better use of insecticides by spraying with fixed-wing aircraft against savanna species of tsetse, and with helicopters against tsetse inhabiting gallery forests and riverine areas. Where control measures using insecticides have been properly implemented, significant success has been achieved, for example in Nigeria, Zimbabwe, Botswana and Zambia (MacLennan 1980, 1981).

Despite the proven efficacy of tsetse control by insecticides, there are severe limitations of this approach on practical, economic and environmental grounds. In Africa at present there is a lack of trained personnel to implement insecticide control programmes. The costs of insecticide control programmes are high. Natural or man-made barriers are required to defend sprayed areas and prevent reinvasion, and constant surveillance for early detection of reinvasion is essential. Finally, there are increasing demands for restricting the use of insecticides because of possible environmental impact on fauna and flora. In this respect it should be emphasized, however, that although non-target organisms can be affected by anti-tsetse spraying and quantities of insecticide remain in the environment, these effects appear to be transitory, rarely lasting for more than a year (Jordan 1986).

Current research involves the development of new potent insecticides with low toxic environmental effect, e.g. synthetic pyrethroids, and the improvement of the technology of aerial spraying, with respect to droplet size, rate of delivery and extent of dispersion (Molyneux 1986).

### Traps and screens

Traps and screens have been used for many years as a means of sampling tsetse populations. However, with recent advances in the design and colour of traps and with the identification of tsetse attractants, increasing attention is being given to the use of traps as a method for tsetse control. Elegant studies carried out in Zimbabwe (Vale 1980, 1982, 1987) and at the Tsetse Research Laboratory in Bristol (Jordan 1986) have culminated in the development of simple insecticide-impregnated visual targets incorporating acetone and 1-octen-3-ol to attract tsetse. Their dramatic effects on tsetse populations and the subsequent drop in incidence of trypanosomiasis in cattle have been reported (Opiyo, Dolan, Njogu, Sayer & Mgutu in press). These developments are of considerable importance with respect to introducing a simple, safe and relatively cheap method of control, although it must be emphasized that they have only proved effective with the 'savanna' tsetse, and suitable olfactory attractants have not as yet been identified for the 'riverine' or 'forest' tsetse.

## Biological methods

A variety of approaches has been considered, including the breeding of tsetse with reduced susceptibility to infection, insect growth regulators, pheromones (sex hormones) as attractants, and predators, parasites or pathogens which might serve to control tsetse populations (Jordan 1986). Most of these, however, are at an early stage of investigation. One method which has reached the developmental stage is release of sterile males, a concept based on the fact that tsetse females copulate only once and if the male is sterile the female will never reproduce. Thus, a series of field trials has been carried out to evaluate the impact of the release of gamma-irradiated male tsetse on tsetse populations. Trial work has been carried out at Mkwaja Ranch in Tanzania and also in Zimbabwe, Upper Volta, Nigeria and Zambia (Jordan 1986). Although under field conditions it has been demonstrated that this approach can significantly reduce tsetse populations, the majority opinion is that sterile male release is not a practical proposition because it is too sophisticated and too expensive. It is estimated that ten sterile males are required per female. In order to reduce the number of sterile males required it is necessary to carry out two to three insecticide sprays and then to release 12 000 sterile males per $km^2$, even in areas where the tsetse density is low. Each species of tsetse must be treated separately, so the cost would increase with the number of species to be controlled.

## Trypanocidal drugs

Escalating costs and other constraints on initiating and maintaining tsetse control campaigns, together with the non-availability of a vaccine, have led to the livestock industries in the vast tsetse-infested areas of Africa having to rely almost entirely on the use of trypanocidal drugs to treat or prevent the disease. Without these drugs, the outcome would be disastrous. Despite the need and demand for effective trypanocides, no new drug has been produced for commercial use in the last 25 years. Currently, there are five drugs available commercially. Samorin/Trypamidium (isometamidium chloride) is a long-acting drug with a prophylactic activity against trypanosomes that can last for several months (Whitelaw, Bell, Holmes, Moloo, Hirumi, Urquhart & Murray 1986). Berenil (diminazene aceturate), Novidium (homidium chloride) and Ethidium (homidium bromide) have a much shorter prophylactic action and are used therapeutically to treat infections. Recently, quinapyramine sulphate and pro-salt have been reintroduced, because of efficacy against *T. evansi*, particularly in camels. There would appear to be no immediate prospects of new drugs for commercial use. Furthermore, the cost of registering a new drug for use in animal trypanosomiasis is regarded by pharmaceutical companies as too high in relation to their estimates of the potential financial return. This is despite the fact that some 120 million cattle, sheep and goats are exposed to infection. Even if animals were treated only twice per year, 240 million doses would be required. This is ten times the number currently used, as estimated by the Food and Agriculture Organisation of the United Nations. Several reasons would appear to account for the lack of use of the trypanocidal drugs:

1. The belief that the cost of trypanocidal drugs and their use is high.

2. The concern that repeated use of trypanocidal drugs might lead to side effects in the host and to drug resistance.

3. The widely held premise that trypanocidal drugs should and can only be used in sophisticated ranch management systems and that control programmes involving trypanocidal drugs should not and cannot be implemented in village-managed systems. This belief has serious implications as currently 90% of livestock in Africa are reared in villages and by small holders.

However, there are an increasing number of reports of the successful use of trypanocidal drugs in both indigenous and imported cattle under ranch or village management. Thus, some 12 000 Boran are maintained in Mkwaja Ranch in Tanzania in an area where Boran cattle rapidly succumb to trypanosomiasis if left untreated. As a result of the strategic use of Samorin in combination with Berenil, the level of productivity achieved is close to that of Boran reared in tsetse-free conditions on ranches in Kenya considered among the best in the world (Trail *et al.* 1985). Application of similar strategic treatment in Galana Ranch in Kenya (Wilson, Njogu, Gatuta, Mgutu & Alushula 1981) resulted in good economic returns. When a similar drug strategy was implemented in East African Zebu cattle (700 head) under village management in Kenya, it resulted in a 20% increase in performance (Maloo *et al.* 1988). In both these situations, the level of tsetse challenge was considered high. On the other hand, where disease risk is low and where it is possible to examine individual animals at regular intervals, therapeutic trypanocidal drug strategies have been successfully employed, e.g. at Kilifi Plantation on the coast of Kenya. This dairy ranch is one of the biggest in Africa, supporting 800 breeding females based on Sahiwal × Ayrshire crosses. The owner was virtually out of business because of trypanosome-induced abortion storms until he successfully introduced a systematic therapeutic drug strategy which he has maintained for over 20 years (Wissocq, Trail, Wilson & Murray 1983).

The fact that these control programmes were carried out on large numbers of cattle over a long period of time with financially successful results and with no evidence of drug resistance offers hope for livestock and socio-economic development programmes in tsetse-infected areas of Africa by demonstrating that the rational use of trypanocidal drugs can be integrated into livestock management, whether in villages or in ranches.

### Prospects for vaccination

The major constraint on the development of a vaccine against trypano-somiasis is the phenomenon of antigenic variation, as it would appear that host protective responses are effected mainly by antibodies directed against surface coat antigens, i.e. the variant-specific glycoprotein (VSG) (Murray & Urquhart 1977). The repertoire of these antigens generated by bloodstream forms of the parasite is large, whereas the repertoire of antigens produced by metacyclic parasites following transmission through the tsetse is much more limited (Crowe, Barry, Luckins, Ross & Vickerman 1983). Thus, it has been possible to immunize cattle and goats against tsetse-transmitted homologous (but not heterologous) strains of *T. congolense* and *T. brucei* (reviewed by Morrison, Murray & Akol 1985), but not *T. vivax* (Emery, Moloo & Murray 1987), by prior exposures to metacyclic parasites which have been grown in tissue culture, or by prior infection via tsetse flies followed by trypanocidal drug treatment. The immunity produced has been shown to last for as long as five months. Nevertheless, the feasibility of production and the efficacy of a vaccine against metacyclic trypanosomes will depend on the relative stability of the metacyclic antigen repertoire for each species of trypanosome and on the number of serodemes which occur in the field. Currently, research is directed towards these objectives. It is thought, however, that the number of different serodemes is likely to be prohibitively large for the production of a cocktail vaccine containing the appropriate metacyclic antigens. As a result the development of a vaccine against African trypanosomiasis has been considered unlikely.

However, most research has concentrated on the VSGs of the trypanosome and data on the subcellular distribution and properties of other antigens are surprisingly sparse. Recently, a flagellar pocket membrane fraction has been identified in *T. rhodesiense* at the parasite surface at the emergence of the flagellum from the flagellar pocket. This would appear to be non-variable between different serodemes and to have protective potential (Olenick, Wolff, Nauman & McLaughlin 1988). At the same time, receptor-mediated endocytosis of low-density lipoprotein (LDL) and transferrin has been demonstrated with *T. brucei* (Coppens, Opperdoes, Courtoy & Baudhuin 1987). While cholesterol is the major sterol in the membrane of the trypanosome, there is no evidence that it can be synthesized by the trypanosome *de novo*. Cholesterol is not freely available in the mammalian bloodstream but is buried within LDL particles. As a result, it was hypothesized and confirmed by *in vitro* studies that the ability to endocytose LDL is essential for optimal trypanosome growth: removal of LDL or addition of antibodies against the purified LDL receptors inhibits growth. The receptor appears to be highly conserved and to be localized to the flagellar pocket membrane and the flagellar membrane, and to be completely absent from the rest of the pellicular membrane (F.R. Opperdoes personal communication). These exciting findings indicate that suitable antigens for immuno-targeting may exist and that the possibilities of vaccine development should be reconsidered.

### Genetic resistance: trypanotolerance

Genetic resistance to the effects of infection by African trypanosomes occurs in certain breeds of domestic livestock (ILCA 1979) and in many species of wildlife (Murray, Grootenhuis *et al.* 1981). The term trypanotolerance is used to describe this trait, which in domestic livestock is best recognized in the N'Dama and West African Shorthorn, indigenous West African taurine breeds of cattle that have been in Africa for 5000 to 7000 years (Payne 1964; Epstein 1971).

The trait has recently been reported in an East African Zebu and the Orma Boran (Wilson *et al.* 1981; Njogu, Dolan, Wilson & Sayer 1985). Confirmation not only that these breeds possess a high degree of genetic resistance to trypanosomiasis (reviewed by Murray, Morrison & Whitelaw 1982; Njogu, Ismael, Dolan, Okech, Sayer, Opiyo & Alushula 1988) but also that they have considerable potential for production (ILCA 1979) has stimulated interest both in making greater use of them as domestic livestock and also in studying them in order to understand the basic mechanism(s) responsible for the genetic resistance.

### Criteria of trypanotolerance

Comparative investigations on the question of trypanotolerance have been carried out on cattle throughout Africa, e.g. in Nigeria (Desowitz 1959; Roberts & Gray 1973a,b), The Gambia (Murray, Clifford, Gettinby, Snow

& McIntyre 1981; reviewed by Murray, Morrison *et al.* 1982), Senegal (Toure, Gueye, Seye, Ba & Mane 1978), Burkina Faso (reviewed by Roelants 1986; Akol, Authie, Pinder, Moloo, Roelants & Murray 1986), Kenya (Njogu, Dolan *et al.* 1985; Ismael, Njogu, Gettinby & Murray 1985; Paling, Moloo & Scott in press) and in the ILCA/ILRAD Trypanotolerance Network (1986a,b). The main breeds studied included Ayrshire, Friesian, Holstein, Hereford, and their crosses, as well as indigenous African breeds such as Zebu, Boran, West African Shorthorn and N'Dama. As described earlier, a limited number of studies have also been performed in wild Bovidae. Irrespective of the mode of infection, the outcome of each study consistently confirmed the superior resistance of the N'Dama (and the West African Shorthorn) and showed that the basis of this trait was associated with the capacity of these animals to develop less severe anaemia. Anaemia is a well recognized and inevitable consequence of trypanosome infections in domestic animals in general and cattle in particular (Hornby 1921; Murray 1974; Morrison, Murray & McIntyre 1981). Maasai herdsmen recognize that cattle in tsetse-infested areas can 'run out of blood'. It has now been definitively established that the measurement of anaemia gives a reliable indication of the disease status (Murray 1979) and productive performance (ILCA/ILRAD 1986a,b) of trypanosome-infected cattle.

Furthermore, resistance of these breeds to anaemia appeared to be correlated with the ability to limit the intensity, prevalence and duration of parasitaemia (Dargie, Murray, Murray, Grimshaw & McIntyre 1979). A similar capacity to control parasitaemia and resist anaemia has been shown by the Orma Boran cattle in Kenya (Ismael *et al.* 1985) and to an even greater extent by several species of wild Bovidae (Murray, Grootenhuis *et al.* 1981). As parasitaemia wave remission is effected by antibodies directed against the surface coat antigens of the trypanosome (reviewed by Murray & Urquhart 1977), it is generally assumed that the superior capacity of trypanotolerant animals to control parasitaemia is associated with a better immune response, although in cattle, at any rate, there are only a few preliminary studies in N'Dama infected with *T. vivax* (Desowitz 1959) and Bauole (West African Shorthorn) infected with *T. congolense* (Akol *et al.* 1986) which suggested that this might be the case.

Thus, trypanotolerance is associated with at least three possibly related characteristics, namely, the ability to control parasitaemia, the ability to develop an effective immune response, and the ability to resist anaemia.

Another possible aspect of 'trypanotolerance' is that it would appear that indigenous African cattle and wild Bovidae are more resistant to environmental constraints because of superior physiological adaptation in terms of food utilization, heat tolerance and water conservation (reviewed by Murray, Morrison *et al.* 1982).

### Susceptibility to other diseases

There is also evidence, sometimes experimental and sometimes anecdotal, that the taurine trypanotolerant breeds of West Africa are resistant not only to trypanosomiasis but also to several other important infectious diseases. Of major significance are the reports that both the N'Dama and the West African Shorthorn are resistant to streptothricosis (Stewart 1937; Coleman 1967; Oduye & Okunaiya 1971). Furthermore, N'Dama appear to be more resistant to tick-borne diseases, including heartwater (*Cowdria ruminantium*), anaplasmosis and babesiosis (Epstein 1971). These observations might indicate a greater resistance to ticks *per se*. N'Dama would also seem to possess some degree of resistance to helminthiasis (A.A. Ilemobade personal communication). In this respect, the Red Maasai sheep of Kenya have been shown to be significantly more resistant than other breeds not only to trypanosomiasis (Griffin & Allonby 1979a,b) but also to haemonchosis (Preston & Allonby 1979). On the other hand, N'Dama and West African Shorthorn are reputed to be much more susceptible to rinderpest than Zebu (Stewart 1937, 1951; Cornell & Evans 1937; Ferguson 1967), and Van Hoeve (1972) noted that Muturu (a West African Shorthorn type), but not N'Dama, were highly susceptible to footrot.

While the foregoing observations require confirmation and more detailed investigation, it would appear that the N'Dama and West African Shorthorn possess a unique genetic advantage that makes them more resistant to trypanosomiasis, as well as to some of the other major infectious diseases of Africa.

### Production potential

It has been widely believed that indigenous breeds of cattle in Africa and, in particular, the smaller taurine trypanotolerant breeds possess low potential for production. The ILCA publication of 1979 has gone a long way in encouraging a reassessment of this conclusion. In a survey of the status of trypanotolerant cattle in 18 countries in West and Central Africa, indices of productivity were computed using all the production data available for the different breeds. A total of 30 trypanotolerant cattle herds and 20 Zebu herds were investigated, with approximately 4000 data per trait. Traits evaluated included reproductive performance, cow and calf viability, milk production, growth and cow body weight. These parameters were used to compute the index of the total weight of calf and liveweight equivalent of milk produced per 100 kg of cow maintained per year. This final index related the production traits back to the actual weight of breeding cow that had to be supported, this being closely connected with maintenance costs. It was found in areas where the tsetse fly risk was low that the productivity of Zebu (38.6 kg per 100 kg of cow maintained per year) was only 4% higher than that of the N'Dama and West African Shorthorn (37.1 kg). Data directly comparable between breeds were not available in many areas because the levels of tsetse challenge was such that breeds other than trypanotolerant ones did not exist. Furthermore, it has been found at the International Trypanotolerance Centre in The Gambia that supplementary feeding of N'Dama with local by-products, which are normally discarded, resulted in growth rates as much as 1 kg per day, producing two-year-old animals weighing over 300 kg. At the same time, little attention has been paid to milk production in N'Dama and it has generally been assumed to be too low to even record. However, preliminary results at the International Trypanotolerance Centre in The Gambia have shown that the average milk yield of 65 N'Dama cows over a practical extraction period of 10 months was 313 kg (K. Ayemang, personal communication). Such a yield is remarkable for animals that had an average bodyweight of 230 kg and were exposed to tsetse challenge.

## Conclusions

Through a process of rigorous natural selection, Africa has provided her own answer to the problem of disease control in domestic livestock. Breeds of cattle are now known to exist that possess a significant degree of resistance to tsetse-transmitted trypanosomiasis, as well as to several other important infectious diseases. What is more, it has been shown that these animals can be productive and, with proper management, have considerable potential. Recognition that the degree of anaemia induced during a trypanosome infection reflects the trypanotolerant status of the host could allow rational breeding programmes to be instituted and estimates of heritability made.

Major research priorities must be to identify the mechanisms that regulate parasite growth, allow the development of an effective immune response, and prevent pathogenic degrees of anaemia. An understanding of the mechanisms responsible might permit their manipulation and lead to novel strategies for the control of trypanosomiasis by therapeutic or immunological means, or even by molecular genetics. At the same time, marker(s) might be identified for genetic resistance, permitting the selection of breeding stock without having to infect animals.

## The way ahead

With the expansion of the human population in Africa and the continuing threat of the tsetse, there is mounting pressure on the tsetse-free pastures and farmlands for increased production. Many of these tsetse-free areas are in the drier regions of Africa where the ecology is fragile and where the problems of overgrazing emphasize the need to bring into production more favourable agricultural areas. The tsetse-infected areas of Africa possess such resources. At the same time, these areas support a unique array of wild animals.

Can the former only be developed at the expense of the latter? The tsetse fly, or rather the infection it transmits, has been described as 'the greatest menace to tropical Africa' (Ormsby-Gore 1925), while Cloudsley-Thompson (1977) believed it to be 'a blessing in disguise', preserving much of Africa from overgrazing by cattle and land degradation. The latter view has resulted in pleas from some sources for postponement of decisions regarding future tsetse eradication until the implications are more fully appreciated. While recognizing some of the reasons for concern, these recommendations can only be considered as unrealistic, as it is inevitable that the growth of Africa's human population will result in increasing encroachment into tsetse-infested areas. Furthermore, we agree with Jordan (1986) that it is immoral to attempt to prevent change by allowing a disease to remain unchecked. As he pointed out, we should always ask ourselves how long such a serious disease would be allowed to remain unchecked in the so-called developed world.

Will the occupation of tsetse-infested areas result in an ecological disaster? Undoubtedly, overgrazing must lead to pasture degradation. However, far from increasing overgrazing, planned tsetse control and eradication, by opening up new areas for agricultural and livestock development, should greatly reduce the pressure on the arid and semi-arid zones north of the main tsetse infestations. It could also be argued that wildlife conservation might be better served by the increased utilization of tsetse-infested areas; the massive food shortages in Africa mean that wildlife are sometimes one of the few sources of meat and protein with the result that in several areas, particularly of West Africa, very few remain.

What of the future? Domestic and wild ungulates must not be regarded as competitors for tsetse-infested Africa but they must be recognized as having an integral role in Africa's socio-economic development: domestic animals providing meat and milk for consumption, draught power and manure for agriculture; the wild animals, man's heritage, providing fascination, beauty, unique genetic resources as well as an additional source of the badly needed protein from game ranches, such as have been established in East Africa (Hopcraft 1982).

That tsetse control programmes with proper land reclamation may have a successful outcome can be seen today in Zululand. Many years ago G. *pallidipes* was eradicated from some 18 000 km$^2$ using insecticides (DuToit 1954). Today, Zululand is an important cattle-producing area integrated with game reserves supporting a rich diversity of wildlife.

Tsetse-infested areas of Africa are being and will be developed. Therefore, it is imperative that all parties concerned ensure that these massive changes are supported by balanced multidisciplinary advice that recognizes the needs of Africa and Africans.

# References

Akol, G.W.O., Authie, E., Pinder, M., Moloo, S.K., Roelants, G.E. & Murray, M. (1986). Susceptibility and immune responses of Zebu and taurine cattle of West Africa to infection with *Trypanosoma congolense* transmitted by *Glossina morsitans centralis*. *Vet. Immun. Immunopath.* **11**:361–73.

Ashcroft, M.T. (1958). An attempt to isolate *Trypanosoma rhodesiense* from wild animals. *Trans. R. Soc. trop. Med. Hyg.* **52**:276–82.

Ashcroft, M.T. (1959). The importance of African wild animals as reservoirs of trypanosomiasis. *E. Afr. med. J.* **36**:289–97.

Ashcroft, M.T., Burtt, E. & Fairbairn, H. (1959). The experimental infection of some African wild animals with *Trypanosoma rhodesiense, T. brucei* and *T. congolense*. *Ann. trop. Med. Parasit.* **53**:147–61.

Awan, M.A.Q. (1979). Identification by the blood incubation infectivity test of *Trypanosoma brucei* subspecies isolated from game animals in the Luangwa Valley, Zambia. *Acta trop.* **36**:343–7.

Awan, M.A.W. & Dillmann, J.S.S. (1973). *Trypanosoma brucei* infection in a lion (*Panthera leo*) in Zambia. *Trop. Anim. Hlth Prod.* **42**:75–8.

Baker, J.R. (1968). Trypanosomes of wild mammals in the neighbourhood of the Serengeti National Park. *Symp. zool. Soc. Lond.* No. 24:147–58.

Baker, J.R., Sachs, R. & Laufer, I. (1967). Trypanosomes of wild mammals in an area northwest of the Serengeti National Park, Tanzania. *Tropenmed. Parasit.* **18**:280–4.

Bruce, D., *et al.* (1914). Reports of the Sleeping Sickness Commission of the Royal Society No. 15: 1–157. The Royal Society, London. (Reports first published in *Proc. R. Soc. Lond.* (B) **85–87**: passim).

Bruce, D., *et al.* (1915). Reports of the Sleeping Sickness Commission of the Royal Society No. 16: 1–221. The Royal Society, London. (Reports first published in *Proc. R. Soc. Lond.* (B) **88**: passim).

Burridge, M.J., Reid, H.W., Pullan, N.B., Sutherst, R.W. & Wain, E.B. (1970). Survey for trypanosome infections in domestic cattle and wild animals in areas of East Africa. II. Salivarian trypanosome infections in wild animals in Busoga District, Uganda. *Br. vet. J.* **126**:627–33.

Carmichael, J. (1934). Trypanosomes pathogenic to domestic stock and their effect in certain species of wild fauna in Uganda. *Ann. trop. Med. Parasit.* **28**:41–5.

Cloudsley-Thompson, J.L. (1977). *Man and the biology of arid zones*. Edward Arnold, London.

Coleman, C.H. (1967). Cutaneous streptothricosis of cattle in West Africa. *Vet. Rec.* **81**:251–4.

Coppens, I., Opperdoes, F.R., Courtoy, P.J. & Baudhuin, P. (1987). Receptor-mediated endocytosis in the bloodstream form of *Trypanosoma brucei*. *J. Protozool.* **34**:465–73.

Cornell, R.L. & Evans, S.A. (1937). On the value and limitations of tissue vaccines against rinderpest. *J. comp. Path.* **50**:122–5.

Cross, G.A.M. (1975). Identification, purification and properties of clone-specific glycoprotein antigens constituting the surface coat of *Trypanosoma brucei*. *Parasitology* **71**:393–417.

Crowe, J.S., Barry, J.D., Luckins, A.G., Ross, C.A. & Vickerman, K. (1983). All metacyclic variable antigen types of *Trypanosoma congolense* identified using monoclonal antibodies. *Nature, Lond.* **306**:389–91.

Dargie, J.D., Murray, P.K., Murray, M., Grimshaw, W.R.I. & McIntyre, W.I. (1979). Bovine trypanosomiasis: the red cell kinetics of N'Dama and Zebu cattle infected with *Trypanosoma congolense. Parasitology* **78**:271–86.

Desowitz, R.S. (1959). Studies on immunity and host–parasite relationships. I. The immunological response of resistant and susceptible breeds of cattle to trypanosomal challenge. *Ann. trop. Med. Parasitol.* **53**:293–313.

Dillmann, J.S.S. & Townsend, A.J. (1979). A trypanosomiasis survey of wild animals in the Luangwa Valley, Zambia. *Acta trop.* **36**:349–56.

Dräger, N. & Mehlitz, D. (1978). Investigations on the prevalence of trypanosome carriers and the antibody response in wildlife in Northern Botswana. *Tropenmed. Parasit.* **29**:223–33.

Duke, H.L., Mettam, R.W.M. & Wallace, J.M. (1934). Observations on the direct passage from vertebrate to vertebrate of recently isolated strains of *Trypanosoma brucei* and *Trypanosoma rhodesiense. Trans. R. Soc. trop. Med. Hyg.* **28**:77–84.

Du Toit, R. (1954). Trypanosomiasis in Zululand and the control of tsetse flies by chemical means. *Onderstepoort J. vet. Res.* **26**:317–87.

Dwinger, R.H., Grootenhuis, J.G., Murray, M., Moloo, S.K. & Gettinby, G. (1986). Susceptibility of buffaloes, cattle and goats to infection with different stocks of *Trypanosoma vivax* transmitted by *Glossina morsitans centralis. Res. vet. Sci.* **41**:307–15.

Emery, D.L., Moloo, S.K. & Murray, M. (1987). Failure of *Trypanosoma vivax* to generate protective immunity in goats against transmission by *Glossina morsitans morsitans. Trans. R. Soc. trop. Med. Hyg.* **81**:611.

Epstein, H. (1971). *The origin of the domestic animals of Africa* **1** and **2**. Africana, New York.

FAO-WHO-OIE (1982).*Animal health yearbook 1981.* (Ed. Kouba, V.). No.18. Food and Agriculture Organization of the United Nations, Rome.

Ferguson, W. (1967). Muturu cattle of Western Nigeria: II. Survivability, reproductive and growth performances in an area of light *Glossina palpalis* density. *J. W. Afr. Sci. Ass.* **12**:37–44.

Ford, J. (1971). *The role of the African trypanosomiases in African ecology: A study of the tsetse fly problem.* Clarendon Press, Oxford.

Gavora, J.S. (1982). Disease resistance. In *Second world congress on genetics applied to livestock production* **5**:143–59. Ministerio de Agricultura Pesca y Alimentacion, Servicio de Publicationes Agraria, Madrid.

Geigy, R., Jenni, L., Kauffmann, M., Onyango, R.J. & Weiss, N. (1975). Identification of *T. brucei*-subgroup strains isolated from game. *Acta trop.* **32**:190–205.

Geigy, R., Kauffmann, M. & Beglinger, R. (1967). A survey of wild animals as potential reservoirs of trypanosomiasis in the Ulanga District (Tanzania). *Acta trop.* **24**:97–108.

Geigy, R., Mwambu, P.M. & Kauffmann, M. (1971). Sleeping sickness survey in Musoma District, Tanzania. IV. Examination of wild mammals as a potential reservoir for *T. rhodesiense. Acta trop.* **28**:211–20.

Griffin, L. & Allonby, E.W. (1979a). Trypanotolerance in breeds of sheep and goats with an experimental infection of *Trypanosoma congolense*. *Vet. Parasit.* 5:97–105.

Griffin, L. & Allonby, E.W. (1979b). Studies on the epidemiology of trypanosomiasis in sheep and goats in Kenya. *Trop. Anim. Hlth Prod.* 11:133–42.

Heisch, R.B. (1952). Presence of trypanosomes in bush babies after eating infected rats. *Nature, Lond.* 169:118.

Heisch, R.B., McMahon, J.P. & Manson-Bahr, P.E.C. (1958). The isolation of *Trypanosoma rhodesiense* from a bushbuck. *Br. med. J.* 1958(2):1203–4.

Hopcraft, D. (1982). Wildlife ranching in perspective. *Tigerpaper* 9:17–20.

Hornby, H.E. (1921). Trypanosomes and trypanosomiasis of cattle. *J. comp. Path.* 34:211–40.

Hornby, H.E. (1952). *Animal trypanosomiasis in Eastern Africa, 1949.* His Majesty's Stationery Office, London.

ILCA (1979). *Trypanotolerant livestock in West and Central Africa* 2. International Livestock Centre for Africa, Addis Ababa, Ethiopia.

ILCA (1984). *Annual Report, 1983. Improving livestock and crop livestock systems in Africa.* International Livestock Centre for Africa, Addis Ababa, Ethiopia.

ILCA/ILRAD (1986a). *The ILCA/ILRAD Trypanotolerance Network. Situation report, December 1985. Proceedings of a Network meeting held at ILCA, Nairobi. June 1986.* International Livestock Centre for Africa, Addis Ababa, Ethiopia.

ILCA/ILRAD (1986b). *The African Trypanotolerant Livestock Network. Indications from results. 1983–1985.* International Livestock Centre for Africa, Addis Ababa, Ethiopia.

Ismael, A.A., Njogu, A.R., Gettinby, G. & Murray, M. (1985). Susceptibility of Orma and Galana Boran cattle to infection with bloodstream forms of *Trypanosoma congolense* and *T. vivax*. International Scientific Council for Trypanosomiasis Research and Control. 18th Meeting. Harare, Zimbabwe, 1985. *OAU/STRC Publ.* No. 113:176–81.

Jordan, A.M. (1986). *Trypanosomiasis control and African rural development.* Longman, London & New York.

Kinghorn, A. & Yorke, W. (1912). Trypanosomes infecting game and domestic stock in the Luangwa Valley, North Eastern Rhodesia. *Ann. trop. Med. Parasit.* 6:301–15.

Losos, G.J. & Gwamaka, G. (1973). Histological examination of wild animals naturally infected with pathogenic African trypanosomes. *Acta trop.* 30:57–63.

McCulloch, B. (1967). Trypanosomes of the *brucei* subgroup as a probable cause of disease in wild zebra, *Equus burchelli. Ann. trop. Med. Parasit.* 61:261–4.

McDowell, R.E. (1977). *Ruminant products: more meat than milk.* Winrock International Livestock and Training Centre, Marilton, Arkansas.

MacLennan, K.J.R. (1980). Tsetse-transmitted trypanosomiasis in relation to the rural economy in Africa: Part 1. Tsetse infestation. *World Anim. Rev.* 36:2–17.

MacLennan, K.J.R. (1981). Tsetse-transmitted trypanosomiasis in relation to the rural economy in Africa: Part 2. Techniques in use for the control or eradication of tsetse infestations. *World Anim. Rev.* 37:9–19.

Maloo, S.H., Chema, S., Connor, R., Durkin, J., Kimotho, P., Maehl, J.H.H.,

Mukendi, F., Murray, M., Rarieya, J.M. & Trail, C.J.M. (1988). The use of chemoprophylaxis in East African Zebu village cattle exposed to trypanosomiasis in Muhaka area, Kenya. In *Livestock production in tsetse infested areas of Africa*: 283–8. International Livestock Centre for Africa, Addis Ababa, Ethiopia.

Molyneux, D.H. (1986). African trypanosomiasis. *Clinics trop. Med. commun. Dis.* 1:535–55.

Morrison, W.I., Murray, M. & Akol, G.W.O. (1985). Immune responses of cattle to African trypanosomes. In *Immunology and pathogenesis of trypanosomiasis*: 103–31. (Ed. Tizard, I.). CRC Press, Inc., Boca Raton, Florida.

Morrison, W.I., Murray, M. & McIntyre, W.I.M. (1981). Bovine trypanosomiasis. *Curr. Top. vet. Med.* 6:469–97.

Mukasa-Mugerwa, E. (1981). *The camel* (Camelus dromedarius): *a bibliographical review*. International Livestock Centre for Africa, Addis Ababa, Ethiopia.

Murray, M. (1974). The pathology of African trypanosomiasis. In *Progress in immunology* 2:181–92. (Eds Brent, L. & Holborow, J.). North Holland Publishing Co., Amsterdam.

Murray, M. (1979). Anaemia of bovine African trypanosomiasis: an overview. In *Pathogenicity of trypanosomes: proceedings of a workshop held at Nairobi, Kenya, 20–23 November 1978*:121–7. (Eds Losos, G. & Chouinard, A.) (*IDRC Monogr.* No. 132e). International Development Research Centre, Ottawa.

Murray, M., Clifford, D.J., Gettinby, G., Snow, W.F. & McIntyre, W.I.M. (1981). Susceptibility to African trypanosomiasis of N'Dama and Zebu cattle in an area of *Glossina morsitans submorsitans* challenge. *Vet. Rec.* 109:503–10.

Murray, M. & Gray, A.R. (1984). The current situation on animal trypanosomiasis in Africa. *Prev. vet. Med.* 2: 23–30.

Murray, M., Grootenhuis, J.G., Akol, G.W.O., Emery, D.L., Shapiro, S.Z., Moloo, S.K., Dar, F., Bovell, D.L. & Paris, J. (1981). Potential application of research on African trypanosomiases in wildlife and preliminary studies on animals exposed to tsetse infected with *Trypanosoma congolense*. In *Wildlife diseases research and economic development*:40–5. (Eds Karstad, L., Nestel, B. & Graham, M.). (*IDRC Monogr.* No. 179e). International Development Research Centre, Ottawa.

Murray, M., Morrison, W.I. & Whitelaw, D.D. (1982). Host susceptibility to African trypanosomiasis: trypanotolerance. *Adv. Parasitol.* 21:1–68.

Murray, M. & Urquhart, G.M. (1977). Immunoprophylaxis against African trypanosomiasis. In *Immunity to blood parasites of animals and man*:209–41. (Eds Miller, L.H., Pino, J.A. & McKelvey, J.J., Jr.). Plenum Publishing Corporation, New York.

Njogu, A.R., Dolan, R.B., Wilson, A.J. & Sayer, P.D. (1985). Trypanotolerance in East African Orma Boran cattle. *Vet. Rec.* 117:632–6.

Njogu, A.R., Ismael, A.A., Dolan, R.B., Okech, G., Sayer, P.D., Opiyo, E.A. & Alushula, H. (1988). Studies on trypanotolerance in East Africa: the Orma Boran. In *Livestock production in tsetse infested areas of Africa*: 447–8. International Livestock Centre for Africa, Addis Ababa, Ethiopia.

Oduye, O.O. & Okunaiya, O.A. (1971). Haemotological studies on the White Fulani and N'Dama breeds of cattle. *Bull. epizoot. Dis. Afr.* 19:213–8.

Olenick, J.G., Wolff, R., Nauman, R.K. & McLaughlin, J. (1988). A flagellar pocket membrane fraction from *Trypanosoma brucei rhodesiense*: immunogold localization and nonvariant protection. *Infect. Immun.* 56:92–8.

Opiyo, E.A., Dolan, R.B., Njogu, A.R., Sayer, P.D. & Mgutu, S.P. (In press). Tsetse control on Galana Ranch. International Scientific Council for Trypanosomiasis Research and Control. 19th Meeting, Lome, Togo, 1987. *OAU/STRC Publ.* No. 114.

Ormsby-Gore, W. (1925). *Report on the East African Commission Cmd. 2387.* His Majesty's Stationery Office, London.

Paling, R.W., Moloo, S.K. & Scott, J.R. (In press). The relationship between parasitaemia and anaemia in N'Dama and Zebu cattle following four sequential challenges with *Glossina morsitans centralis* infected with *Trypanosoma congolense*. International Scientific Council for Trypanosomiasis Research and Control. 19th Meeting, Lome, Togo, 1987. *OAU/STRC Publ.* No. 114.

Payne, W.J.A. (1964). The origin of domestic cattle in Africa. *Emp. J. exp. Agric.* **32**:97–113.

Preston, J.M. & Allonby, E.W. (1979). The influence of breed on the susceptibility of sheep of *Haemonchus contortus* infection in Kenya. *Res. vet. Sci.* **26**: 134–9.

Rickman, L.R. (1984). The blood incubation infectivity test (BIIT) as an epidemiological tool in the study of African human trypanosomiasis (sleeping sickness). In *New approaches to the identification of parasites and their vectors*:199–216. (Eds Newton, B.N. & Michal, F.). Schwabe & Co., Basel.

Roberts, C.J. & Gray, A.R. (1972). Trypanosome infections in captive antelope. *Trans. R. Soc. trop. Med. Hyg.* **66**:335.

Roberts, C.J. & Gray, A.R. (1973a). Studies on trypanosome-resistant cattle. I. The breeding and growth performance of N'Dama, Muturu and Zebu cattle maintained under the same conditions of husbandry. *Trop. Anim. Hlth Prod.* **5**:211–9.

Roberts, C.J. & Gray, A.R. (1973b). Studies on trypanosome-resistant cattle. II. The effect of trypanosomiasis on N'Dama, Muturu and Zebu cattle. *Trop. Anim. Hlth Prod.* **5**:220–33.

Roelants, G.E. (1986). Natural resistance to African trypanosomiasis. *Parasite Immun.* **8**:1–10.

Rurangirwa, F.R., Musoke, A.J., Nantulya, V.M., Nkonge, C., Njuguna, L., Mushi, E.Z., Karstad, L. & Grootenhuis, J.G. (1986). Immune effector mechanisms involved in control of parasitaemia in *Trypanosoma-brucei*-infected wildebeest (*Connochaetes taurinus*). *Immunology* **58**:231–7.

Sachs, R., Schaller, G.B. & Baker, J.R. (1967). Isolation of trypanosomes of the *T. brucei* group from lion. *Acta trop.* **24**:109–12.

Stewart, J.L. (1937). The cattle of the Gold Coast. *Vet. Rec.* **49**:1289–97.

Stewart, J.L. (1951). The West African Shorthorn cattle. Their value to Africa as trypanosomiasis resistant animals. *Vet. Rec.* **63**:454–7.

Toure, S.M., Gueye, A., Seye, M., Ba, M.A. & Mane, A. (1978). A comparison between the pathology of a natural trypanosome infection in Zebu and N'Dama cattle. *Rev. Elev. Med. vet. Pays trop.* **31**:293–313.

Trail, J.C.M., Sones, K., Jibbo, J.M.C., Durkin, J., Light, D.E. & Murray, M. (1985). *Productivity of Boran cattle maintained by chemoprophylaxis under trypanosomiasis risk.* International Livestock Centre for Africa, Addis Ababa, Ethiopia. (*ILCA Res. Rep.* No. 9).

Vale, G.A. (1980). Field studies of the responses of tsetse flies (Glossinidae) and other Diptera to carbon dioxide, acetone and other chemicals. *Bull. ent. Res.* 70:563–70.

Vale, G.A. (1982). The improvement of traps for tsetse flies (Diptera: Glossinidae. *Bull. ent. Res.* 72:95–106.

Vale, G.A. (1987). Prospects for tsetse control. *Int. J. Parasitol.* 17:665–70.

Van Hoeve, K. (1972). Some observations on the performance of N'Dama and Muturu cattle under natural conditions in Northern Nigeria. International Scientific Council for Trypanosomiasis Research and Control, 13th Meeting, Lagos. *OAU/STRC Publ.* No. 105:103–6.

Van Meirvenne, N., Magnus, E. & Vervoort, T. (1977). Comparisons of variable antigenic types produced by trypanosome strains of the sub-genus *Trypanozoon*. *Ann. Soc. Belge Méd. trop.* 57:409–23.

Weitz, B. (1963). The feeding habits of *Glossina*. *Bull. Wld Hlth Org.* 28:711–29.

Whitelaw, D.D., Bell, I.R., Holmes, P.H., Moloo, S.K., Hirumi, H., Urquhart, G.M. & Murray, M. (1986). Isometamidium chloride (Samorin) prophylaxis against experimental metacyclic *Trypanosoma congolense* challenge and its relationship to the development of immune responses in Boran cattle. *Vet. Rec.* 118:722–6.

Wijers, D.J.B. (1969). The history of sleeping sickness in Yimbo location (Central Nyanza, Kenya), as told by the oldest inhabitant of the location. *Trop. geogr. Med.* 21:323–39.

Willett, K.C. (1970). Epizootiology of trypanosomiasis in livestock in East and Central Africa. In *The African trypanosomiases*:766–73. (Ed. Mulligan, H.W.). Allen & Unwin, London.

Wilson, A.J., Njogu, A.R., Gatuta, G., Mgutu, S.P. & Alushula, H. (1981). An economic study of the use of chemotherapy to control trypanosomiasis in cattle in Galana Ranch, Kenya. International Scientific Council for Trypanosomiasis Research and Control, 17th Meeting, Arusha, Tanzania, 1981. *OAU/STRC Publ.* No. 112:306–17.

Wissocq, Y.J., Trail, J.C.M., Wilson, A.D. & Murray, M. (1983). Genetic resistance to trypanosomiasis and its potential economic benefits in cattle in East Africa. International Scientific Council for Trypanosomiasis Research and Control, 17th Meeting. Arusha, Tanzania, 1981. *OAU/STRC Publ.* No. 112:361–4.

*Symp. zool. Soc. Lond.* (1989) No. 61:241–252

# Men, elephants and competition

I.S.C. PARKER
and A.D. GRAHAM

*P.O. Box 15093*
*Nairobi, Kenya*

*B.P. 1444*
*Bangui*
*Central African Republic*

## Synopsis

Elephant decline has occurred widely across Africa over the past two centuries in concert with that of other large mammals and of forests—not all of which carried or produced valuable trophies or timber. The trends parallel accelerating human increase. The events are so in keeping with the theory of competitive exclusion that we suggest that this is the primary cause of decline to which all other influences are secondary.

The only conservation solution in keeping with theory is to provide elephants and those species with which man competes completely with sufficiently large zones of non-overlap; e.g. national parks from which human competition is excluded. Such zones can only exist at cost to society and many parks in Africa fail because that cost is not met. Where it is, exclusive competition can be suppressed and elephants, other large mammals and trees conserved successfully.

## Men, elephants and competition

'Complete competitors cannot coexist' is how Hardin (1960) summarized the competitive exclusion principle. Embodied in the catch phrase 'survival of the fittest', this is—as pointed out by Krebs (1972) among others— central to the concept of Darwinian evolution. Much in the palaeontological record is interpreted in this light; e.g. hyracoids, proboscids and anthra- cotheres were once dominant herbivores, but were out-competed and supplanted by artiodactyls and perissodactyls (Kingdon 1971).

Man has caused extinctions—among them the passenger pigeon, the dodo and Stellar's sea cow and, one supposes, his close relatives—the australopithecines. Conventionally these events are attributed to cupidity, stupidity and lack of foresight. But one man's cupidity may be another's commonsense; stupidity in hindsight may have appeared wise at the time and apparent lack of vision may merely arise from looking in different directions. All are subjective. While they reveal prevailing emotions, they

ZOOLOGICAL SYMPOSIUM No. 61
ISBN 0–19–854009–4

contribute little to elucidating the biological causes of human-induced extinctions.

If competitive exclusion has been a major factor in extinction through evolution, it is common sense to first consider man's role as part of the process. If the competitive exclusion principle proves adequate, alternative and unique explanations may be unnecessary.

Overall, the African elephant has declined numerically and in range (Brooks & Buss 1962; Stewart & Stewart 1963; Laws, Parker & Johnstone 1975; Douglas-Hamilton 1979; Kingdon 1979; Burrill & Douglas-Hamilton 1987, among many) during this century. The trend began even further back; Skead (1980) reported that it had been under way in the Cape from the 17th century; Parker (1979) and Largen & Yalden (1987) documented it in West Africa and Ethiopia respectively in the 19th century. Other examples are given by numerous observers and historians.

Hardin's (1960) contention that where the demands of two sympatric species were sufficiently similar, competition would lead to the extinction of one, raises the question of whether human and elephant demands are so similar as to bring them into competition. Mayr (1963) held that competition for any essential resource—food, a place to live, or to hide or to breed—would be sufficient to bring about extinction.

Parker & Graham (in press) make the point that there are no human food plants that African elephants do not eat. Further, there are no plants used by domestic herbivores that they will not consume. Their dietary catholicism even extends into omnivorism, for elephants close to human dwellings rummage through refuse and also ingest animal protein—a fact well-known in zoos. In dietary terms they compete completely with man for his food plants and indirectly by using the same food resource as his domestic stock. Even in the Sahel, herdsmen lopping *Acacia* branches to bring the foliage into their goats' reach, or camels browsing such trees, compete directly with elephants for food.

By virtue of man's technique of producing food by replacing wild animals and plants with a limited range of domesticants, the two species also compete for land, for every acre under crops is an acre denied to elephants.

The only places in which elephants can hide from man are forest and, to a limited degree, swamps. As pointed out in Parker (1984), elephants in forest are difficult to exterminate. This was demonstrated in Uganda's Budongo forest where sustained Government hunting over 35 years to eliminate elephants failed, even though the area was only 474 km² (Laws *et al.* 1975). Roth's (1984) illustration of elephant distribution in West Africa and his observation that many of the relict populations occurred in remnant forest patches where there was no deliberate elephant conservation policy but, on the contrary, free-for-all hunting, makes the same point.

Yet man competes with forest trees for access to land (Harrison 1987) and in doing so, competes with elephants for places in which they can hide. It seems, therefore, that man competes directly with elephants, not for only one essential resource, but for several. In fact, there is nowhere an elephant can live that man does not covet. The stage is set for Mayr's (1963) predicted extinction of one by the other.

Mayr (1963) also wrote that 'competition becomes more acute as the population of either species increases'. Man has increased throughout this century and currently does so at a faster rate than ever before. The African population's doubling time is now down to 18 years—and it is naive to argue that the observed increase does not reflect a corresponding expansion of agricultural and pastoral activity.

This could, no doubt, be expressed in biometric terms. The rising human biomass must cause a net decrease in wild biomass from the areas coming under human domination. However, given the innate inefficiencies of primate digestion, the trade-off cannot be 1:1. A large proportion of man's crops are consumed by vermin and an even larger proportion wasted by inefficient storage and distribution systems. Indeed it would be surprising if the trade-off was not greater than an order of 1 new human biomass unit to 100 wild units. And, as Africa exports more and integrates with economies outside the continent, so this ratio will further worsen. But the salient point is that Africa's great human increase must, through competitive exclusion, have been matched by a decline in so complete a competitor as the elephant.

Parker & Graham (in press) have explained elephant decline in terms of the competitive exclusion principle. They note that in East Africa, over a 46-year span (1925–71) it corresponded closely to trends in the expansion of human range, human population growth and rising ivory exports (as an index of elephant deaths). Figure 1, derived from Parker & Graham (in press), illustrates the lack of correlation between trends in elephant decline (ivory exports) and the price of ivory (an index of cupidity). Subsequent to 1971 the price of ivory did increase substantially and may have accelerated elephant decline, but only as an additional factor in the process already well established.

In Fig. 2, we show how East African human distribution maps as the negative of elephant distribution at the same time. Compelling evidence that these maps reflect competitive exclusion is provided by J. Kingdon (pers. comm.) who witnessed an almost instantaneous expansion of elephant range when, under ex-President Nyerere's programme of 'Ujamma' communes, people in Tanzania's Rukwa area were forced away from long-settled land. The rapidity with which elephants occupied it strongly suggests that human presence had been the single influence keeping them away.

Parker & Graham (in press) illustrate the two species' preferences for the same rainfall regimes and the inverse relationship between increasing humans and diminishing elephants, predicting extinction at a human

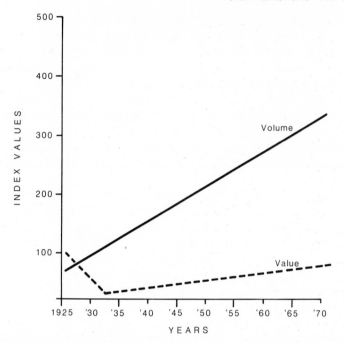

**Fig. 1.** Trends in East African ivory exports and average annual values between 1925 and 1971 expressed as indices with 1925 = 100 (from Parker and Graham in press). The export volume regression equation is $y = -81.372 + 5.8297x$, $r = 0.79$ ($p = 0.001$). The two value equations are:– 1925–33: $y = 340.27 - 9.54x$, $r = -0.97$ ($p = 0.001$) and 1934–71: $y = -7.496 + 1.214x$, $r = 0.84$ ($p = 0.001$).

density of 82.5/km$^2$ on fertile East African soils, but at lower densities on infertile soils in Zimbabwe (Figs 3, 4). The wealth of evidence is so in keeping with competitive exclusion that we accept it as the underlying process in the African elephant's decline.

This immediately raises the question of whether it is also the cause of the declines noted in many other species (Kingdon, 1979; Sidney 1965; Stewart & Stewart 1963, among many). Notably these, too, are most severe where human population densities are greatest. Figure 5 illustrates declines in the East African ranges of elephant, black rhinoceros, the common zebra and giraffe. While the four patterns of contraction show differences, they also indicate that the greatest declines have taken place about the foci of human concentration.

With browsers like the black rhinoceros, competition was particularly severe because they cannot eat the grasses with which humans replaced the thickets that were their best food resource. A much-publicized case was the 996 black rhinoceros shot in Kenya's Makueni location of Machakos District in the late 1940s to make way for settlement (Hunter 1952). The

A.  1949  Human densities  > 25Km$^2$  (■)            B.  1950  Elephant distribution  (■)

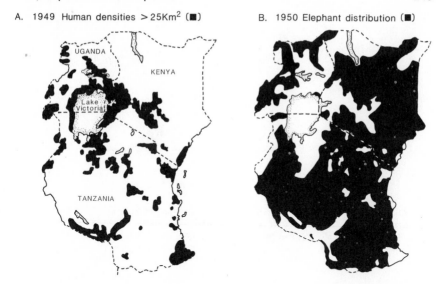

**Fig. 2.** East African elephant range in 1950 (from Kingdon 1979) and the distribution of human densities in excess of 25/km$^2$ in 1948 (from Goldthorpe & Wilson 1960), to show how one is largely the negative of the other.

annual sale of rhinoceros horn by the Kenya Government obtained from 'control' measures between 1930 and 1970 reflected the same process over 40 years (Parker & Martin 1979). Zebra compete directly for cattle grazing and also for planted wheat and they were shot in such numbers in the first

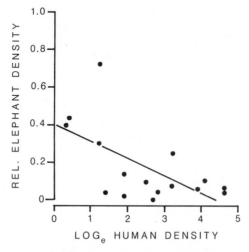

**Fig. 3.** A regression illustrating the relationship between human and elephant densities in Kenya's agricultural districts (from Parker & Graham in press). the regression is $y = -0.925x$, $+ 0.4093$, $r = 0.82$, and predicts elephant extinction at a human density of 82.5/km$^2$.

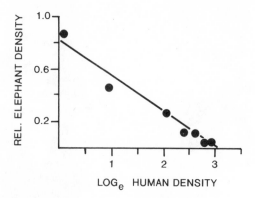

**Fig. 4.** A regression illustrating the relationship between human and elephant densities in seven Zimbabwe districts (from Parker & Graham in press). The regression is $y = -0.268x + 0.787$, $r = -0.99$, and predicts elephant extinction at human densities of 18.9/km². The difference between the Kenya (Fig. 3) and Zimbabwe data is attributed principally to geological factors.

four decades of this century that at one time their hides were used as grain sacks instead of jute or hessian bags. Giraffe, like black rhinoceros, suffered from destruction of browse, though not to the same degree, since by exploiting foliage up to 6 m above ground they are able to make use of residual trees that survive the opening of thickets by humans.

The evidence from these other species, while not as complete as that assembled for elephants, is nonetheless entirely consistent with the competitive exclusion principle. And nowhere is the point made more forcefully than in the results of aerial surveys over > 200,000 km² of eastern Africa which show that > 90% of large wild herbivores occurred where there were no people (M. Norton-Griffiths pers. comm.). It is so in keeping with theory that we suggest competitive exclusion as the primary cause of the observed decline in African wildlife.

Our suggestion does not imply that other factors have not influenced the trends. In some species a valuable commodity—ivory and rhinoceros horn for instance—may add impetus to the rate of change. An analogy might be declining virility in the human male. The outcome, impotence, is the pre-ordained consequence of inexorably increasing age. Drink, like ivory's value, may hasten the flaccid end and provide titillating scuttlebut, but it has scant bearing on the underlying process and what may be done to reverse it.

Competitive exclusion brings Mayr's (1963) second consequence of competition to the fore. He concluded that the coexistence of competing species was only possible if there was 'a sufficiently large zone of non-overlap.' National parks provide such zones: indeed, while it has never been stated in so many words, the prohibition of human settlement in national parks implicitly acknowledges competitive exclusion. They are artificial

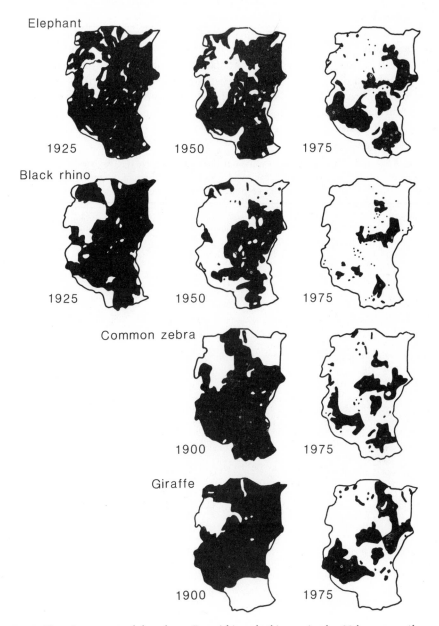

**Fig. 5.** Changing ranges of four large East African herbivores in the 20th century (from Kingdon 1979).

zones of non-overlap, examples, to paraphrase Hardin (1960), of conservation prostheses.

Resistance to as powerful a process as human expansion can be expected to be costly. Bell & Clarke (1985) recognized this and indicated that successful conservation in Africa entails ± \$200/km² in recurrent costs. We have analysed the 75 national parks listed in *The Directory of Afrotropical Protected Areas* (IUCN 1987) for which some evidence of annual expenditure is given. Table 1 lists them with their sizes and the apparent annual expenditure in \$/km².

**Table 1.** A list of those African national parks for which IUCN (1987) has published annual recurrent or total costs allocated for their management, their sizes and whether or not they experience poaching or trespass as major problems.

| Country | National park | Size (km²) | \$/km² | Success | Failure |
|---------|---------------|-----------|--------|---------|---------|
| Ivory coast | Azagny | 170 | 0 | | + |
| | Banco | 300 | 0 | | + |
| | Mt Peko | 340 | 0 | | + |
| | Tai | 3300 | 121 | | + |
| Ethiopia | Abijatta-Shalla | 800 | 7 | | + |
| | Awash | 720 | 116 | | + |
| | Bale Mts | 2200 | 36 | − | |
| | Mago | 1500 | 63 | | + |
| | Nechisar | 700 | 92 | | + |
| | Omo | 3450 | 2 | | + |
| | Simen | 225 | 67 | | + |
| | Yanguda Rassa | 2000 | 70 | | + |
| Kenya | Aberdare[a] | 766 | 175 | | + |
| | Amboseli | 392 | 231 | − | |
| | Lake Nakuru | 200 | 362 | − | |
| | Lambwe Valley | 120 | 139 | − | |
| | Meru | 870 | 120 | | + |
| | Mt Elgon | 169 | 365 | | + |
| | Nairobi | 117 | 1564 | − | |
| | Ol Doinyo Sabuk | 18 | 1444 | − | |
| | Tsavo | 20821 | 26 | | + |
| | Kisite Mpungwe | 39 | 1162 | − | |
| | Malindi Watamu | 16 | 189 | − | |
| Liberia | Sapo | 1307 | 51 | | + |
| Malawi | Kasungu | 2316 | 34 | − | |
| | Lake Malawi | 94 | 532 | − | |
| | Lengwe | 887 | 32 | | + |
| | Liwonde | 548 | 75 | − | |
| | Nyika | 3134 | 14 | | + |
| Mauritania | Banc d'Arguin | 11730 | 15 | | + |
| Namibia | Etosha | 22270 | 46 | − | |

| | | | | |
|---|---|---|---|---|
| Niger | Parc W | 2200 | 5 | + |
| Nigeria | Kainji Lake | 5431 | 3 | + |
| South Africa | Kruger | 19485 | 590 | − |
| | Kalahari Gemsbok | 9591 | 37 | + |
| | Karoo | 270 | 370 | − |
| | Augrabies Falls | 94 | 2335 | − |
| | Addo | 89 | 3090 | − |
| | Mountain Zebra | 65 | 5760 | − |
| | Golden Gates | 62 | 7258 | − |
| | Tsitsikama | 33 | 15152 | − |
| | Bontebok | 28 | 1786 | − |
| | Pilanesburg | 500 | 3234 | − |
| | Borakalalo | 74 | 10730 | − |
| Tanzania | Arusha | 137 | 350 | + |
| | Gombe | 52 | 98 | + |
| | Katavi | 2253 | 8 | + |
| | Kilimanjaro | 756 | 96 | + |
| | Lake Manyara | 320 | 80 | − |
| | Mikumi | 3230 | 33 | + |
| | Ruaha | 12950 | 6 | + |
| | Serengeti | 14763 | 15 | + |
| Zaire | Kundelunga | 7600 | 2 | + |
| | Maiko | 10830 | 19 | + |
| | Selonga | 36560 | 9 | + |
| | l'Upemba | 11730 | 42 | + |
| Zambia | Blue Lagoon | 450 | 111 | + |
| | Isangano | 840 | 18 | + |
| | Kafue | 22400 | 13 | + |
| | Kasanka | 390 | 64 | + |
| | Liuwa Plain | 3660 | 10 | + |
| | Lochinvar | 254 | 122 | + |
| | Lower Zambezi | 4140 | 4 | + |
| | Luambe | 254 | 118 | − |
| | Lukusuzi | 2720 | 33 | + |
| | Lusenga | 880 | 6 | + |
| | Mosi-oa-Tunya | 66 | 1364 | − |
| | Mweru Wantipa | 3134 | 14 | + |
| | North Luangwa | 4636 | 26 | + |
| | Nsumbu | 2020 | 50 | + |
| | Nyika | 80 | 100 | − |
| | Sioma Ngwezi | 5276 | 7 | + |
| | South Luangwa | 9050 | 44 | + |
| | West Lunga | 1684 | 18 | + |
| Zimbabwe | Matobo | 432 | 162 | − |

+, poaching or trespass reported and the park arbitrarily deemed to be failing.

−, no poaching or trespass reported or only considered a minor problem and the park deemed a success.

[a] The Directory indicates no serious poaching or trespass in the Aberdare National Park, but the loss of black rhinoceros there indicates that poaching does take place and is serious.

The success of many national parks depends, in large part, on preventing human-induced changes in plant or animal populations. *The Directory of Afrotropical Protected Areas* indicates whether parks listed suffer from poaching or trespass, both of which are inimical to their basic purpose. We have taken the *Directory*'s lack of comment on either poaching or trespass, or statement that they are slight, to be an arbitrary indication that the park concerned is successfully managed. Conversely, where the *Directory* (or our own knowledge) indicates that either is common or out of control, we have taken it as evidence of failure. The outcome is also given in Table 1 and in Fig. 6.

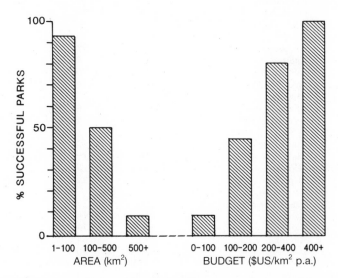

**Fig. 6.** Two histograms to illustrate the relationship between the successful suppression of trespass and poaching in African national parks and (a) their size and (b) their annual recurrent budgets in $US/km$^2$.

The data are too subjective for a rigorous analysis, but even the simple approach we have taken makes two vital points:

1. Park size has a direct relationship to success or failure. The smaller a park, the greater its chance of preventing poaching or trespass and, conversely, the larger it is, the lower its chances. More than 90% of those that are $< 100$ km$^2$ are successful; 50% of those between 200 and 500 km$^2$ succeed; and $< 10\%$ of those over 500 km$^2$ succeed.

2. Money spent on management also has a direct influence on ability to prevent poaching or trespass. Over 90% of parks which spend $< \$100$ km$^2$ annually fail; this drops to 55% in parks which spend $\$100–200/$km$^2$ annually; it drops further to 40% where $\$200–400/$km$^2$ are spent annually

and failure falls to zero where outlay exceeds $400/km^2$ annually. The data complement the Bell & Clarke (1985) finding.

As would be expected, size and budget are correlated with the higher expenditures per km$^2$ linked to the smaller parks. Budgets of > $400/km$^2$ are only reported for parks of < 200 km$^2$ (with two exceptions—Kruger and Pilanesburg). Thus while large size, *per se*, does not necessarily predispose to failure—in fact larger parks could be expected to be more resilient—nearly all large parks have a poor prognosis as long as the imperative of a minimum appropriate budget is unrecognized or side-stepped.

The analysis has profound implications for conservation. Small park size and/or adequate funding appear to be vital, correlated, ingredients in conserving Africa's large mammals and plants successfully. Failure to grasp that this is so results in money and manpower being spent where human competition cannot be effectively inhibited, or applied too thinly in the parks to achieve success.

Either the authorities concerned have been unaware of the fundamental ecological processes (to which man is integral) that they are dealing with and the real cost of manipulating those processes, or they have not considered the price worth paying. Yet it is here, in the realms of basic biological law and socio-economics, that Africa's widely reported conservation failures are rooted.

It seems to us naive, if not perverse, to disregard the process of competitive exclusion and, instead, to promote alternative explanations for the failure to conserve elephants, forest trees, etc. To continue to interpret emotionally derived factors such as cupidity as primary rather than secondary processes can only cloud the basic issues and expose the only places where species can be kept free of human competition to an increasing risk of obliteration. Even on reserved land, competition from man is an active and intensifying process throughout sub-Saharan Africa. Conservation needs to be elevated to an objective plane based on ecological and socio-economic principles.

Where elephants are concerned, they decline in parks where funds are insufficient to stop competition through poaching. Where adequate funds are available, they are not declining. And if this stark writing on the wall is not read, then the first alternative proposed by Mayr (1963)—extinction—can be the only outcome.

## References

Bell, R.H.V. & Clarke, J.E. (1985). Funding and financial control. In *Conservation and wildlife management in Africa: the proceedings of a U.S. Peace Corps workshop held in Kasungu National Park, Malawi*: 543–55. (Eds Bell, R.H.V. & McShane-Cahuzi, E.). U.S. Peace Corps, Washington. D.C.

Brooks, A.C. & Buss, I.O. (1962). Past and present status of the elephant in Uganda. *J. Wildl. Mgmt* **26**:38–50.

Burrill, A. & Douglas-Hamilton, I. (1987). *African elephant database project: final report.* UNEP, Nairobi.

Douglas-Hamilton, I. (1979). *African elephant action plan.* IUCN, Gland, Switzerland.

Goldthorpe, J.E. & Wilson, F.B. (1960). *Tribal maps of East Africa and Zanzibar.* East African Institute of Social Research, Kampala, Uganda.

Hardin, G. (1960). The competitive exclusion principle. *Science, N.Y.* **131**:1291–7.

Harrison, P. (1987). *The greening of Africa.* Paladin, London.

Hunter, J.A. (1952). *Hunter.* Hamish Hamilton, London.

IUCN (1987). *The directory of Afrotropical protected areas.* IUCN, Gland, Switzerland.

Kingdon, J. (1971). *East African mammals: an atlas of evolution in Africa.* I. Academic Press, London, New York.

Kingdon, J. (1979). *East African mammals: an atlas of evolution in Africa.* **IIIB** (*Large mammals*). Academic Press, London, New York, San Francisco.

Krebs, C.J. (1972). *Ecology:· the experimental analysis of distribution and abundance.* Harper & Row, New York, London.

Largen, M.J. & Yalden, D.W. (1987). The decline of elephant and black rhinoceros in Ethiopia. *Oryx* **21**:103–6.

Laws, R.M., Parker, I.S.C. & Johnstone, R.C.B. (1975). *Elephants and their habitats: the ecology of elephants in North Bunyoro, Uganda.* Clarendon Press, Oxford.

Mayr, E. (1963). *Animal species and evolution.* Harvard University Press, Cambridge (Mass.).

Parker, I.S.C. (1979). *The ivory trade,* Report to the United States Department of the Interior, U.S. Government, Washington D.C. (Mimeo. 900pp.)

Parker, I.S.C. (1984). Rainfall, geology, elephants and men. In *Proceedings of a symposium on the extinction alternative*:137–77. (Ed. Mundy, P.). Endangered Wildlife Trust, Johannesburg.

Parker, I.S.C. & Graham, A.D. (In press). Elephant decline: an hypothesis. *Int. J. envir. Stud.*

Parker, I.S.C. & Martin, E.B. (1979). Trade in African rhino horn. *Oryx* **15**:153–8.

Roth, H.H. (1984). Distribution and status of elephants in West Africa in relation to vegetation zones. In *The status and conservation of Africa's elephants and rhinos*:186–7. (Eds Cumming, D.H.M. & Jackson, P.). IUCN, Gland, Switzerland.

Sidney, J. (1965). The past and present distribution of some African ungulates. *Trans. zool. Soc. Lond.* **30**:1–397.

Skead, C.J. (1980).*Historical mammal incidence in the Cape Province* **1**. Department of Nature and Environmental Conservation of the Provincial Administration, Cape Town.

Stewart, D.R.M. & Stewart, J. (1963). The distribution of some large mammals in Kenya. *J. E. Afr. nat. Hist. Soc.* **24**:1–52.

Symp. zool. Soc. Lond. (1989) No. 61:253–265

# A survey of wildlife populations in Tanzania and their potential for research

K.N. HIRJI

*Serengeti Wildlife Research Institute*
*P.O. Box 661*
*Arusha, Tanzania*

## Synopsis

From 1985 to 1988, periodic aerial surveys of wildlife were conducted in the Ngorongoro Conservation Authority Area and in the major national parks such as Serengeti, Tarangire, Manyara, Rusha, Katavi and Mikumi. Similar counts were also carried out in the major game reserves, namely the Selous, Mkomazi and Rungwa-Kizigo. Invariably, in all the wildlife conservation areas studied, the population of elephants showed significant, and in most cases alarming, reduction. For instance, in the Selous which is one of the largest known depositories of elephants in the world, the 1976 population estimate of 109 419 was reduced to 55 153 in 1986. In the Ruaha National Park, the 1977 population estimate of 24 625 elephants was reduced to 3478 in 1987 whilst that of Rungwa-Kizigo Game Reserve was reduced form 14 528 to 8662 for the same period. Elsewhere in Serengeti, Manyara, Mkomazi, Mikumi and probably in the rest of the conservation areas in Tanzania, there has been a significant decline in elephant population. Nationally the elephant population has gone down from 198 700 in 1977 to 89 100 in 1987 and the trend indicates that the decline will continue in the years to come. With reference to buffalo, a significant increase in the estimated population was noted in the Selous, from 78 893 in 1976 to 157 536 in 1988. However, in the Serengeti National Park, the population decreased from 74 237 in 1975 to 43 456 in 1986.

In the rest of the areas, the population has remained steady with minor fluctuations. In the Serengeti, the population of wildebeest doubled between 1971 and 1977 when the population reached 1.4 million. Thereafter, the population was estimated at 1 337 849 in 1984 and 1 146 340 in 1986. The decline in 1986 may be due to the dry spell in 1985 and possibly to poaching. However, comparison of long-term data shows that the population is stable around 1.2 million. The zebra population has stabilized in all the areas visited; and in all these areas, particularly in the Selous and Katavi, the population of hippopotamus was observed to be higher than before. The estimates of wildlife populations in Tanzania indicate that, with the exception of species such as the rhinoceros and elephant, the situation is non-critical.

ZOOLOGICAL SYMPOSIUM No. 61
ISBN 0–19–854009–4

However, it is acknowledged that in order to maintain the present populations of species, additional conservation effort and strategies are needed now and in the future.

Without any doubt, Tanzania, through its research organization the Serengeti Wildlife Research Institute, offers excellent opportunities for wildlife research. Through its Research Centres in the Selous, Mahale, Gombe and Serengeti the Institute co-ordinates research throughout the National Parks, the Game Reserves and the Ngorongoro Conservation Area as well as in the rest of the country.

## Introduction

Before 1985, most of the aerial surveys for the purpose of estimating wildlife populations were limited to the Serengeti National Park and the Ngorongoro Conservation Area Authority, particularly to the Crater. Some surveys were also carried out in the Selous (Douglas-Hamilton 1976), Ruaha National Park (Norton-Griffiths 1975; Barnes & Douglas-Hamilton 1982) and Tarangire (Ecosystems Ltd 1980). Populations of wildlife were estimated in particular areas and at specific times at irregular intervals since 1961. However, little effort was made to cover all National Parks and Game Reserves and the surveys were not standardized. Until 1984, a population estimate of wildlife for the whole country was not available. Upgraded estimates for Serengeti, Ngorongoro, Selous, Ruaha and Mikumi were not available and it was not possible to evaluate the population trends in these areas. The Institute was therefore asked by the Department of Wildlife to carry out frequent population surveys in as many areas as possible, starting with the major conservation areas, the objective being that by the year 1989 all National Parks, Game Reserves and some Controlled Areas should be surveyed.

In order to carry out this function, the Serengeti Ecological Monitoring Programme (SEMP) was established under the auspices of the Institute with financial support from the World Wildlife Fund, the Frankfurt Zoological Society and the Serengeti Wildlife Research Institute. The main objectives of the SEMP were as follows:–

1. To monitor all ecological parameters, specifically wildlife populations, of the Serengeti Ecosystem. This in fact was to be a continuation of a previous programme existing in Serengeti. Every effort has been made to collaborate with the Masai-Mara programme and we have already conducted two counts together.

2. To conduct wildlife population surveys in all major National Parks and Game Reserves and other areas of special interest.

3. To standardize methodology of aerial surveys so that continuous comparisons can be made in order to determine population trend for each area.

4. To train local wildlife scientists and wildlife management officers in current ecological monitoring techniques. Specifically, to train a team of local personnel to conduct aerial surveys.

To date, SEMP has carried out aerial surveys in the Serengeti, Ngorongoro, Tarangire, Ruaha, Katavi, Rungwa-Kizigo and Manyara. It has also collaborated in the Selous, Mkomazi, Mikumi and Manyara surveys.

This paper attempts to draw some conclusions from the population estimate of wildlife thus far obtained from different conservation areas and establish trends for some key species. The paper will also attempt to identify the potential and critical areas of research.

## Methodology

All aerial surveys were conducted in single-engine four-seater aircraft using standard systematic reconnaissance flight (SRF) methods (Norton-Griffiths 1978). In summary, flight lines spaced at intervals of 5 or 10 km were flown either in an east-west-east or north-south-north direction. Census strips were delineated by rods attached to the wing struts and were set to be 150 m at 350 feet above ground level. Strip width calibrations were carried out before and after the census and resulted in a combined calculated strip width of 340 m at a flying height of 330 feet (calculated averages of four surveys). Sub-units were determined by 30 or 60 s of flying time. Radar altimeter readings were recorded at the start of each sub-unit by the Front Seat Observer (FSO), who was also responsible for calling out the start of each new sub-unit. The FSO also recorded the presence or absence of required ecological parameters such as water availability, soil erosion, burns, vegetation type and cover, human activities, roads and tracts, tree damage, etc. within each sub-unit. Two Rear Seat Observers (RSOs) recorded on tape recorders all wildlife, livestock and human settlements, including poaching camps, seen within the defined counting strip width at sequential sub-units called out by the FSO. No oblique photographs were taken by the RSOs during counting.

When a photographic survey was required, an aircraft equipped with radar altimeter and a vertically mounted and remotely controlled Nikon 35 mm camera with 250 exposure magazine was used. Vertical sample photographs (transparencies) were taken at intervals of 20 or 30 s simultaneously with the radar altimeter reading by the FSO along the determined transect. These photographs are analysed for vegetation cover and pattern (Norton-Griffiths 1979) and form baseline data for some selected areas such as Tarangire, Katavi, Northern Selous, Serengeti and Ngorongoro.

## Results and discussion

For most areas, wet- and dry-season surveys are being conducted in order to obtain population estimates for comparison. Where possible, previous counts are duplicated, in order to determine population trends.

### Serengeti National Park (SNP)

Population surveys have been conducted in the SNP since 1961. In particular, records of the populations of migrating species, specifically the wildebeest, were kept from 1961 until 1977. From 1977 to 1983 there were no records for SNP. In 1984 and 1986, the monitoring of the wildebeest population was resumed. To date, three major counts have been undertaken in SNP. The wildebeest were counted in 1984 and 1986, the brown animal count of 1986 covered the northern SNP and the Masai-Mara, whilst those of 1985 and 1988 covered the whole SNP. Also in 1986, a total count of elephants and buffalo was carried out in SNP and Masai-Mara. Summarizing the above counts and taking into account the previous counts, the following conclusions can be drawn.

1. The wildebeest population has remained stable over the decade. The population, which was estimated at 0.7 million in 1971, doubled to 1.4 million in 1977 and gradually declined to 1.14 million in 1986 (Sinclair 1987). The drop from 1977 to 1986 may be due to severe dry-season mortality especially in 1984, to diseases or to increased poaching.

2. With regard to brown animals, there has been a significant increase in the populations of topi, kongoni and impala from 27 000, 9000 and 57 000 in 1971 to 53 000, 13 000 and 69 000 in 1985 respectively. There has been a slight decrease in the populations of giraffe and eland, whilst zebra have maintained their population at 200 000 (Borner 1985). The population of Thomson's gazelle has levelled out at around 572 920 ($\pm$129 431, 95% C.L.) in 1986 from 700 000 in 1972. Previously, it was shown that the population may have been reduced to approximately 250 000 in 1985 (Borner 1985). The brown animal count will estimate the correct population of Thomson's gazelle which is the second most numerous species in SNP.

3. The 1986 total count of elephants and buffalo for the whole of SNP has shown that the population of elephants in the Park has declined from approximately 2200–2500 animals (1966–1976 census) to 467 in 1986. This is a decrease of 80% of the original population. The population of buffalo in SNP has also been reduced from 74 237 in 1975 to 43 456 in 1986, a decrease of 59%. Both elephants and buffalo have been subjected to intensive and widespread poaching between 1977 and 1985. Population counts and estimates of predators and endangered species such as hunting-dogs are continuing and we hope that by the end of 1989 we will have a preliminary list of the major predators.

## Ngorongoro Conservation Area Authority (NCAA)

In 1987, wet (April) and dry (October) season counts were conducted in order to estimate the population and distribution of wildlife and livestock in NCAA. In this exercise, the Crater was considered separately. Results presented in Table 1 highlight the extremely seasonal distribution of Ngorongoro wildlife populations. It is obvious that species such as eland, Thomson's gazelle, impala, kongoni, ostrich and zebra migrate away from NCAA during the dry season. Wildebeest and zebra migrate annually and the figures given for the SNP apply to some extent to NCAA.

**Table 1.** Comparison of wet (April) and dry (October) season estimates within Ngorongoro stratum (1987).

| Species | Wet season estimates (April) | Dry season estimates (October) | Value of B |
|---|---|---|---|
| Buffalo | 388 | 629 | 0.40 |
| Eland | 5372 | 161 | 12.95 |
| Giraffe | 1666 | 1226 | 1.37 |
| Grant's gazelle | 6715 | 6452 | 0.11 |
| Thomson's gazelle | 145373 | 1484 | 7.36 |
| Impala | 3301 | 452 | 2.62 |
| Kongoni | 275 | 48 | 1.87 |
| Ostrich | 2217 | 613 | 3.46 |
| Zebra | 59832 | 2855 | 4.06 |

Results presented in Table 2 summarize the population estimates and trends of key animal species in the Ngorongoro Crater from 1964 to 1987. The population of wildebeest in the Crater has decreased yearly whilst that of zebra has remained stable. Buffalo population has increased from 600 in 1964–68 to over 2000 in 1987. Eland, on the other hand, have decreased drastically and in future may be altogether absent from the Crater. The rhinoceros population shows an upward trend and is currently estimated at over 15 animals.

Table 3 gives a summary of human and livestock population trends in the NCAA from 1957 to 1987. Whereas the population of people has risen steadily from 10 633 in 1957 to 22 637 in 1987, that of cattle has declined from 161 034 to 136 292 during the same period.

NCAA is an area of multiple land use. The major issue here is to achieve a desired balance between conservation of wildlife, tourism, pastoralism and eventually agriculture. The potential for research along these lines with direct input towards achieving the desired balance is almost unlimited.

## The Selous Game Reserve

Table 4 summarizes the population estimates of key species of animals for 1986 and compares these estimates with those of 1976. Invariably, all key

**Table 2.** Historical data on population estimates in the Ngorongoro Crater.

| | Wildebeest | Zebra | Buffalo | Gazelle | Thomson's gazelle | Kongoni | Eland | Warthog | Ostrich | Rhinoceros |
|---|---|---|---|---|---|---|---|---|---|---|
| 1964–1968 | | | | | | | | | | |
| Wet season | 13764 | 4026 | 661 | 1599 | 3090 | 140 | 387 | 59 | 38 | 31 |
| Dry season | 16535 | 4499 | 228 | 2578 | 3657 | 145 | 214 | 31 | 29 | 14 |
| Long-term averages | | | | | | | | | | |
| 1967–1974 | 15112 | 4116 | 369 | 1531 | 3277 | 57 | 338 | 57 | | |
| 1975–1983 | 12695 | 4108 | 942 | 1364 | 2709 | 158 | 217 | 21 | | |
| Serengeti Ecological Monitoring Programme 1986 | | | | | | | | | | |
| Dry season | 11847 | 4297 | 1455 | 2136 | 3392 | 72 | 59 | 19 | 26 | 2 |
| Ngorongoro Ecological Monitoring Programme 1987 | | | | | | | | | | |
| Wet season | 9011 | 3127 | 2714 | 3588 | 4342 | 70 | 64 | 25 | 23 | 9 |
| Dry season | 7415 | 4332 | 2855 | 1135 | 4677 | 112 | 7 | 31 | 40 | 8 |

**Table 3.** Human and livestock population trends in the Ngorongoro Conservation Area (1957–1987).

| Year | | People | Cattle | Sheep & Goats |
|---|---|---|---|---|
| 1957[a] | | 10633 | | |
| 1960 | | | 161034 | 100689 |
| 1962 | | | 142230 | 83120 |
| 1963b Dec. | | | 116870 | 66320 |
| 1964 June | | | 132490 | 82980 |
| 1966 Feb. | | 8728 | 94580 | 68590 |
| 1968 | | | 103568 | 71196 |
| 1974 | | 12645 | 123609 | 157568 |
| 1978 | | 17982 | 107838 | 186985 |
| 1980 | | 14645 | 118358 | 144675 |
| 1987[c] | Bomas | People | Cattle | Sheep/Goats |
| (a) | 1070 | 22637 | 137398 | 137389 |
| (b) | 2904 | – | 136292 | 121272 |

[a] Data for 1957–1980 from J. Kayera, paper contributed to the IUCN Workshop—'Towards a regional conservation strategy for the Serengeti'.
[b] Data from Ngorongoro's annual report for 1967.
[c] Preliminary reports by the IUCN Ngorongoro Conservation and Development Project and SEMP:
(a) Ground census carried out from 23 April to 27 June 1987.
(b) Systematic reconnaissance flight (SRF) from 14–16 April.
Merged wet- and dry-season estimates.

**Table 4.** Selous census zone (area 74 000km$^2$) comparison of dry-season estimates.

| Species | 1976 | 1986 | % Change | D-factor |
|---|---|---|---|---|
| Baboon | 5932 | 1367 | −77 | −2.17 |
| Buffalo | 78893 | 157536 | 100 | 1.79 |
| Eland | 11009 | 4303 | −61 | −1.84 |
| Elephant | 109419 | 55153 | −50 | −5.50 |
| Elephant (dead) | 6493 | 11390 | 75 | 3.78 |
| Giraffe | 1332 | 2000 | 50 | 0.70 |
| Greater kudu | 1635 | 380 | 77 | −2.31 |
| Hartebeest | 34507 | 15288 | −56 | −4.07 |
| Hippopotamus | 18505 | 16857 | −9 | −0.32 |
| Impala | 43891 | 16731 | −62 | −5.46 |
| Rhinoceros | 2541 | 51 | 98 | −5.87 |
| Sable | 9728 | 4379 | −55 | −2.74 |
| Warthog | 20232 | 6657 | −67 | −4.51 |
| Waterbuck | 12060 | 4936 | −59 | −2.70 |
| Wildebeest | 69044 | 63202 | −8 | −0.30 |
| Zebra | 44421 | 22299 | −50 | −3.40 |

species have declined during the ten-year interval, with the exception of buffalo, which shows a 100% increase. The most important change is that of elephants which declined from the estimated population of 109 419 animals in 1976 to 55 153 in 1986.

The Selous is the major tourist hunting area in Tanzania. The decline of the populations of key species, especially trophy animals, must be linked with this industry. Poaching has been the major problem and the elephant population decline is due to very intensive and widespread poaching. The population of rhinoceros has been reduced from the estimated figure of 2500 animals in 1976 to less than 50 in 1986.

### Other game reserves and national parks

Population estimates of key species were determined for Rungwa-Kizigo Game Reserve and Ruaha, Katavi and Tarangire National Parks during the dry season of 1987. Results of the surveys are represented in Tables 5, 6, 7 and 8 respectively. Duplicated wet-season counts are being conducted in May 1988. The major objective of these surveys is to establish a standardized sampling procedure for future comparisons. The wet-season counts of May 1988 will as far as possible replicate those of October 1987 in order to get comparisons between the dry and the wet season.

Previously, it has not been possible to compare population estimates from different counts. With the exception of elephants, no meaningful conclusions can be drawn from these estimates.

### Trends in elephant populations in Tanzania

The total estimated elephant population for 1977 was 180 700. Over 150 000 of the elephants were confined to the Selous (109 419), Ruaha

**Table 5.** Population estimates and confidence limits (C.L.): Rungwa and Kizigo Game Reserves wildlife census, October 1987.

| Species | Estimate of numbers | 95% C.L. |
|---|---|---|
| Elephant | | |
|   Live | 9049 | 16.6 |
|   Skeletons | | |
|     Grey | 719 | 30.1 |
|     White | 3463 | 50.0 |
|     Total | 4182 | – |
|   Carcase ratio | 31.6% | – |
| Buffalo | 13393 | 121.9 |
| Duiker | 490 | 37.0 |
| Eland | 3103 | 115.0 |
| Greater kudu | 163 | 140.5 |
| Giraffe | 1993 | 18.5 |
| Impala | 2646 | 89.9 |
| Roan antelope | 2809 | 63.9 |
| Sable antelope | 98 | 105.0 |
| Warthog | 1797 | 14.5 |
| Zebra | 8036 | 43.5 |
| Kongoni | 1927 | 66.3 |

**Table 6.** Population estimates and confidence limits (C.L.): Ruaha National Park Wildlife Census, October 1987.

| Species | Estimate of numbers | 95% C.L. |
|---|---|---|
| Elephant | | |
|   Live | 3478 | 103.0 |
|   Dead | | |
|     Grey skeleton | 3445 | 30.4 |
|     White skeleton | 1950 | 24.5 |
|     Rotten | 1365 | 145.1 |
|     Fresh | 33 | 206.1 |
|     Total | 6793 | − |
|   Carcase ratio | 66.1% | |
| Buffalo | 30423 | 134.2 |
| Duiker | 1268 | 154.3 |
| Eland | 2470 | 129.8 |
| Greater kudu | 975 | 156.2 |
| Giraffe | 3705 | 23.8 |
| Impala | 5396 | 13.3 |
| Ostrich | 195 | 79.1 |
| Roan antelope | 813 | 94.3 |
| Warthog | 1040 | 81.0 |
| Zebra | 9328 | 9.7 |

**Table 7.** Population estimates and confidence limits (C.L.): Katavi wildlife survey, October 1987.

| Species | Estimates of numbers | | | Totals and 95% C.L. | |
|---|---|---|---|---|---|
| | Katavi N. Park | Northern area | Western area | Total | C.L. (%) |
| Elephant | | | | | |
|   Live | 1284 | 1317 | 458 | 3059 | 102.2 |
|   Skeletons | | | | | |
|     Grey | 121 | 75 | 0 | 196 | 37.4 |
|     White | 76 | 30 | 44 | 150 | 36.5 |
|     Recent | 0 | 30 | 0 | 30 | 129.9 |
|   Carcase ratio | 13.3% | 9.3% | 9.6% | 10.9% | − |
|   Poacher's camps | 45 | 60 | 15 | 120 | 44.2 |
| Zebra | 2312 | 973 | 0 | 3285 | 61.3 |
| Warthog | 227 | 150 | 59 | 436 | 41.8 |
| Eland | 0 | 15 | 207 | 222 | 199.8 |
| Giraffe | 287 | 45 | 30 | 362 | 77.6 |
| Buffalo | 93473 | 299 | 0 | 93792 | 93.6 |
| Duiker | 121 | 60 | 44 | 225 | 20.0 |
| Roan antelope | 453 | 45 | 0 | 498 | 177.6 |
| Kongoni | 121 | 105 | 0 | 226 | 106.9 |
| Reedbuck | 136 | 0 | 0 | 136 | 108.1 |
| Waterbuck | 212 | 0 | 0 | 212 | 172.1 |
| Topi | 1375 | 75 | 0 | 1450 | 104.9 |
| Hippopotamus | 1481 | 0 | 0 | 1481 | 154.6 |
| Bushbuck | 30 | 0 | 15 | 45 | 111.2 |

**Table 8.** Population estimates and confidence limits (C.L.): Tarangire wildlife survey, October 1987.

| Species | Estimate of numbers | 95% C.L. |
|---|---|---|
| Cattle | 96047 | 63.5 |
| Sheep and goats | 56188 | 76.8 |
| Donkeys | 152 | 111.4 |
| Elephant | | |
|   Live | 6110 | 51.5 |
|   Dead | | |
|     Grey skeleton | 152 | 53.5 |
|     White skeleton | 258 | 32.2 |
|     Fresh | 30 | 142.8 |
|   Carcase ratio | 6.7% | |
| Buffalo | 23818 | 102.1 |
| Zebra | 20089 | 1.8 |
| Wildebeest | 24197 | 71.8 |
| Impala | 7217 | 33.7 |
| Grant's gazelle | 1486 | 20.3 |
| Eland | 1592 | 62.5 |
| Giraffe | 2441 | 13.8 |
| Ostrich | 1395 | 14.5 |
| Kongoni | 2365 | 15.2 |
| Lesser kudu | 61 | 97.4 |
| Oryx | 637 | 88.5 |
| Reedbuck | 212 | 142.0 |
| Waterbuck | 106 | 199.0 |
| Warthog | 212 | 60.3 |

National Park (22 852) and Rungwa-Kizigo Game Reserve (19 575). In 1987, the total population was estimated at 89 100 (Selous—55 153, Ruaha—3673 and Rungwa-Kizigo—9049) giving a reduction rate of 50% during the ten-year period. Table 9 shows that if the current trend continues, there will be an estimated population of only 62 300 animals in 1990 (Serengeti Ecological Monitoring Programme 1988). Tables 10 and 11 show the decline of the elephant population in both the major and minor dispersal areas.

The elephant population decline in Ruaha National Park from 22 852 in 1977 to 3673 in 1987 (11 929 in 1983) is the most dramatic (Table 10).

**Table 9.** Total elephant population numbers and trends.

| Year | Total | Annual decrease % |
|---|---|---|
| 1977 | 180 700 | — |
| 1987 | 89 100 | 7.7 |
| 1990 projection | 62 300 | 11.3 |

**Table 10.** Estimates of numbers from elephant populations in Tanzania: major populations.

|  | Year | Ruaha Nat. Park | Rungwa and Kizigo G.R. | Selous Zone |
|---|---|---|---|---|
| Census data: | 1972 | 14816 | – | – |
|  | 1976 | – | – | 109419 |
|  | 1977 | 22852 | 13482 | – |
|  | 1983 | 11929 | 19575 | – |
|  | 1986 | – | – | 55153 |
|  | 1987 | 3673 | 9049 | – |
| Projections: | 1987 | – | – | 51513 |
|  | 1988 | 2736 | 7465 | 50740 |
|  | 1989 | 2039 | 6159 | 45666 |
|  | 1990 | 1519 | 5081 | 41100 |

**Table 11.** Estimates of numbers from elephant populations in Tanzania: minor populations.

| Year | Serengeti Nat. Park | Manyara N. Park | Mkomazi G.R. | Kilombero Valley | Tarangire N. Park | Katavi N. Park |
|---|---|---|---|---|---|---|
| 1961 | 470 | – | – | – | – | – |
| 1965 | 2058 | 421 | – | – | – | – |
| 1966 | 2216 | – | – | – | – | – |
| 1967 | 1967 | 380 | – | – | – | – |
| 1968 | 2168 | – | 3000 | – | – | – |
| 1970 | 2460 | 380 | 2067 | – | – | – |
| 1975 | 2352 | – | – | – | – | – |
| 1976 | – | 453 | 667 | 5848 | – | – |
| 1980 | – | – | – | – | – | – |
| 1981 | – | 485 | – | – | – | – |
| 1984 | – | 373 | – | – | – | – |
| 1985 | – | 434 | – | – | – | – |
| 1986 | 467 | – | – | 2230 | – | – |
| 1987 | – | 181 | – | – | 6110 | 3058 |
| Forward projections: |  |  |  |  |  |  |
| 1987 | – | – | – | 2047 | – | – |
| 1988 | 374 | – | 93 | 2007 | 6171 | 2997 |
| 1989 | 299 | – | – | 1806 | 6233 | 2937 |
| 1990 | 239 | 150 | 193 | 1625 | 6295 | 2878 |

The annual decrease is calculated at 25.5% for Ruaha, 17.5% for Rungwa-Kizigo, and 6.6% for the Selous. If this trend continues, the population is projected to 1519 (Ruaha), 5081 (Rungwa-Kizigo) and 41 100 (Selous) in 1990. Elsewhere, the elephant populations have crashed, notably in Serengeti and Manyara National Parks and in Mkomazi Game Reserve (Table 11). Without doubt, the major factor contributing towards this alarming decline in elephant population is poaching.

**Research potential**

In addition to the various research activities being conducted in the Serengeti, some of which have direct bearing on certain of the issues raised, additional research needs to be conducted in the following important areas.

1. Long-term studies on the population dynamics and the ecology of the wildebeest should be re-instated. This study should be closely co-ordinated with that in Masai-Mara. Preferably, there should be a single large programme to cover the whole of the Serengeti Masai-Mara ecosystem.

2. Research on the biology and ecology of the Thomson's gazelle in the Serengeti is a priority. In particular, resource utilization and competition for food with wildebeest and zebra in the short-grass plains need to be assessed.

3. Studies on the primary and secondary productivity of the Serengeti grasslands should be expanded.

4. Studies on the impact of tourists on wildlife populations and vegetation in the Ngorongoro Crater need to be initiated.

5. Previous studies on the biology and ecology of the rhinoceros in the Crater have been re-initiated and are being upgraded.

6. Long-term studies on the interaction between wildlife and livestock are a priority in Ngorongoro. These studies will supplement those being conducted by the Ngorongoro Ecological Monitoring Programme.

7. In the Selous Game Reserve, a long-term research programme on resource utilization, especially hunting by tourists, and its impact on wildlife populations should be initiated.

8. Ecological Monitoring and Training Programmes (EMTP) for the major conservation areas should be established. Already, two such programmes have been established for the Serengeti and Ngorongoro. Proposals for similar programmes for Ugalla Game Reserve and Ruaha/ Rungwa have been submitted for financial assistance. The main objective of the EMTPs is to provide scientific information on a continuous basis to the Park or Game Reserve management.

8. Single-species studies in the conservation areas in southern Tanzania are also a priority.

9. The effect of human activities such as encroachment, poaching, cropping, etc. on wildlife populations in the different conservation areas needs to be assessed.

# Acknowledgements

I am grateful to the Zoological Society of London and the organizers of the Symposium, Professor G.M. Ole Maloiy and Professor P.A. Jewell, for their kind invitation. I am also grateful to the Commonwealth Foundation for their grant which enabled me to attend the Symposium and also to meet and discuss aspects of co-operation in wildlife research with colleagues in Britain.

# References

Barnes, R.F.W. & Douglas-Hamilton, I. (1982). The numbers and distribution patterns of large mammals in the Ruaha-Rungwa area of southern Tanzania. *J. appl. Ecol.* **19**:411–25.

Borner, M. (1985). *Brown animal census, Serengeti National Park. A summary of results.* Report submitted to Serengeti Wildlife Research Institute, Arusha, Tanzania. (Mimeo.)

Douglas-Hamilton, I. (1976). *Aerial census of wildlife in the Selous Game Reserve.* World Wildlife Fund Project 3173 Report.

Ecosystems Ltd. (1980). *The status and utilization of wildlife in the Arusha Region, Tanzania.* Ecosystems Ltd., Nairobi.

Norton-Griffiths, M. (1975). The numbers and distribution of large mammals in Ruaha National Park, Tanzania. *E. Afr. Wildl. J.* **13**:121–40.

Norton-Griffiths, M. (1978). *Counting animals.* (2nd edn). African Wildlife Leadership Foundation, Nairobi.

Norton-Griffiths, M. (1979). The influence of grazing, browsing, and fire on the vegetation dynamics of the Serengeti. In *Serengeti: dynamics of an ecosystem*: 310–52. (Eds Sinclair, A.R.E. & Norton-Griffiths, M.). University of Chicago Press, Chicago & London.

Serengeti Ecological Monitoring Programme (1988). *Programme report.* Serengeti Research Institute, Arusha, Tanzania.

Sinclair, A.R.E. (1987). *Long-term monitoring in the Serengeti-Mara. Trends in wildebeest and gazelle populations.* Report No. 5. Serengeti Ecological Monitoring Programme, Serengeti, Tanzania.

Symp. zool. Soc. Lond. (1989) No. 61:267–298

# Development of research on large mammals in East Africa

Richard M. LAWS

*St Edmund's College
Cambridge, CB3 0BN UK*

## Synopsis

East Africa has great potential to contribute to ecological theory and practice. This review lays stress on the recent origin of research on large mammals in Africa, which initially stemmed mainly from the work of Game Departments in conservation and control. Modern research began in the 1950s and some of the more important contributions are mentioned. Up to 1961, no well-found research centre existed primarily for research on large wild mammals, but in the 1960s three in particular were established. These were the Nuffield Unit of Tropical Animal Ecology, the Serengeti Research Institute and the Tsavo Research Project; they were funded largely by contributions from foundations in developed countries.

Research on large mammals is broadly reviewed and some highlights are described. The development of methodologies for aerial counting led to important advances in the collection of data. Research on population ecology has involved the development of age-determination methods to establish age structures and age-specific parameters of populations. Such information is particularly needed in studies of growth, condition, mortality and survival, reproduction, comparative life histories, population dynamics and modelling of populations. These topics are discussed, giving examples from work in East Africa. Progress on investigations of food and feeding, especially in relation to ecological separation, is summarized; different mechanisms appear to operate in wet and dry seasons and between regions of high and low rainfall, also in grazers, browsers and generalists. Social organization and behaviour is a topic which has attracted great attention and a comprehensive synthesis has related it to feeding behaviour. The East African contribution to research on large mammals has also provided examples and data towards the understanding of mammalian mating systems and the conditions under which the different types occur. With regard to conservation theory and management the need for basic research as well as management-orientated research is emphasized. Some examples of problems are presented, including elephant, giraffe and hippopotamus; the development of conservation theory as affecting East Africa is discussed. Unfortunately, theory seems to have limited application there, because political and economic policies dominate in the region rather than principles of ecological design.

This review emphasizes the importance of long-term studies and much has been achieved at relatively low cost. However, the promise of the 1960s appears not to have been realized by the 1980s. Limited funding and the difficulties in obtaining research permits since independence mean that there are still few career scientists in Africa to provide continuity and the field is dominated by young scientists on short-term projects. Much prestige could accrue to African governments following a relatively modest increase in their investment in research to provide core support for national research institutes. There is a growing international community dedicated to conservation which is a source of additional support once the basic structures are in place. A suggestion is advanced as to how this might be promoted.

## Introduction

My involvement in African research began in 1961 with the Arusha Conference, which was a turning point for research on large African mammals, marking a change from essentially descriptive and anecdotal studies towards sampling, experiment and management-orientated research. My direct involvement in this field ended in the 1970s with the publication of a book describing the results of elephant research in Uganda (Laws, Parker & Johnstone 1975).

Africa has the potential to contribute greatly towards ecological and behavioural research on mammals and to general theory, because of the abundance, diversity, accessibility and ease of study compared with work in other regions of the globe. There is a wide range of environments and habitats. Much has been achieved with remarkably limited resources, but inadequate resources and political instability have prevented the full realization of the potential for science. Logistics are simple in Africa and mammal research is not expensive, compared with my other main research interests—marine mammals and Antarctic science. The continuity of research in those fields is one of the main contrasts with research in Africa.

I am aware of the danger of superficiality in a review of this nature, because of the great diversity of topics to be covered, so I have sought to narrow the scope. The papers in this Symposium are mainly concerned with large herbivores; there are none on primates and only one dealing with a carnivore. Therefore I confine the review to research on large herbivores. Their ecophysiology is not covered. The review can also be narrowed geographically, by excluding forests and concentrating primarily on East African savannas. Because of space limitations I have not reviewed wildlife disease, nor tsetse control research, both of which have relevance to studies of large African mammals. This restricted focus is also more in accord with my own first-hand experience. More detailed information and fuller bibliographies can be found in the accounts of Eltringham (1979), Laws *et al.* (1975) and W. Leuthold (1977).

There are three important consequences of the development of investigations over time which lead to changes in approach. First, the accumulation of knowledge and data becomes increasingly valuable with the passage of time, providing progressively greater opportunities for synthesis and comparative studies which did not exist at the beginning. Also the theoretical background has been changing all the time as new concepts are introduced, for example life history strategies, r- and K-selection, ecological separation, mating strategies, kinship theory, foraging theory, island biogeography and conservation theory.

Secondly, the early studies were necessarily cross-sectional and the value of longitudinal studies increases year by year, with the accretion of data. Examples are the Serengeti monitoring programme now extending over 25 years (Sinclair & Norton-Griffiths 1979) and the 16 years of behavioural data on the Amboseli elephants (Moss 1988). A particular advantage of longitudinal studies is that the effects of environmental variation, such as climate, space (including the effects of human population increase), vegetation changes and release from disease (e.g. rinderpest control) can be assessed.

Thirdly, new techniques, often developed in other regions, are becoming available. The exploitation of microchip technology has led to the almost universal application of computers and the development of mathematical modelling; drug immobilization facilitates handling and marking of animals; radio- and satellite-tagging and telemetry are making remarkable advances; other techniques include auto-analysis for biochemical assay, monoclonal antibodies and DNA fingerprinting, to mention a few.

## Origins of large mammal research in East Africa

The beginnings of such research can probably be seen in the setting up of Game Departments in the then Colonial Territories. They were concerned with the control and conservation of wild populations. Control operations were seen to be needed 'to keep wildlife within reasonable bounds' and conservation 'implies the maintenance in perpetuity of a reasonable quantity of wildlife' (Worthington 1959). There was a conflict between these objectives from the beginning.

The authorities in the Belgian Congo pioneered the concept of strict reserves from 1925 onwards. An independent Board of Trustees in Brussels directed a well-found organization for administration and research and initiated a prestigious publication series. Similar systems were founded in the French and Portuguese territories.

The research carried out remained essentially descriptive and inventorial in nature, although experimental ecology was begun. The first African National Park, the Kruger National Park, set up in 1898, included a small research unit. The East African Game Departments employed some biologists; one thinks particularly of Alan Brooks in Uganda and Hugh Lamprey in Tanganyika. Several museums in the region also specialized in natural history. A number of societies promoted research, including the Zoological Society of London, the precursor of the present Fauna and Flora Preservation Society, the Comité de Chasse Colonial Français, the East African Natural History Society, the American Committee for Wildlife Protection, the International Union for the Conservation of Nature and Natural Resources, and others. There were relevant international conferences in 1933, 1938, 1953 and 1961.

Research began as: (1) the conservation and control of game animals; (2) investigation of the diseases of domestic animals and man, e.g. rinderpest; (3) studies of the tsetse fly in relation to trypanosomiasis, including control shooting schemes; (4) population dynamics, initially academic studies on small mammals; (5) counts to assess abundance, for example in the Kruger National Park and Parc National Albert; (6) the newly developing subject of ethology (Hediger 1950). In an article in *Nature* Harrison Matthews (1954) illustrated how little had been done at that time. He made a plea for serious work to be begun before it was too late, remarking that the 'scientists actually engaged on such problems in Africa are very few and isolated'. The Zoological Society, of which he was Scientific Director, supported a literature survey (Sidney 1965).

Great opportunities to obtain material from control shooting schemes were lost, particularly those involving elephant and tsetse control shooting. John Perry (1953) made a brave try which led to the first significant paper on elephant reproduction and formed the basis for subsequent work on elephant biology. A pioneer study by Hubert (1947) was on the effects of exclusion of fire in savanna ecosystems; he demonstrated the consequent change to bush vegetation with an associated change in mammal communities.

## Start of the modern era

In Kenya the Tsavo National Park was created in 1948; in Tanzania the Serengeti National Park was set up in 1948 and extended in 1951; in Uganda the Queen Elizabeth and Murchison Falls National Parks were established in 1952. Up to 1961, however, there existed no well-found research centre for this kind of research, with resident staff and appropriate facilities.

### Uganda National Parks

Concern had been felt by the Trustees of the Uganda National Parks for several years about the increase of the hippopotamus population in the Queen Elizabeth National Park, where habitat deterioration due to overgrazing by hippopotamus was becoming evident (Bere 1959). Two American Fulbright Fellows, G.A. Petrides and W.G. Swank, recommended a culling programme to reduce the population which began in 1957. W.H. Longhurst, another Fulbright Fellow, arrived to conduct research in conjunction with the culling programme, which was carried out by National Parks staff under the supervision of F. Poppleton. At the same time research was being carried out in Uganda by I.O. Buss and H.K. Buechner, Fulbright Fellows, who also identified an elephant problem, involving the destruction of woodland in the Murchison Falls National Park.

Discussions were held between the Zoological Society of London, the Nature Conservancy and the Bureau of Animal Population at Oxford, which led to an initiative by Cambridge University, Makerere University College and the Uganda National Parks. They obtained support from the Nuffield Foundation to set up a research station at Mweya in the Queen Elizabeth National Park in 1961, named the Nuffield Unit of Tropical Animal Ecology (NUTAE) (Beadle 1974). Subsequently contributions were received from other organizations and collaboration was extended to University College, Nairobi, EAFFRO and EAVRO.

In 1962, with C.R. Field, I started a series of intensive studies of the ecology of large African mammals which was to produce scientific knowledge about them of a quality and quantity never before achieved. The programme began with a detailed investigation of the hippopotamus. The existing culling operation was replanned on an experimental basis, around a series of study areas in which the hippopotamus population grazing densities were maintained at different levels, by culling, from zero to 23 per km$^2$. At the same time population densities of other herbivores were monitored by regular counts and vegetation change was monitored by studies involving quadrats, transects and exclosures (Lock 1972).

The programme expanded with the participation of research students and visiting workers to include the biology of other large mammals: elephant, buffalo, waterbuck, Uganda kob, topi, warthog and lion. The vegetation was mapped and studied and soils and climate also received attention. In 1965 the work of the unit spread to incorporate research on elephant and hippopotamus in the Murchison Falls National Park, including the first systematic scientific culling of elephants (Laws & Parker 1968). Dr S.K. Eltringham was the second Director of NUTAE and in 1971 it became the Uganda Institute of Ecology under the direction of Dr E. Edroma. Workers involved in the initial programme included Clough, Field, Grimsdell, Kangwage, Lock, Neal and Spinage.

### Serengeti National Park

The first modern studies in the Serengeti were conducted by B. and M. Grzimek (1960a,b) and by L. Talbot (Talbot & Talbot 1963) on the plains game. The

Grzimeks were the first to use light aircraft in extensive studies of wildlife populations and the vegetation.

In 1961 the Serengeti Research Project based on Banagi began work. Initially the workers were R.M. Watson, one of my research students, R. Bell, J. Verschuren (who had earlier worked in the Congo National Parks with F. Bourlière and others) and H. Kruuk. Watson added greatly to knowledge of plains game ecology, particularly wildebeest (Watson 1967, 1969, 1970) and Bell's initial work on topi developed into a very influential study of ecological separation through diet (Bell 1969, 1970a,b; Gwynne & Bell 1968). In 1966 the Serengeti Research Institute was founded at Seronera, with H.P. Lamprey as Director. Well-found laboratories and living accommodation were established, capital funding being provided by the Fritz Thyssen Foundation, FRG, and recurrent funds by the Ford Foundation. The objectives were threefold; research on habitats and primary production; research on the plant, herbivore, predator system; and an investigation of wildlife disease. A long-term (30-year) programme was envisaged including the use of quadrats, aerial photography, and an experimental approach to studies of vegetation and fire. Workers involved in the initial programme included Lamprey, Bell, Sinclair, Jarman, Douglas-Hamilton, Croze and Hendricks (Lamprey 1969). Success of the programme is clear from Sinclair & Norton-Griffiths' synthesis (1979).

### Tsavo National Park

There had been concern over elephant damage to the woody vegetation in Tsavo from 1959 and this was enhanced by the consequences of the 1961 drought. In 1962 a wide-ranging aerial count produced an estimate of some 16 000 elephants in the 8000 square miles of Tsavo National Park (East). At the same time investigations were made on elephant-carrying capacity (Glover 1963) and on elephant diet (Bax & Sheldrick 1963): and Schenkel & Schenkel-Hulliger (1969) studied the black rhinoceros population which had been severely affected by the drought. In 1964 the Aruba Research Conference on elephant behaviour and control was held, representing a wide variety of experience; the general opinion was that concurrent research and management culling of elephants was desirable. A further aerial count revealed a total of 20 300 elephants in Tsavo National Park (East) in 1965, which meant that the solution advanced by Glover (1963) would be inadequate. In 1965 too, I had put a proposal to the Ford Foundation for a programme of research on elephants throughout East Africa, directed from NUTAE in Uganda.

The Ford Foundation was interested and agreed to support it, but for political reasons it was not possible. In the event the grant was divided between NUTAE, with funds channelled through Makerere University, and Kenya, funded through the Ministry of Tourism and Wildlife. The Kenya proposal was for a three-year programme of research and management on elephant biology in the Tsavo National Park. Before this was implemented I had at the invitation of the Trustees of the Kenya National Parks conducted an experimental culling of 300 elephants, in the Koito area of the Tsavo National Park (East) with the assistance of I.S.C. Parker (Laws & Parker 1968). In February 1967 the Tsavo Research Project began its work. Laboratories and living accommodation were constructed under the Ford Foundation grant and a research programme conducted by myself and R.M. Watson began. This was planned from the beginning as an inter-disciplinary programme over an ecological unit extending over some 17 000 square miles, including Tsavo National Park (East and West) and the surrounding area occupied by elephant populations.

It was planned to include previous land-use, drawing on studies of environmental archaeology and recent human history, current dynamics of the ecosystem, including climate, soils and vegetation (30 fixed aerial photographic transects were selected), animal populations and elephant research. This research quickly showed that there were 30–40 000 elephants in the region, apparently in ten populations. Monthly aerial surveys provided the first regular recordings of distribution of large mammals. Some further sampling was undertaken by the culling of 300 elephants from each of two populations in the Mkomazi Game Reserve, Tanzania, which was part of the ecological unit. New findings emerged on elephant group sizes and social organization. Detailed studies were made of reproduction, growth, age structure, tusk weights and age, and for the first time the age structure of elephant populations was investigated by aerial photographic measurements of elephant back length. The age structure was also studied by means of a collection of found lower jaws of elephants that had died naturally, made at my request by the Warden, D.E. Sheldrick. J. Goddard joined us to work on the black rhinoceros (Goddard 1968, 1970). I made recommendations for concurrent management and research, but these were not accepted by the National Park Trustees and I resigned in 1968; Watson had left in 1967 (Laws 1969a,b).

## Review of research in East Africa

### General topics

Analysis of titles of papers in the *African Journal of Ecology* (formerly *East African Wildlife Journal*) over the 35 years 1963–87 reveals that, of 378 titles relevant to this review of large herbivore biology, a quarter dealt with aspects of elephant biology, half were studies of one or more other species and a quarter were mixed or general papers on community or environmental interactions. The proportion of elephant papers increased from 21% in 1963–79 to 34% in 1980–87; papers on other species decreased from 54%

in 1963–79 to 42% in 1980–87 and mixed subjects represented 25% in the former period and 24% in the latter. Over the last three volumes 20 out of 44 relevant titles are on elephant biology. This emphasizes the growing importance of the elephant in research as a keystone species posing management and conservation problems.

A selective survey of titles in a recently published book on African ungulates (Kingdon 1982) indicates that of 807 relevant titles listed, 18.8% were published before 1960, 15.5% in 1960–64, 24.5% in 1965–69, 27.6% in 1970–74, and 12.8% in 1975–79. Many of the pre-1960 titles are very descriptive notes and anecdotes and the increase to the mid-1970s supports the view that modern wildlife research in Africa began in the 1960s. In fact the decade of the 1960s was a particularly productive period, as represented by publications up to the mid-1970s; research fell off in the 1970s due to political instability and reduction of support for research in Africa. I hope that may be changing again now.

Two central problems that characterize ecological studies are distribution and abundance and this is particularly true in East African studies. In order to make progress in understanding these topics one has to conduct research on individuals, their age, growth, behaviour and physiology; on populations, their dynamics, factors limiting population size (particularly births and deaths, immigration and emigration movements) feeding strategies and social behaviour; on communities, interactions and changes of animal populations in their habitats; and on environmental change, that is, climate, soils, nutrient cycling, energy flow, habitats and vegetation and the human factor.

### Distribution and abundance

There have been several useful regional surveys of geographical distribution (Brooks & Buss 1962; Sidney 1965; Stewart & Stewart 1963; Kingdon 1971–82).

There has been a remarkable number of total counts, reflecting the urge to find a justification for field work on animals; when unable to think of something better to do the solution has often been to go out and make a count! Examples of early counts that made a significant contribution include those by Grzimek & Grzimek (1960b) in the Serengeti, and later aerial counts in the same area (Stewart & Zaphiro 1963; Watson 1967); by Buechner, Buss, Longhurst & Brooks (1963) of elephants in the Murchison Falls area, by Watson and Laws in the Tsavo ecological unit (Laws 1969b), by Watson, Parker & Allan (1969) in the Mkomasi Game Reserve, and of hippopotamus in the Nile at Murchison Falls National Park (Laws *et al.* 1975). On the ground, Lamprey (1963) had put well-organized transect counts by Game Scouts to good use in his study of ecological separation in the Tarangire Game Reserve. Field & Laws (1970) describe the result of repeated ground counts in ten different study areas in the Queen Elizabeth National Park. Hippopotamus counts have been carried out from river banks and lake shores and from boats.

It is now accepted that the minimum level of precision needed to achieve the purpose of a count should be defined before the start and certainly before expensive surveys are mounted. The turning point in this respect was the Workshop on the use of light aircraft in research (Swank, Watson, Freeman & Jones 1969). It dealt with aerial photography, sampling theory and methods, and adopted an experimental approach to observer error and variability; problems due to clumping were addressed.

Norton-Griffiths (1975) published a useful handbook on techniques of counting from the air and on the ground which standardized methods of conducting sample counts, transects and block or quadrat counts, survey design and analysis of results. It arose from his experience on the long-term monitoring programme in the Serengeti, which is the most significant work in this field (Sinclair & Norton-Griffiths 1979).

## Population ecology

Population ecology involves research on age determination to establish age structures and age-specific parameters, growth and condition, mortality and survivorship, reproduction and natality, comparative life histories, population dynamics and population modelling.

### Age determination

Spinage (1973) has reviewed methods of age determination for African mammals. A problem to be overcome before ageing methods can be used is the assignment of chronological ages to relative age categories. This can be resolved by the use of known-age animals, from marking experiments or longitudinal studies of known individuals. Spinage (1967a) took tooth impressions from individual waterbuck, recaptured at intervals, to establish the chronology of replacement and wear. Vital staining of cementum with a dye such as tetracycline is another check that has been used.

### Tooth replacement and wear

Growth, eruption, replacement (particularly in young animals) and wear of teeth are all related to age. Incisor, premolar and molar replacement have been used. Examples are for zebra (Klingel & Klingel 1966) buffalo (Grimsdell 1969), hippopotamus (Laws 1968b), black rhinoceros (Goddard 1970), white rhinoceros (Hillman-Smith, Owen-Smith, Anderson, Hall-Martin & Selaladi 1986), waterbuck (Spinage 1967a), elephant (Laws 1966; Sikes 1966, 1968). Spinage (1972) compared decreasing molar height in zebra, in a wear model with the visual assessment of incisor changes (Klingel & Klingel 1966) and found good agreement. Hall-Martin (1975) used the wear patterns of maxillary first molars in giraffe.

### Internal layers in teeth

Laws (1952) drew attention to the use of internal growth layers, seasonally laid down in teeth, as a measure of age and this has come into wide application. It has been largely neglected in Africa, partly because seasonality is less pronounced there than in temperate and polar regions. Examples of its application to African mammals are the establishment that lines in the cementum of the first incisor of waterbuck are correlated with seasonal changes (Spinage 1967a, 1982), a similar finding for buffalo, using the first molar (Grimsdell 1969), and for several teeth of white rhinoceros (Hillman-Smith *et al.* 1986).

### Methods for live animals

Aerial photographic measurements which can be converted to age provide an extremely valuable tool, because animal populations are widespread and often clumped in distribution, and large samples can be obtained by this method. Watson (1967) used body length to distinguish yearlings and the size and shape of the horns to sex adult wildebeest. Laws (1969b) developed a method of ageing elephants from measurements of back length on aerial photographs, relating these to age/length keys obtained from post-mortem samples. The method was refined by Croze (1972). On the ground photographs can be used to measure shoulder heights of individuals in order to estimate age (Douglas-Hamilton 1972; Jachmann 1986). It may not be valid to compare populations from different regions in this way, without calibration by means of samples obtained from post-mortem; this is another reason why such samples are valuable. Jachmann (1986) has also used measurements of droppings to obtain an age structure, and foot impressions, by virtue of their correlation with shoulder height, can also be applied (Western, Moss & Georgiadis 1983).

### Collection of found jaws and skulls

Found jaws or skulls, representing natural deaths, may make an important contribution to understanding population structures and dynamics. This was a feature of the research programme at NUTAE. The method also proved to be particularly valuable for elephants in the Tsavo ecological unit. Laws (1969b) analysed a collection of found jaws, representing the period 1960–67, and Corfield (1973) repeating the work in a later period, 1970–72, was able to show how age- and sex-specific mortality rates had changed as a result of the 1970–72 drought.

### Tusk dimensions and weight as a measure of sex and age

Laws (1969b) drew attention to the use of elephant tusk dimensions and weights to estimate age and sex composition of samples. This method was later used by Parker (1979), and Pilgram & Western (1986) on commercial ivory samples. Barnes (1982) used it to study seasonal mortality of elephants in Tanzania. Again, an important qualification is that it is probably very misleading uncritically to apply age-specific tusk weights established by detailed post-mortem study of one geographical population to another.

## Growth and condition

Age criteria have been applied to measurements and weights of a number of species in order to obtain growth curves; these are commonly expressed as length, shoulder height or weight. Examples are buffalo (Grimsdell 1969; Sinclair 1974a), waterbuck (Spinage 1970), hippopotamus (Laws 1968b), elephant (Laws *et al.* 1975; Hanks 1972, 1979), but such growth data are also available now for a number of other species. The von Bertalanffy growth equation was used by Laws (1968b) and Laws *et al.* (1975) to fit curves to the data, demonstrating that in the case of the male elephant at puberty social maturity was delayed until growth was complete. McCullagh (1969) studied instantaneous growth rates in two populations of elephants by means of the analysis of hydroxyproline/creatinine ratios, a biochemical index which reflects the rate of breakdown of collagen and is therefore a measure of the current growth rate. Condition indices are revealing in ecological studies because they reflect the animal's relation to its environment. Hanks (1981) has reviewed the use of such methods; those most widely used are based on the kidney fat index or bone marrow fat. They have been applied, among other species, to impala and elephant (Dunham & Murray 1982; Laws *et al.* 1975).

## Age-specific mortality

Again, using methods of estimating the age of living or dead animals, a number of studies have investigated mortality and survivorship. One of the first, by Petrides & Swank (1966), constructed a survivorship curve for a living elephant population, assuming a stationary age distribution and based on field estimates of numbers in broad age groupings and the proportion of calves. Spinage (1970) drew up life tables for waterbuck based on found skulls; Laws (1968b, 1969b) produced similar analyses for hippopotamus and elephant, and Laws *et al.* (1975) gave age structures for several elephant populations. Sinclair (1974b) presented similar data for buffalo and Goddard (1970) for black rhinoceros. In their study of elephant age structures Laws *et al.* (1975) showed that for several separate populations adult mortality rates remained fairly constant, while juvenile mortality appeared to vary according to environmental change; Sinclair (1974b) concluded that, in buffalo, adult mortality was more important in population regulation.

### Reproduction and natality

In many species the breeding season and reproductive cycle have been studied directly by field observation of matings and births. Population fertility can be studied from the proportion of calves of the year in samples counted on the ground or from the air (e.g. Sinclair 1974a). But the best results have come from sampling by shooting (e.g. Laws *et al.* 1975) or by longitudinal studies (e.g. Moss 1988).

Thus, post-mortem data have made it possible to establish, for several elephant populations, the age of maturity, seasonality of reproduction (conceptions and births), age-specific fecundity (including the presence of a post-reproductive segment in the female population), and mean calving interval (from proportions pregnant and from placental scar counts (Laws 1969b; Laws *et al.* 1975). All these parameters have been shown to be remarkably plastic in response to habitat and climatic changes.

### Comparative life histories

In recent years there has been a development of interest in co-adapted traits of growth and life histories (Calder 1983; Peters 1983; Stearns 1983). (I had published on aspects of this in marine mammals in the 1950s: Laws 1956, 1959.) The concept relies on biochemical and physiological processes being scaled to size by allometry. Western (1979) has shown that, for a number of African mammals, life history parameters such as growth rates, age at first reproduction, gestation period, lifespan, intrinsic rate of natural increase, birth rate, net reproductive rate, and litter weight, are allometrically scaled to size and interrelated. Size accounts for most of the variation in life history parameters between species and so size scaling is important for comparative studies. He distinguished between first-order strategies where these parameters are due to size alone, and second-order strategies where they vary between populations and according to environmental circumstances. Size is a central theme of Jarman's (1974) comparative study, discussed below. Georgiadis (1985) has considered sexual dimorphism as a special case.

### Population dynamics

I have dealt with some aspects of population dynamics already. Ecologists have given much attention to factors influencing population regulation and in the 1950s and 1960s the emphasis was upon group selection (Wynne-Edwards 1962) and behavioural regulation; selection by food availability was considered to be less important. This has now changed. Most studies have had no clearly defined objectives at the start and few studies have specifically set out to investigate the natural regulation of a population. An exception was the outstanding study by Sinclair (1974a,b,c), which used a natural experiment—the release of a buffalo population from the depressive effect of rinderpest—to explore and model natural regulation. He studied the subject by a range of techniques (already mentioned) including collected samples and comparison with two populations studied in Uganda (Grimsdell 1969).

The regular aerial photographic samples indicated large annual fluctuations in fertility and recruitment which were unrelated to density and therefore not regulatory; they were not correlated with rainfall and were possibly random. The strong seasonality of births was correlated with food quality and rainfall. A positive relation between annual rainfall and buffalo density in different areas of East Africa indicated that regulation was taking place, with nutrition probably an important determining factor. Sinclair could determine three causes of population reduction each year—due to fertility, to juvenile mortality and to adult mortality. The latter was the only detectable change which acted as a negative feedback and the only one needed to regulate the population, as indicated by observed population trends. A composite life table was constructed and showed constant age-specific mortality until old age. It was concluded that predation and disease acted by hastening the response to changes in the food supply, which was the main regulatory factor, acting through adult mortality caused ultimately by inadequate food in the dry season. Intra- and inter-specific competition with wildebeest was also implicated. The hypothesis that regulation by social factors could maintain the population at a density lower than the ceiling allowed by the food supply was rejected.

Apart from these studies of natural regulation of buffalo populations, I have not included detailed reference in this review to the outstanding long-term programme on the dynamics of the Serengeti ecosystem because it would be difficult to do it justice. A full account is given by Sinclair & Norton-Griffiths (1979). These investigations took advantage of the natural perturbation of wildlife populations by the great rinderpest epizootic beginning in 1890, and the release of the ungulates following its final control in 1962 (wildebeest) and 1963 (buffalo); zebra were unaffected. Large population increases of the ruminants followed. Sinclair & Norton-Griffiths (1979) include chapters on the environment, grassland-herbivore dynamics, the eruption of the ruminants, the migration and grazing succession, feeding strategies and the pattern of resource partitioning, foraging behaviour, social organization, the influence of grazing, browsing and fire—among other topics.

Although wildebeest in this area increased as expected, zebra remained constant at the 1970–80 numbers. Senzota (1988) drew up a conceptual model and suggested that the constraint was due to social organization and/ or mortality due to disease or predation, but this remains to be tested.

Populations of other species have been studied and modelled. In elephants, Hanks & McIntosh (1973) also emphasized that mortality was of greater importance than reproductive parameters. From their model they estimated the maximum possible annual rate of increase at 4%. However, Jachmann (1986) has presented evidence for an 11% annual increase in Malawi elephants, and in the Addo National Park an annual increase of 7% has been recorded from 1953 to 1976, and of 6% over the longer period 1953 to 1986 (Bosman & Hall-Martin 1986). Models are only as good as their assumptions and do need to be tested by input of factual data.

Laws *et al.* (1975) modelled elephant populations centred on the Murchison Falls National Park. They concluded that adult natural mortality appeared not to have altered, but calf mortality had increased by at least 50% since 1946. Together with deferred maturity and reduced fecundity this had led to a massive decline in recruitment. The population decrease was estimated at 64% between 1946 and 1971, less than half of it due to control shooting operations. It was also estimated that natural regulatory mechanisms alone caused a 51% decrease, in response to habitat changes, probably mediated by food quality and quantity.

## Food and feeding

The availability of food and its quality has a profound effect on ecology and behaviour. Ecological separation is also promoted by differing food requirements.

### Dry regions

Lamprey (1963) made an important observational study in the Tarangire Reserve, Tanganyika. He showed that ecological separation was probably achieved in one or more of six ways: through habitat, food, the occupation of different areas in the same season, of the same area in different seasons, feeding at different height levels, and by occupation of different dry-season refuges. This impressive quantitative study, which indicated how competition between species was avoided, was the first of its kind.

In the western Serengeti, Bell (1969, 1970a,b), working in a relatively dry area, mapped the position of five species in his study area for a year, relating their positions to levels of the catena. He found that species entered and left the upper levels in order of their body weights, the lightest being first up and last down. In the wet season grass growth is vigorous on top of the catena and can stand heavy grazing; the forage then is highly nutritious. In the dry season the food on the tops is scarcer and animals move down to the sumps. Species which tolerate long grass (little protein, much fibre), such as zebra and buffalo, move down first. Small animals, such as gazelles, need more protein, move up first in the wet season and are more selective feeders.

The diets were studied for wildebeest, topi, zebra and Thomson's gazelle. The gazelle is a browser/grazer, even taking berries; the other species take differing proportions of leaf, sheath and stem of grasses (Gwynne & Bell

1968). The largest, zebra, takes more stem, therefore less protein, but by virtue of its post-gastric digestive system, it can cope.

This leads to a mutually beneficial grazing succession. The species associate in the wet season, when their combined grazing pressure keeps the grass growing and maximizes protein production. As the zebras leave the summit in the dry season their trampling and coarse feeding opens up and stimulates grass growth, increasing the leaf available to others. Topi and wildebeest next remove much grass and stimulate the dicotyledons, which the gazelle, last in the succession, prefers. Vesey-Fitzgerald (1965) described a similar seasonal grazing succession in the Rukwa Valley. Tanganyika. See also the excellent contribution by Bell (1982). One may conclude that these studies show that ecological separation is achieved because species eat different organs of the same plant, often at different stages of growth.

Similar studies have been carried out on browsing species, but here there is less evidence of ecological separation through food. Goddard (1968) showed that black rhinoceros had a marked preference for legumes which have a high protein content. Leuthold carried out studies on gerenuk and lesser kudu (W. Leuthold 1970, 1971) and on giraffe (B.M. Leuthold & Leuthold 1972), showing that in the latter much of the browsing was in fact near the ground.

### Wetter areas
In the wetter region of western Uganda, Field (1972) showed that the frequencies of different plants in the stomachs of shot animals varied between species. Stewart (1965) obtained similar results in the Nairobi National Park. In individual species there was also a seasonal change in proportions of different plants eaten, except for the hippopotamus; in that species the large mouth made for very unselective feeding. These studies employed a technique based on the microstructure of the plant cuticle (Storr 1961) which was further developed in East Africa by Stewart (1965) and Field (1972); the hairs, papillae and silica bodies vary between plant species and are diagnostic. In the wet season there is more overlap, because the grasses are more palatable then and the abundance of food makes for less competition. In the dry season ecological separation is enhanced by food selection and behavioural differences. Complementary diurnal feeding behaviours of hippopotamus and warthog enable them to eat the same food species. As in the Serengeti, there are mutual benefits (e.g. the coarse-grazing buffalo promoting the formation of the hippopotamus 'lawns').

### Elephants
Several studies have demonstrated that elephants tend to select food in proportion to what is available (e.g. Bax & Sheldrick 1963). The percentage of browse is greatest in the dry season—and it can be critical in semi-arid

regions to see them through. Problems develop in drought years. Barking is another prominent activity of elephants which has been suggested as related to meeting a need for calcium. Weir (1973) concluded that there was enough calcium in normal food and that sodium was likely to be more critical. Jachmann & Bell (1985) have demonstrated a significant correlation with phosphorus and sodium levels and utilization, not with crude fibre. Pushing over trees is part of the feeding strategy and improves food availability during the dry season. Heavy browsing leads to coppicing and increased stem density. Jachmann & Bell (1985) consider that the elephant has different feeding strategies in moist oligotrophic woodlands and in dry eutrophic savannas.

Olivier (1978) believed that elephants are specialized grass feeders, from consideration of their dentition; but because there is great seasonal variation in grass availability they are forced to switch to browse in the dry season. Also, since they have no rumen to synthesize amino acids and vitamins by means of rumen bacteria, they supplement their diet with a wide range of plants. This diversity is necessary to avoid concentration of toxic plant secondary chemicals. In the wet season, however, the bulk of the diet is grass, which has low toxin concentrations.

It is generally accepted that the protein/fibre ratio determines acceptability of food, but it is now being realized that other factors are also involved. Fuller understanding of the constraints on animals must await understanding of the distribution and role of plant secondary compounds and of minerals.

### Other factors
It is generally accepted that the protein/fibre ratio determines acceptability grasses) may also have a part to play in developing new concepts (Tieszen & Imbamba 1980). Finally, M.G. Murray (pers. comm.) is examining new hypotheses related to distinctive differences in rates of movement. Hartebeest and wildebeest when walking require less energy than topi and waterbuck, but expend more energy per unit weight when resting. Differences in the quality of the food ingested appear to result from the differences in rates of movement of the foraging animals, even in the same habitat type.

## Social organization and behaviour
### Related to feeding behaviour
One of the more exciting syntheses based on work in East Africa related feeding behaviour to social organization. Jarman (1974) examined the types of social organization found in 73 species of African antelopes and the buffalo. These species were assigned to five classes based on ecology and on social and behavioural characteristics. They are distinguished by the mating strategies of the reproductively active males and the effect of these strategies on other social castes. Different feeding styles, related to food selection and

home range coverage, are also related to the maximum group size of feeding animals through the influence of the dispersal of food items on group cohesion. As mentioned above, feeding styles are also related to body size and habitat choice, both of which in turn influence anti-predator behaviour and possible minimum group size. It follows that the group size and pattern of movement over the annual home range affects the likelihood of females being in a given place at a given time. This in turn determines the male mating strategy and so the social organization of the species. Effects of these different male mating strategies are seen in such aspects of the species' biology as sexual dimorphism, adult sex ratios, and differential distribution of the sexes.

## Mammalian mating systems

Subsequently W. Leuthold (1977) comprehensively reviewed behavioural ecology of African ungulates in much more detail. More recent information generally conforms quite well with both of these major reviews. However, new concepts have become current since these earlier studies. Mammalian mating systems are now seen as the outcome of the reproductive strategies of individuals, rather than as the evolved characteristics of species. Behaviour is usually closely adapted to maximizing reproductive success and in fact most mammalian mating systems represent different forms of mate guarding. Thus, the reproductive behaviour of males is adjusted to the spatial and temporal distribution of females. In contrast to the males, reproductive success of females is not dependent on multiple matings and their distribution is related to resources, safety from predators and the costs of social living. Clutton-Brock (in press) describes eight main types of mammalian mating systems and the conditions in which they occur. These are: facultative monogamy/polygyny (e.g. some small ungulates); roving males (e.g. elephants); obligate monogamy (e.g. klipspringer); polyandry; resource-based polygyny (e.g. waterbuck, puku, Grevy's zebra); group-based polygyny (e.g. Burchell's zebra, eland, roan); multi-male breeding groups (e.g. buffalo); and leks, where local density of females is high (e.g. Uganda kob, white-eared kob, topi). The last of these mating systems is particularly interesting, the males occupying the lek area being visited by females alone or in groups, the females exercising mate choice (Buechner & Roth 1974). It should be noted that different breeding sysems often occur in different populations of the same species. For example the leks of Uganda kob are related to local peaks in density and in topi according to whether the population is resident or migratory.

This is a rapidly developing field for which East Africa has contributed data and insights and has much to offer in opportunities for further investigation. Field experiments are badly needed and the spectrum of ungulate species should be studied in terms of their social organization, mating systems, lifetime reproductive success and DNA fingerprinting (to determine relationships), among other approaches.

### Elephant social organization and behaviour

Until the 1960s knowledge of elephant behaviour was superficial and anecdotal. Few biologists had worked specifically on elephants and among the first were I.O Buss (Buss 1961; Buss & Smith 1966), and S.K. Sikes (1971). Subsequently two main lines of approach developed. The first, initiated by Laws (Laws & Parker 1968; Laws 1969b; Laws *et al.* 1975) depended primarily on aerial surveys and post-mortem examination of elephants in groups taken in culling operations. The second, begun by Douglas-Hamilton (1972, 1973), involved detailed observations of the movements and behaviour of individual elephants, recognized by natural markings, and followed closely over several years. C. Moss participated in Douglas-Hamilton's study, in the Lake Manyara National Park, Tanzania, and subsequently embarked on her own longitudinal study, in the Amboseli, which has continued without a break for 16 years (Moss 1983, 1988; Western & Lindsay 1984).

*Comprehensive cross-sectional studies*

Laws *et al.* (1975) in the Northern Bunyoro, Uganda, elephant range of 3200 km$^2$ determined that the average elephant density was 3.5/km$^2$ in 1967. However, two density strata existed, a central zone averaging 1.9 elephants/km$^2$ and a smaller peripheral area averaging 3.8/km$^2$. At any time the distributions were clumped. In these two strata the overall mean group size was 12.0, but group size in the central, medium density zone was 6.6 as against 22.5 in the peripheral high density zone. Of 800 elephants examined in the culling operations 14.6% were males, with a mean group size of 3.1; the age distribution of solitary males was the same as in bull herds. Some 85% of those cropped were in family units, the simplest structure averaging 2.75 individuals, including a single adult female accompanied by her offspring, and an average of two or three of these mother-offspring groups were usually combined as a matriarchal group of mother, mature daughters and their calves. Some 74% of groups were matriarchal groups. The age distribution of bulls in 26 bull herds was reported and also the structure by age and reproductive status of 59 herds of mixed sex, using this information and placental scar counts to establish presumed relationships within the groups. The association of adult males with groups was shown to be temporary and there was no evidence of relationship. Further analysis of another 177 family units from other populations in East Africa gave a similar picture. In the peripheral zone of the North Bunyoro elephant range very large herds (containing up to 1200 elephants) were recorded. It was suggested that they formed because of the breakdown of the normal social organization due to the shooting of the matriarchs in zones of conflict with man, by poaching and crop protection shooting. Laws *et al.* (1975) also made other behavioural observations and presented data on reproductive biology, which showed variability between populations.

*Comprehensive longitudinal studies*
Douglas-Hamilton (1972, 1973) in the course of a four-and-a-half year study investigated elephant numbers and density, social organization and behaviour, the reproductive rate, interactions with the habitat and population dynamics of the Lake Manyara National Park population. In this 320 km$^2$ area the average density year-round was 5/km$^2$, the highest ever recorded. Some 350 elephants were individually recognized and a photographic file was set up. The family groups were found to be confirmed as the basic stable social unit of elephant society. A less stable association of two to four family units, numbering up to 50 animals, was believed to represent a kinship group. In all, 48 separate family units and eight separate bulls were distinguished. Home ranges of family units and bulls varied from 14 to 52 km$^2$ and overlapped. There was no evidence of territorial behaviour nor of a high level of intra-specific aggression even at these very high densities. When disturbed the elephants bunched to form larger units. All births, deaths and departures were recorded. The number of births fluctuated from year to year in relation to rainfall levels, the mean calving interval was 4.7 years, the mean age of female puberty was 11 years, mortality rates were calculated and during the study period the population increased at 3–4% a year. The unpublished thesis contains much detail and it is matter of regret that it has not been published, although a popular book describes the study (Douglas-Hamilton & Douglas-Hamilton 1975).

Moss (1983, 1988) and her colleagues, particularly Poole (1982), have made an outstanding long-term study of the behaviour and dynamics of an elephant population in the Amboseli National Park, Kenya. This is a population of about 680 individuals all of which can be recognized by the workers involved; 160 are males over 10 years and 225 are adult females. It is quite impossible to do justice to this body of work in this review, but the quality and quantity of the data recorded can be judged from their paper in this symposium, describing exciting new work on very low-frequency vocal communication over long distances by elephants (Poole & Moss 1989). Moss (1988) has given a full bibliography of the 24 papers so far published. This work has extended the pioneering work at Lake Manyara. Among the topics addressed are habitat selection, social structure and group dynamics, population dynamics, social and reproductive behaviour, musth and male–male competition (Poole & Moss 1981), oestrous behaviour and female choice, maternal investment and allomothering.

The average family unit size was ten. Females were shown not to be territorial and actively to search for a male when in oestrus, which lasts for three to six days. An average of 75 females come into oestrus each year, the number varying with the rainfall (in the bad drought of 1976 very few came into oestrus). Natality therefore varies with rainfall as Laws (1969b) and Laws *et al.* (1975) described. By photographic measurements of shoulder height Moss and her colleagues were also able to confirm the growth curves presented by Laws *et al.* (1975) and, by implication, the validity of the age criteria described by Laws (1966). I therefore see no reason to modify these age criteria in the light of Jachman's (1988) paper.

## Other studies

The work on elephant group sizes and unit populations in the Tsavo ecological unit (Laws 1969b) has already been mentioned. W. Leuthold & Sale (1973) showed by radio-tracking four males and four females that the home range was 350 km$^2$ in Tsavo National Park (West) and 1580 km$^2$ in the drier Tsavo National Park (East). They suggested that long-distance movements, to 80 km or more, were in direct response to localized rainfall, and that the subdivision into unit populations described by Laws (1969b) did not exist in the wet season although it may occur in the dry season. Laws's observations of movements of 20–30 km were held to represent a temporary stable situation in which food was plentiful. However, it is their study which is unlikely to represent normal conditions, because it followed a very severe drought in which at least 6000 elephants died in the area, particularly females and calves (Corfield 1973), presumably entailing a breakdown in the social organization of the populations. Thus, Caughley & Goddard (1975) described a mean annual range of movement in the Luangwa Valley, Zambia, of 25 km. Home ranges in other elephant populations have been established as 94–262 km$^2$ (Dunham 1986), 15–52 km$^2$ (Douglas-Hamilton 1972), and radio-collared elephants at Amboseli moved up to 15 km in the dry season and up to 40 km in the wet season (Western & Lindsay 1984); this is also in line with information from Uganda.

## Conservation and management

Soulé, Wilcox and Holtby (1979) concluded that in the absence of reserve management—'benign neglect'—isolation of conservation areas would lead to long-term losses of up to 90% of large mammal species in East African reserves. Other workers, however, have predicted much lower extinction rates. The relevance of species-area and equilibrium theory (MacArthur & Wilson 1967; Connor & McCoy 1979; Gilbert 1980) in relation to reserve design has been discussed widely (e.g. Diamond 1975; Wilcox 1980; Higgs 1981), but in practice these theoretical considerations have had little practical effect. East (1984) and Western & Ssemakula 1981) have discussed the application of species-area curves to ecological islands or faunal enclaves. The latter authors have examined the relation of the number of ungulate species to the size of each reserve, compared with predictions from earlier biogeographical studies. They found no significant relationship for savanna reserves overall, although when re-analysed for reserves of similar habitat types it was significant. Most of the variation in the number of species in areas that can practically be established as reserves was explained by habitat and landscape diversity. Grassland savanna communities appear to be richer in species than bush, woodland and forest. In real life, they concluded, political and economic policies and practices will override the principles of ecological design. They also suggested that the point of diminishing returns, in terms of the number of species conserved in reserves, is reached at the relatively small area of about 200 km$^2$. East (1984) came to intermediate conclusions but believed that it would probably require intensive management to conserve species diversity.

Problems are greater when the communities include dominant species— keystone species—like elephant and hippopotamus that can have a spectacular impact on their environment. Another keystone species is, of course, man—whether in competition for land or in poaching (Parker & Graham 1989). The usual dilemma is the need for management decisions when scientific information is inadequate, and there is a need for ongoing basic research as well as focused management-orientated research when a problem is perceived to be developing. Krebs (1988) has commented: 'one of the paradoxes of ecology is that the largest and most dramatic ecological phenomena are simultaneously the easiest to observe and the hardest to study . . . and the time scale of the experiments we need to do on the elephant problem exceeds our human life span. The issues involved in the elephant problem are sufficiently important to the long-term conservation of wildlife in Africa that we must realize that *the larger the animal, the larger the problem and the longer the time frame we must adopt.*'

While setting up the long-term research we also have to address immediate problems, as will be discussed in the next section.

### The elephant problem

Many authors have drawn attention to habitat changes caused by elephants (e.g. Buechner & Dawkins 1961; Laws 1969b, 1970; Laws *et al.* 1975; Harrington & Ross 1974; Croze 1974; Barnes 1980, 1985; Spence & Angus 1971; Hatton & Smart 1984). In these studies evidence was given of deaths of large trees and bushes, and the effect of heavy browsing on regeneration. In some national parks (e.g. Murchison Falls National Park and Tsavo National Park) the vegetation has changed dramatically from forest, woodland or bush to open grassland, with consequential effects on the biota associated with wooded vegetation. These are the classic case studies. Other workers have held that fire is more important than elephants in producing the observed results (e.g. Harthoorn 1966). Tree deaths may be due to other causes (Western & Van Praet 1973).

Four hypotheses have been proposed (Caughley 1976). First, the equilibrium hypothesis, which holds that elephant/forest, woodland, bush systems possess a stable equilibrium point. There is little if any support for this view today. Second, the compression hypothesis, which explains the problem as caused by the displacement of elephants by expanding human populations, a situation which has been described as a sea of elephants including islands of humans that is reversed so that islands of elephants are now surrounded by a sea of humans (Parker & Graham 1989). Elephants are progressively compressed into these diminishing reserves which creates artificially high densities. A time-lag in the response of elephant populations (increased juvenile mortality, deferred maturity and reduced pregnancy rate leading to reduced recruitment), results in overpopulation, high densities in excess of carrying capacity and tree destruction (Laws 1969b, 1970; Laws *et al.* 1975). Third, there is the intrinsic eruption hypothesis, which Caughley (1976) attributes to Riney. This holds that elephant eruptions have occurred on a similar time-scale in widely separate regions of Africa. They are considered to be a response to the release of vegetation occasioned by the rinderpest panzootic at the end of the last century, essentially by relaxing competition between buffalo and elephant. Caughley (1976) has advanced reasons for not accepting this hypothesis. Instead, he has proposed a fourth hypothesis as a result of his experience in the Luangwa Valley, Zambia. This assumes that there is no natural equilibrium between elephants and wooded habitats, but there is a stable limit cycle involving a simple density-dependent relationship between elephants and trees. Elephants increase while thinning the forest and decline until they reach a low density that allows regeneration of woody vegetation, which in turn triggers an increase in elephant numbers, when the cycle repeats. By taking cores of baobab trees in the Luangwa Valley to establish their age distribution he showed that there had been a cycle in their abundance (*c.* 200 years) which he postulated was related to a cycle in elephant numbers. (Phillipson (1975), however, has proposed climatic cycles of *c.* 50 years as the causative factor of elephant cycles.) Cyclicity is proposed as integral to the relationship between elephants and forests, not a pathological displacement from equilibrium. Caughley considered that if densities of elephants or forest are displaced artificially from their cyclic values, the cycle will re-establish itself when the external forces causing displacement are relaxed.

It may be that there is no general explanation for the elephant problem that will apply to all areas. My own view is that a combination of the stable limit cycle and compression can explain most if not all situations that have been described. A critical factor is a consideration of the spatial shifts in elephant populations on the larger geographical scale, as they modify the woody vegetation by their activities. Thus, in any one locality the vegetation element of the stable limit cycle could proceed—e.g. thinning and resurgence of forest—while the elephant population overall could remain stable or slightly fluctuating, as one would expect for such an extremely K-selected species. Once the human population increase reversed the density relationship between people and elephants, compression would come into play because the elephant population movements would be prevented. The elephant/tree cycle would then be artificially held at the trough where woody vegetation is suppressed by elephant browsing. The history of the changes in North Bunyoro (Laws *et al.* 1975) and the Tsavo ecological unit (Laws 1969b; Corfield 1973) support this interpretation.

## Other species

### Giraffe

Pellew (1983) studied the interaction between giraffe, elephants and the environment in the Serengeti National Park. He showed that in this region there had been a marked increase in the amount of browse as a result of elephant activity destroying mature *Acacia* trees, together with the effects of a fire protection policy. The giraffe population, as described by aerial counts, fecundity and mortality data, had been increasing at 5–6% a year. When compared with the Nairobi National Park giraffes, the Serengeti population had a larger sub-adult component, faster juvenile growth rates, shorter calving intervals, and attained maturity at earlier ages. These properties resulted in higher fecundity, higher calf survival and lower adult mortality. He suggested that the eradication of rinderpest in the early 1960s, together with an increased dry season rainfall in the 1970s, had caused an eruption of the grazing ruminant populations in the Serengeti (Sinclair & Norton-Griffiths 1979). By reducing the amount of standing grass in the dry season, less of the park burned, thus promoting the regeneration of trees and shrubs, especially in areas where elephant damage had reduced the woodland canopy. The giraffe, previously food-limited, had expanded greatly. He suggested that a stable limit cycle might be implicated and made some suggestions for management action. Reducing giraffe browsing would have a greater immediate effect than removing elephant, the primary effect of which would be to arrest the decline in the number of mature trees. Thus, for the most rapid recovery of the woodland the recommended policy was to shoot giraffe and give the areas of regenerating woodland protection from fire. However, because the giraffe were retarding, rather than suppressing, the woody regeneration, fire control might be sufficient to promote the renewal of the *Acacia* woodland.

*Hippopotamus*

Reference has been made earlier to the hippopotamus problem in the Queen Elizabeth National Park, Uganda. This was one of the first overpopulation problems to be recognized and the first example of effective management in East Africa. Large areas of grassland within 5 km of lake shores had been overgrazed and turned into compacted earth: massive, deep erosion gullies had formed. Culling hippopotamus led to habitat recovery, accelerated by improved rainfall, and, in the area where hippopotamus were most heavily culled, an increase in the total numbers of animals of 50% over a decade. At the same time there were large increases in total mammal biomass and energy levels. The increase in species diversity was beneficial and in line with the policy for the National Park, because the trend towards reduction in diversity, by extreme single-species dominance, was halted and reversed. An experimental approach was adopted in the 1960s, as described earlier, with parallel studies of climate, vegetation and other grazing mammals. Considerable advances were made in knowledge of the biology of hippopotamus and other mammals in relation to their environments. Unfortunately the experimental culling programme ended in 1966, so the opportunity was lost for a long-term experimental investigation over more than one rainfall cycle. Some key papers are by Eltringham (1974, 1979), Field & Laws (1970), Laws (1968a,b), Laws & Clough (1966), Lock (1972), Thornton (1971).

## The future

This review reinforces the importance of long-term studies, which are necessary to take account of climatic and environmental changes, population dynamics and their influence on social behaviour; also for studies of such topics as lifetime reproductive success and population cycling. Change is seen to be the norm, not stability. The value of an experimental approach is also clear. The sustained programmes in the Serengeti ecosystem and the Amboseli basin provide good models. Much has already been achieved at minimal cost.

Unfortunately there has been a decline in the East African research institutions since the 1960s and a great opportunity has been lost. By now there could have been a cadre of career scientists in this field of research, in each country, if the appropriate structures had been created and maintained. (The comparison with Antarctic science and marine mammal research is instructive.) In Africa able people tend to go into administration where they are badly needed at an early stage in their career. As a result there is still a very small nucleus of African research workers and few of any seniority.

African governments are not looking for basic science, but for advice on policy and resources; this is not a task for the inexperienced. The rational approach would be to define the problem, decide the scientific approach, generate testable hypotheses, undertake manipulative experiments and observations and reach conclusions. The scientists' role is to make sure that the managers and politicians have the necessary information to enable them to appreciate the consequences of adopting various options; they make the decisions, not the scientists.

East African wildlife research has largely been funded by overseas foundations. Research tends to take the form of short field projects, funded by grants, consultancies or contracts, and the essentially *ad hoc* approach is encouraged by the practice since independence of issuing two-year research permits, each of which may take 11 months to obtain. This leads to the work being undertaken increasingly by graduate students. Their supervisors in the home countries may influence their policy contributions, but young scientists should not be entering wildlife politics as so many have. There is also a temptation for them to produce a startling result, which may help to get their next job.

In addition, peer review and criticism is often lacking, compared with the situation in their home country. Too many results in recent years have been confined to unpublished theses (more or less inaccessible to criticism, but quoted), or as unpublished reports to governments or international organizations; these too are relatively inaccessible and outside the normal refereed journal system. Even when properly published in refereed journals the work is dispersed through the literature. The *African Journal of Ecology* and symposia such as this are valuable in bringing the results together. But in general the result has been a decline in the overall quality and quantity of work and a corresponding decline in support from the funding bodies.

Much prestige could accrue to African governments supporting research in this field, where Africa offers pre-eminent opportunities. Research in African mammals is not 'big science' and is relatively inexpensive, an obvious area in which African countries could make a scientific reputation. A relatively modest investment could contribute to basic ecological knowledge and theory and there would be applications as well. In the longer term, for example, as well as contributing to conservation and management, there is potential in wildlife for biotechnology to improve domestic stock. There would be support from a growing international movement dedicated to wildlife and environmental protection.

If governments were to provide long-term core support for research on large mammals and their environments, by regenerating their national institutes, the probability of obtaining matching funding from overseas would be enhanced. Stronger links with overseas universities could also be developed, given the presence of well-found research facilities in Africa. Governments should also make it easier for overseas scientists to carry out field research by relaxing present restrictions (e.g. work permits and research permits) and they could then help to train African biologists.

As a first step it is suggested that a workshop might be organized to address these ideas and opportunities. The main objective would be to demonstrate to African governments the importance of supporting biological research on their uniquely valuable ecosystems and animal communities. It would not be a conventional symposium, but a working meeting, with a title such as 'Mammals in Africa: a regional challenge'. A good precedent is the FAO Consultation on marine mammals, 'Mammals in the Sea', held in Bergen, Norway, in 1976 (FAO 1978–82), which had important implications for the development of research on marine mammals. Large mammal research in Africa needs a similar refocusing.

## References

Barnes, R.F.W. (1980). The decline of the baobab tree in Ruaha National Park, Tanzania. *Afr. J. Ecol.* **18**:243–52.

Barnes, R.F.W. (1982). A note on elephant mortality in Ruaha National Park, Tanzania. *Afr. J. Ecol.* **20**:137–40.

Barnes, R.F.W. (1985). Woodland changes in Ruaha National Park (Tanzania) between 1976 and 1982. *Afr. J. Ecol.* **23**:215–21.

Bax, P.N. & Sheldrick, D.L.W. (1963). Some preliminary observations on the food of elephant in the Tsavo Royal National Park (East) of Kenya. *E. Afr. Wildl. J* **1**:40–53.

Beadle, L.C. (1974). The Nuffield Unit of Tropical Animal Ecology (1961–1971). *J. Zool., Lond.* **173**:539–48.

Bell, R.H.V. (1969). *The use of the herb layer by grazing ungulates in the Serengeti National Park, Tanzania.* PhD thesis: University of Manchester.

Bell, R.H.V. (1970a). The use of the herb layer by grazing ungulates in the Serengeti. In *Animal populations in relation to food resources*:111–24. (Ed. Watson, A.). Blackwells Scientific Publications, Oxford. (*Symp. Br. ecol. Soc.* No. 10.)

Bell, R.H.V. (1970b). A grazing ecosystem in the Serengeti. *Scient. Am.* **225**:86–93.

Bell, R.H.V. (1982). The effect of soil nutrient availability on community structure in African ecosystems. In *Ecology of tropical savannas*:193–216. (Eds Huntley, B.J. & Walker, B.H.). Springer Verlag, Berlin. (*Ecol. Stud. Anal. Synth.* 42.)

Bere, R.M. (1959). Queen Elizabeth National Park, Uganda: the hippopotamus problem and experiment. *Oryx* **5**:116–24.

Bosman, P. & Hall-Martin. A. (1986). *Elephants of Africa.* C. Struik, Cape Town.

Brooks, A.C. & Buss, I.O. (1962). Past and present status of the elephant in Uganda, *J. Wildl. Mgmt* **26**:38–50.

Buechner, H.K., Buss, I.O., Longhurst, W.M. & Brooks, A.C. (1963). Numbers and migrations of elephants in Murchison Falls National Park, Uganda. *J. Wildl. Mgmt* **27**:145–62.

Buechner, H.K. & Dawkins, H.C. (1961). Vegetation change induced by elephants and fire in Murchison Falls National Park, Uganda. *Ecology* **42**:752–66.

Buechner, H.K. & Roth, H.D. (1974). The lek system in Uganda kob antelope. *Am. Zool.* **14**:145–62.

Buss, I.O. (1961). Some observations on food habits and behavior of the African elephant. *J. Wildl. Mgmt* **25**:131–48.

Buss, I.O & Smith, N.S. (1966). Observations on reproduction and breeding behaviour of the African elephant. *J. Wildl. Mgmt* **30**:375–88.

Calder, W.A., III. (1983). Body size, mortality, and longevity. *J. theor. Biol.* **102**:135–44.

Caughley, G. (1976). The elephant problem—an alternative hypothesis. *E. Afr. Wildl. J.* **14**:265–83.

Caughley, G. & Goddard, J. (1975). Abundance and distribution of elephants in the Luangwa Valley, Zambia. *E. Afr. Wildl. J.* **13**:39–48.

Clutton-Brock, T.G. (In press). Mammalian breeding systems. *Proc. R. Soc.* (B).

Connor, E.F. & McCoy, E.D. (1979). The statistics and biology of the species-area relationship. *Am. Nat.* **113**:791–833.

Corfield, T.F. (1973). Elephant mortality in Tsavo National Park, Kenya. *E. Afr. Wildl. J.* **11**:339–68.

Croze, H. (1972). A modified photogrammetric technique for assessing age-structures of elephant populations and its use in Kidepo National Park. *E. Afr. Wildl. J.* **10**:91–115.

Croze, H. (1974). The Seronera bull problem. II. The trees. *E. Afr. Wildl. J.* **12**:29–47.

Diamond, J.M. (1975). The island dilemma: lessons of modern biogeographic studies for the design of natural reserves. *Biol. Conserv.* **7**:129–45.

Douglas-Hamilton, I. (1972). *On the ecology and behaviour of the African elephant.* PhD thesis: University of Oxford.

Douglas-Hamilton, I. (1973). On the ecology and behaviour of the Lake Manyara elephants. *E. Afr. Wildl. J.* **11**:401–3.

Douglas-Hamilton, I. & Douglas-Hamilton, O. (1975). *Among the elephants.* Collins, London.

Dunham, K.M. (1986). Movements of elephant cows in the unflooded Middle Zambezi Valley, Zimbabwe. *Afr. J. Ecol.* **24**:81–7.

Dunham, K.M. & Murray, M.G. (1982). The fat reserves of impala, *Aepyceros me–lampus. Afr. J. Ecol.* **20**:81–7.

East, R. (1984). Rainfall, soil nutrient status and biomass of large African savanna mammals. *Afr. J. Ecol.* **22**:245–70.

Eltringham, S.K. (1974). Changes in the large mammal community of Mweya Peninsula, Rwenzori National Park, Uganda, following removal of hippopotamus. *J. appl. Ecol.* **11**:855–65.

Eltringham, S.K. (1979). *The ecology and conservation of large African mammals.* Macmillan Press Ltd., London.

FAO (1978–82). *Mammals in the seas. Report of the FAO Advisory Committee on Marine Resources Research Working Party on Marine Mammals* **1–4**. Food and Agriculture Organization of the United Nations, Rome. (*FAO Fish Ser.* No. 5.).

Field, C.R. (1972). The food habits of wild ungulates in Uganda by analyses of stomach contents. *E. Afr. Wildl. J.* **10**:17–42.

Field, C.R. & Laws, R.M. (1970). The distribution of the larger herbivores in the Queen Elizabeth National Park, Uganda. *J. appl. Ecol.* 7:273–94.

Georgiadis, N. (1985). Growth patterns, sexual dimorphism and reproduction in African ruminants. *Afr. J. Ecol.* 23:75–87.

Gilbert, F.S. (1980). The equilibrium theory of island biogeography: fact or fiction? *J. Biogeogr.* 7:209–35.

Glover, J. (1963). The elephant problem at Tsavo. *E. Afr. Wildl. J.* 1:30–9.

Goddard, J. (1968). Food preferences of two black rhinoceros populations. *E. Afr. Wildl. J.* 6:1–18.

Goddard, J. (1970). Age criteria and vital statistics of a black rhinoceros population. *E. Afr. Wildl. J.* 8:105–21.

Grimsdell, J.J.R. (1969). *Ecology of the buffalo,* Syncerus caffer, *in western Uganda.* PhD thesis: University of Cambridge.

Grzimek, M. & Grzimek, B. (1960a). Census of plains animals in the Serengeti National Park, Tanganyika. *J. Wildl. Mgmt* 24:27–37.

Grzimek, M. & Grzimek, B. (1960b). A study of the game of the Serengeti plains. *Z. Säugetierk.* 25:1–61.

Gwynne, M.D. & Bell, R.H.V. (1968). Selection of vegetation components by grazing ungulates in the Serengeti National Park. *Nature, Lond.* 220: 390–3.

Hall-Martin, A.J. (1975). *Studies on the biology and productivity of the giraffe* Giraffa camelopardalis. DSc thesis: University of Pretoria.

Hanks, J. (1972). Growth of the African elephant (*Loxodonta africana*). *E. Afr. Wildl. J.* 10:251–72.

Hanks, J. (1979). *A struggle for survival: the elephant problem.* Country Life Books, London.

Hanks, J. (1981). Characterization of population condition. In *Dynamics of large mammal populations*:47–73. (Eds Fowler, C.W. & Smith, T.D.). John Wiley & Sons, New York, Chichester etc.

Hanks, J. & McIntosh, J.E.A. (1973). Population dynamics of the African elephant (*Loxodonta africana*). *J. Zool., Lond.* 169:29–38.

Harrington, G.N. & Ross, I.C. (1974). The savanna ecology of Kidepo Valley National Park. I. The effects of burning and browsing on the vegetation. *E. Afr. Wildl. J.* 12:93–105.

Harthoorn, A.M. (1966). The Tsavo elephants. *Oryx* 8:233–6.

Hatton, J.C. & Smart, N.O.E. (1984). The effect of long-term exclusion of large herbivores on soil nutrient status in Murchison Falls National Park, Uganda. *Afr. J. Ecol.* 22:23–30.

Hediger, H. (1950). *Wild animals in captivity.* Butterworth, London.

Higgs, A.J. (1981). Island biogeography theory and nature reserve design. *J. Biogeogr.* 8:117–24.

Hillman-Smith, A.K.K., Owen-Smith, N., Anderson, J.L., Hall-Martin, A.J. & Selaladi, J.P. (1986). Age estimation of the White rhinoceros (*Ceratotherium simum*). *J. Zool., Lond.* 210:355–79.

Hubert, E. (1947). *La faune des grands mammifères de la plaine Rwindi-Rutshuru (Lac Édouard), son évolution depuis sa protection totale.* Institut des Parcs Nationaux du Congo Belge, Brussels.

Jachmann, H. (1986). Notes on the population dynamics of the Kasungu elephants. *Afr. J. Ecol.* **24**:215–26.

Jachmann, H. (1988). Estimating age in African elephants: a revision of Laws' molar evaluation technique. *Afr. J. Ecol.* **26**:51–6.

Jachmann, H. & Bell, R.H.V. (1985). Utilization by elephants of the *Brachystegia* woodlands of the Kasungu National Park, Malawi. *Afr. J. Ecol.* **23**:245–58.

Jarman, P.J. (1974). The social organisation of antelope in relation to their ecology. *Behaviour* **48**:215–67.

Kingdon, J. (1971–82). *East African mammals: an atlas of evolution in Africa* **1–3**. Academic Press, London.

Klingel, H. & Klingel, U. (1966). Tooth development and age determination in the plains zebra (*Equus quagga boehmi* Matschie). *Zool. Gart., Lpz.* **33**:34–54.

Krebs, C.J. (1988). *The message of ecology.* Harper & Row, New York.

Lamprey, H.F. (1963). Ecological separation of the large mammal species in the Tarangire Game Reserve, Tanganyika. *E. Afr. Wildl. J.* **1**:63–92.

Lamprey, H.F. (1969). Ecological research in the Serengeti National Park. *J. Reprod. Fert. Suppl.* No.6:487–93.

Laws, R.M. (1952). A new method of age determination for mammals. *Nature, Lond.* **169**:972–3.

Laws, R.M. (1956). Growth and sexual maturity in aquatic mammals. *Nature, Lond.* **178**:193–4.

Laws, R.M. (1959). Accelerated growth in seals, with special reference to the Phocidae. *Norsk Hvalfangsttid.* **48**:425–52.

Laws, R.M. (1966). Age criteria for the African elephant, *Loxodonta a.africana*. *E. Afr. Wildl. J.* **4**:1–37.

Laws, R.M. (1968a). Interactions between elephant and hippopotamus populations and their environments. *E. Afr. agric. For J.* **33** (special issue): 140–7.

Laws, R.M. (1968b). Dentition and ageing of the hippopotamus. *E. Afr. Wild. J.* **6**:19–52.

Laws, R.M. (1969a). Aspects of reproduction in the African elephant, *Loxodonta africana*. *J. Reprod. Fert. Suppl.* No. 6: 193–217.

Laws, R.M. (1969b). The Tsavo Research Project. *J. Reprod. Fert. Suppl.* No. 6:495–531.

Laws, R.M. (1970). Elephants as agents of habitat and landscape change in East Africa. *Oikos* **21**:1–15.

Laws, R.M. & Clough, G. (1966). Observations on reproduction in the hippopotamus *Hippopotamus amphibius* Linn. *Symp. zool. Soc. Lond.* No. 15: 117–40.

Laws, R.M. & Parker, I.S.C. (1968). Recent studies on elephant populations in East Africa. *Symp. zool. Soc. Lond.* No. 21:319–59.

Laws, R.M., Parker, I.S.C. & Johnstone, R.C.B. (1975). *Elephants and their habitats: the ecology of elephants in North Bunyoro, Uganda.* Clarendon Press, Oxford.

Leuthold, B.M. & Leuthold, W. (1972). Food habits of giraffe in Tsavo National Park, Kenya. *E. Afr. Wildl. J.* **10**:129–41.

Leuthold, W. (1970). Preliminary observations on food habits of gerenuk in Tsavo National Park, Kenya. *E. Afr. Wildl. J.* **8**:73–84.

Leuthold, W. (1971). Studies on the food habits of lesser kudu in Tsavo National Park, Kenya. *E. Afr. Wildl. J.* 9:35–45.

Leuthold, W. (1977). *African ungulates. A comparative review of their ethology and behavioral ecology.* Springer Verlag, Berlin. (*Zoophysiology Ecol.* 8.)

Leuthold, W. & Sale, J.B. (1973). Movements and patterns of habitat utilization of elephants in Tsavo National Park, Kenya. *E. Afr. Wildl. J.* 11:369–84.

Lock, J.M. (1972). The effects of hippopotamus grazing on grasslands. *J. Ecol.* 60:445–67.

MacArthur, R.H. & Wilson, E.O. (1967). *The theory of island biogeography.* Princeton University Press, Princeton.

McCullagh, K.G. (1969). The growth and nutrition of the African elephant. I. Seasonal variations in the rate of growth and the urinary excretion of hydroxyproline. *E. Afr. Wildl. J.* 7:85–90.

Matthews, L.H. (1954). Research on the mammals of Africa. *Nature, Lond.* 174:670–1.

Moss, C. (1983). Oestrous behaviour and female choice in the African elephant. *Behaviour* 86: 167–96.

Moss, C. (1988). *Elephant memories. Thirteen years in the life of an elephant family,* Elm Tree Books, London.

Norton-Griffiths, M. (1975). *Counting animals.* African Wildlife Leadership Foundation, Nairobi.

Olivier, R.C.D. (1978). *On the ecology of the Asian elephant.* PhD thesis: University of Cambridge.

Parker, I.S.C. (1979). *The ivory trade.* Report to the Department of the Interior, U.S. Government, Washington, D.C.

Parker, I.S.C. & Graham, A.D. (1989). Humans, elephants and Gause's hypothesis. *Symp. zool. Soc. Lond.* No. 61:241–52.

Pellew, R.A. (1983). The impacts of elephant, giraffe and fire upon the *Acacia tortilis* woodlands of the Serengeti. *Afr. J. Ecol.* 21:41–74.

Perry, J.S. (1953). The reproduction of the African elephant, *Loxodonta africana. Phil. Trans. R. Soc.* (B) 237:93–149.

Peters, R.H. (1983). *The ecological implications of body size.* Cambridge University Press, Cambridge, England.

Petrides, G.A. & Swank, W.G. (1966). Estimating the productivity and energy relations of an elephant population. *Proc. int. Grassld Congr.* 9:831–42.

Phillipson, J. (1975). Rainfall, primary production and 'carrying capacity' of Tsavo National Park (East), Kenya. *E. Afr. Wildl. J.* 13:171–201.

Pilgram, T. & Western, D. (1986). Inferring hunting patterns on African elephants from tusks in the international ivory trade. *J. appl. Ecol.* 23:503–14.

Poole, J.H. (1982). *Musth and male–male competition in the African elephant.* PhD thesis: University of Cambridge.

Poole, J.H. & Moss, C.J. (1981). Musth in the African elephant, *Loxodonta africana. Nature, Lond.* 292:830–1.

Poole, J.H. & Moss, C.J. (1989). Elephant mate searching: group dynamics and vocal and olfactory communication. *Symp. zool. Soc. Lond.* No. 61: 111–25.

Schenkel, R. & Schenkel-Hulliger, L. (1969). *Ecology and behaviour of the black* rhinoceros (Diceros bicornis L.): a field study. Paul Parey, Hamburg & Berlin. (Mammalia Depicta 5.)

Senzota, R.B.M. (1988). Further evidence of exogenous processes regulating the population of zebra in the Serengeti. Afr. J. Ecol. 26:11–6.

Sidney, J. (1965). The past and present distribution of some African ungulates. Trans. zool. Soc. Lond. 30:1–397.

Sikes, S.K. (1966). The African elephant, Loxodonta africana: a field method for the estimation of age. J. Zool., Lond. 150:279–95.

Sikes, S.K. (1968). The African elephant, Loxodonta africana: a field method for the estimation of age. J. Zool., Lond. 154:235–48.

Sikes, S.K. (1971). The natural history of the African elephant. Weidenfeld and Nicholson, London.

Sinclair, A.R.E. (1974a). The natural regulation of buffalo populations in East Africa. II. Reproduction, recruitment and growth. E. Afr. Wildl. J. 12:169–83.

Sinclair, A.R.E. (1974b). The natural regulation of buffalo populations in East Africa. III. Population trends and mortality. E. Afr. Wildl. J. 12:185–200.

Sinclair, A.R.E. (1974c). The natural regulation of buffalo populations in East Africa. IV. The food supply as a regulating factor, and competition. E. Afr. Wildl. J. 12:291–311.

Sinclair, A.R.E. & Norton-Griffiths, M. (Eds). (1979). Serengeti. Dynamics of an ecosystem. University of Chicago Press, Chicago and London.

Soulé, M.E., Wilcox, B.A. & Holtby, C. (1979). Benign neglect: a model of faunal collapse in the game reserves of East Africa. Biol. Conserv. 15:259–72.

Spence, D.H.N. & Angus, A. (1971). African grassland management—burning and grazing in Murchison Falls National Park, Uganda. In Scientific management of animal and plant communities for conservation:319–31. (Eds Duffey, E. & Watt, A.S.). Blackwells Scientific Publications, Oxford etc. (Symp. Br. ecol. Soc. No. 11.)

Spinage, C.A. (1967a). Ageing the Uganda defassa waterbuck Kobus defassa ugandae Neumann. E. Afr. Wildl. J. 5:1–17.

Spinage, C.A. (1967b). The autecology of the Uganda waterbuck Kobus defassa ugandae) with special reference to territoriality and population control. PhD thesis: University of London.

Spinage, C.A. (1970). Population dynamics of the Uganda defassa waterbuck (Kobus defassa ugandae Neumann) in the Queen Elizabeth National Park, Uganda. J. anim. Ecol. 39:51–78.

Spinage, C.A. (1972). Age estimation of zebra. E. Afr. Wildl. J. 10:273–7.

Spinage, C.A. (1973). A review of the age determination of mammals by means of teeth, with especial reference to Africa. E. Afr. Wildl. J. 11:165–87.

Spinage, C.A. (1982). A territorial antelope. The Uganda waterbuck. Academic Press, New York.

Stearns, S.C. (1983). The influence of size and phylogeny on patterns of covariation among life-history traits in the mammals. Oikos 41:173–87.

Stewart, D.R.M. (1965). The epidermal characters of grasses, with special reference to East African plains species. Bot. Jb. 84:63–116.

Stewart, D.R.M. & Stewart, J. (1963). The distribution of some large mammals in Kenya. Jl E. Africa nat. Hist. Soc. 24:1–52.

Stewart, D.R.M. & Zaphiro, D.R.P. (1963). Biomass and density of wild herbivores in different East African habitats. *Mammalia* 27:483–96.

Storr, G.M. (1961). Microscopic analysis of faeces, a technique for ascertaining the diet of herbivorous mammals. *Aust. J. biol. Sci.* 14:157–64.

Swank, W.G., Watson, R.M., Freeman, G.H. & Jones, T. (Eds) (1969). Workshop on the use of light aircraft in wildlife management in East Africa. *E. Afr. agric. For. J.* 34 (Spec. issue):1–111.

Talbot, L.M. & Talbot, M.H. (1963). The wildebeest in Western Masailand, East Africa. *Wildl. Monogr.* No.12:1–88.

Thornton, D.D. (1971). The effect of complete removal of hippopotamus on grassland in the Queen Elizabeth National Park, Uganda. *E. Afr. Wildl. J.* 9:47–55.

Tieszen, L.L. & Imbamba, S.K. (1980). Photosynthetic systems, carbon isotope discrimination and herbivore selectivity in Kenya. *Afr. J. Ecol.* 18:237–42.

Vesey-Fitzgerald, D.F. (1965). The utilization of natural pastures by wild animals in the Rukwa Valley, Tanganyika. *E. Afr. Wildl. J.* 3:38–48.

Watson, R.M. (1967). *The population ecology of the wildebeest* (Connochaetes taurinus albojubatus *Thomas) in the Serengeti,* PhD thesis: University of Cambridge.

Watson, R.M. (1969). Reproduction of wildebeest, *Connochaetes taurinus albojubatus* Thomas, in the Serengeti region, and its significance to conservation. *J. Reprod. Fert. Suppl.* No. 6:287–310.

Watson, R.M. (1970). Generation time and intrinsic rates of natural increase in wildebeeste (*Connochaetes taurinus albojubatus* Thomas). *J. Reprod. Fert.* 22:557–61.

Watson, R.M., Parker, I.S.C. & Allan, T. (1969). A census of elephant and other large mammals in the Mkomazi region of northern Tanzania and southern Kenya. *E. Afr. Wildl. J.* 7:11–26.

Weir, J.S. (1973). Exploitation of water soluble soil sodium by elephants in Murchison Falls National Park, Uganda. *E. Afr. Wildl. J.* 11:1–7.

Western, D. (1979). Size, life history and ecology in mammals. *Afr. J. Ecol.* 17:185–204.

Western, D. & Lindsay, W.K. (1984). Seasonal herd dynamics of a savanna elephant population. *Afr. J. Ecol.* 22:229–44.

Western, D., Moss, C.J. & Georgiadis, N. (1983). Age estimation and population age structure of elephants from footprint dimensions. *J. Wildl. Mgmt* 47:1192–7.

Western, D. & Ssemakula, J. (1981). The future of the savannah ecosystems: ecological islands or faunal enclaves? *Afr. J. Ecol.* 19:7–19.

Western, D. & Van Praet, P.C. (1973). Cyclical changes in the habitat and climate of an East African ecosystem. *Nature, Lond.* 241:104–6.

Wilcox, B.A. (1980). Insular ecology and conservation. In Conservation biology: an evolutionary—ecological perspective: 95–117. (Eds Soulé, M.E. & Wilcox, B.A.). Sinauer Associates, Sunderland, Mass.

Worthington, E.B. (1959). *Science in the development of Africa.* Commission for Technical Cooperation in Africa South of the Sahara and Scientific Council for Africa South of the Sahara.

Wynne-Edwards, V.C. (1962). *Animal dispersion in relation to social behaviour.* Oliver & Boyd, Edinburgh.

# Index